# Ecological Studies, Vol. 93

Analysis and Synthesis

Edited by

W.D. Billings, Durham, USA
F. Golley, Athens, USA
O.L. Lange, Würzburg, FRG
J.S. Olson, Oak Ridge, USA
H. Remmert, Marburg, FRG

# Ecological Studies

Volume 74
**Inorganic Contaminants in the Vadose Zone** (1989)
B. Bar-Yosef, N.J. Barnow, and J. Goldshmid (Eds.)

Volume 75
**The Grazing Land Ecosystems of the African Sahel** (1989)
H.N. Le Houérou

Volume 76
**Vascular Plants as Epiphytes: Evolution and Ecophysiology** (1989)
U. Lüttge (Ed.)

Volume 77
**Air Pollution and Forest Decline: A Study of Spruce *(Picea abies)* on Acid Soils** (1989)
E.-D. Schulze, O.L. Lange, and R. Oren (Eds.)

Volume 78
**Agroecology: Researching the Ecological Basis for Sustainable Agriculture** (1990)
S.R. Gliessman (Ed.)

Volume 79
**Remote Sensing of Biosphere Functioning** (1990)
R.J. Hobbs and H.A. Mooney (Eds.)

Volume 80
**Plant Biology of the Basin and Range** (1990)
B. Osmond, G.M. Hidy, and L. Pitelka (Eds.)

Volume 81
**Nitrogen in Terrestrial Ecosystem: Questions of Productivity, Vegetational Changes, and Ecosystem Stability** (1990)
C.O. Tamm

Volume 82
**Quantitative Methods in Landscape Ecology: The Analysis and Interpretation of Landscape Heterogeneity** (1990)
M.G. Turner and R.H. Gardner (Eds.)

Volume 83
**The Rivers of Florida** (1990)
R.J. Livingston (Ed.)

Volume 84
**Fire in the Tropical Biota: Ecosystem Processes and Global Challenges** (1990)
J.G. Goldammer (Ed.)

Volume 85
**The Mosaic-Cycle Concept of Ecosystems** (1991)
H. Remmert (Ed.)

Volume 86
**Ecological Heterogeneity** (1991)
J. Kolasa and S.T.A. Pickett (Eds.)

Volume 87
**Horses and Grasses: The Nutritional Ecology of Equids and Their Impact on the Camargue** (1992)
P. Duncan

Volume 88
**Pinnipeds and El Nino: Responses to Enviromental Stress** (1992)
F. Trillmich and K.A. Ono (Eds.)

Volume 89
**Plantago: A Multidisciplinary Study** (1992)
P.J.C. Kuiper and M. Bos (Eds.)

Volume 90
**Biogeochemistry of a Subalpine Ecosystem: Loch Vale Watershed** (1992)
J. Baron (Ed.)

Volume 91
**Atmospheric Deposition and Forest Nutrient Cycling** (1992)
D.W. Johnson and S.E. Lindberg (Eds.)

Volume 92
**Landscape Boundaries** (1992)
A.J. Hansen and F. di Castri (Eds.)

B.W. van Wilgen D.M. Richardson
F.J. Kruger H.J. van Hensbergen (Eds.)

# Fire in South African Mountain Fynbos

## Ecosystem, Community and Species Response at Swartboskloof

With 98 Figures and 72 Tables

Springer-Verlag
Berlin Heidelberg New York
London Paris Tokyo
Hong Kong Barcelona
Budapest

Dr. Brian W. van Wilgen
Division of Forest Science and Technology, CSIR
Jonkershoek Forestry Research Centre
Private Bag X5011, Stellenbosch 7600, South Africa

Dr. David M. Richardson
Division of Forest Science and Technology, CSIR
Jonkershoek Forestry Research Centre
Private Bag X5011, Stellenbosch 7600, South Africa

Dr. Frederick J. Kruger
Division of Forest Science and Technology, CSIR
PO Box 395, Pretoria 0001, South Africa

Dr. Hubertus J. van Hensbergen
Department of Nature Conservation
Forestry Faculty, University of Stellenbosch
Stellenbosch 7600, South Africa

Library of Congress Cataloging-in-Publication Data
Fire in South African mountain fynbos: ecosystem, community, and species response at Swartboskloof / B.W. van Wilgen . . . [et al.], eds. p. cm. — (Ecological studies; v. 93)
Includes bibliographical references and index.

ISBN-13: 978-3-642-76176-8 e-ISBN-13: 978-3-642-76174-4
DOI: 10.1007/978-3-642-76174-4

1. Fynbos ecology—South Africa—Swartboskloof River Watershed. 2. Fire ecology—South Africa—Swartboskloof River Watershed. I. Van Wilgen, B.W. (Brian W.) II. Series.
QK396.F47 1992 91–39427 CIP



Softcover reprint of the hardcover 1st edition 1992



Typesetting: Best-set Typesetter Ltd., Hong Kong
31/3145-5 4 3 2 1 0 – Printed on acid-free paper

# Preface

Ecologists are increasingly being drawn into the task of addressing problems of environmental degradation. They are expected to find solutions that will lead to sustainable resource use throughout the world. In doing so, the robustness of the science becomes increasingly important, and the problem of extrapolating the results of research conducted within what is usually a relatively limited geographical scope is increasingly highlighted.

One approach to developing a globally robust ecology involves more or less formal intercontinental comparative studies, usually focused on the question of ecological convergence. These studies are directed at testing the prediction that similar physical and other environmental factors in different parts of the world, through their selective influences, will give rise to ecosystems which share common structural and functional features. Should this be true, the predictive power of ecology developed within such a framework should be sufficient to solve similar problems elsewhere in such biomes. There is a long history of such an approach in mediterranean-type ecosystems, documented in a series of volumes and their accompanying scientific papers beginning with that of Di Castri and Mooney (1973).

These studies have varied in their focus, ranging from global syntheses (Mooney 1977; Miller 1981a; Conrad and Oechel 1982; Kruger et al. 1983; Dell et al. 1986; Tenhunen et al. 1987; Specht 1988; Kruger et al. 1989; Orshan 1989; Groves and Di Castri in press) to detailed site studies (Specht and Rayson 1957 and subsequent papers; Mooney and Parsons 1973; Thrower and Bradbury 1977). Both approaches are necessary to build the body of knowledge required for the progressive development of a predictive ecology. The power of this approach lies in the fact that the mediterranean-climate regions of the world are geographically separated and clearly similar in many of their physical environmental features. They offer a ready-made global experiment for testing the predictions of ecology. Theory developed within such a framework should contribute not only to the solution of problems of resource use in mediterranean-climate regions but also to a broader understanding of ecology, and indeed of the robustness of ecology as a whole.

One factor that pervades mediterranean ecosystems is the frequent occurrence of intense fires in the vegetation. This phenomenon and its consequences has been well documented in many studies and syntheses (e.g. Mooney and Conrad 1977; Gill et al. 1981; Booysen and Tainton 1984). Our understanding of the role of fire in these ecosystems, as well as its use for management for the conservation of ecosystems (and to the benefit of the populations within the regions concerned) has grown rapidly over the past several decades. Even so, there are many substantial gaps in our understanding of the nature of the ecological effects of fire. The policies for fire management have continued to evolve in the search for more effective approaches appropriate to each circumstance. Questions surrounding the issue of fire include that of the optimum fire regime for maintaining biotic diversity, the fire regime which will allow for conservation while at the same time ensuring that protection needs are met, and the sustainable utilization of ecosystems subject to recurrent fire.

In South Africa these issues are of special importance. The funds available for intensive management of fynbos ecosystems, though limited, were relatively freely available until recently. Changing national priorities will dictate a diversion of funds in other directions in the future. The continued conservation of the unique fynbos ecosystems and their contribution to the regional economy through utilization, recreation, education and tourist opportunities, will in future depend on our contributions to a proper understanding and perception of these values, on the one hand, and to cost-effective systems for the management of these ecosystems, on the other hand. Innovative ways of treating the problems of fire management and all the issues surrounding it will be needed if supportable management policies and assistance are to be found.

This book incorporates some of the results of research conducted in the Swartboskloof catchment of the Jonkershoek Valley near Stellenbosch, South Africa. The research was conducted under the auspices of the Fynbos Biome Project, an enterprise within the National Programme for Environmental Sciences (Jarman 1982; Huntley 1987). The work was initiated to provide a detailed understanding of certain aspects of the functioning of a mountain fynbos ecosystem, through comparison with the results obtained from a lowland fynbos ecosystem at Pella (Jarman 1988; see also Brown et al. 1984; Mitchell et al. 1984) as well as with results obtained from detailed studies on other continents. The focus of the research in Swartboskloof was on the ecological effects of a single fire with three clear objectives in view. These were: (1) to document fully the regeneration and recolonization strategies of the biota of the ecosystems concerned, with a view to developing our capability to predict the processes that follow fire in this respect; (2) to study the stability or otherwise of the boundaries between vegetation communities, and to examine the effects of a fire and other factors which determine those boundaries; and (3) to examine the effects of a fire on resource availability, particularly with respect to mineral nutrient cycling and water yield at a catchment level.

The objectives of this volume are to consolidate and document the information on the effects of the fire in Swartboskloof, to place these findings in context with analogous findings from similar site studies elsewhere in the world, and then to draw conclusions from the research regarding future research needs, policies and practices for the management of fire-prone ecosystems. The book addresses the interaction between climate and structural features of the vegetation as it determines the incidence and behaviour of fire. It examines the fire in Swartboskloof, and places this into perspective through comparison to fire in fynbos in general. It examines the composition and structure of vegetation, the effects of fire on this, the responses of the individual species (both indigenous and introduced), and synthesizes this information in such a way as to draw general conclusions about the effects of a fire regime on fynbos communities. The interactions between fire and faunal groups are also examined, with emphasis on the consequences for the regenerating vegetation. Finally, it examines the effects of fire on the water and mineral nutrient balances of the system, and in so doing it addresses questions on the sustainability of fire management in fynbos ecosystems.

In compiling this volume, we recognized the limitations and constraints which operate in these kind of studies. First, any biome (and especially the fynbos biome) is complex in its spatial and temporal variation (Cowling 1991). Clearly, no single site can accommodate this complexity. Studies of this kind can merely provide benchmark information, but as such this information is valuable in that the depth of understanding provided allows more critical comparative analysis than would be available from more superficial replicated or regional studies. The second major constraint arises from the fact that the resources of expertise and finance limit severely the scope of any study of this kind. This, together with the practical problems arising from the heterogeneity of the system addressed at the scale of this study, mean necessarily that the picture which emerges can be no more than the first approximation of the structure and functioning of the ecosystem examined. Nevertheless, this approximation is an important one on the path towards an adequate understanding that is needed to support the formulation of objectives with regard to both fynbos ecosystems and ecology for southern Africa and elsewhere.

Except where authors names are given, taxonomic nomenclature in this volume follows Bond and Goldblatt (1984) for plants, Maclean (1985) for birds, Smithers (1983) for mammals and the South African Museum, Cape Town, for ants.

Pretoria and Stellenbosch, April 1992

F.J. Kruger
B.W. van Wilgen

# Acknowledgements

This volume has come about through the combined efforts of many people and institutions. First, we acknowledge the financial support of the South African Department of Environment Affairs who funded the bulk of the work as part of their Conservation Forestry Research Programme. Part of the work was also funded by the Fynbos Biome Project of the CSIR, the Core Programme of the Foundation for Research Development of the CSIR, the Division of Forest Science and Technology of the CSIR, and the Universities of Cape Town and Stellenbosch.

Thanks are due to the following persons who, with the editors, served on the Swartboskloof Working Group: William Bond, Johan Fourie, Mike Fraser, Margie Jarman, Jan Lampbrechts, Pat Manders, and Robert Prŷs-Jones. Tisha Greyling and Diane Stafford helped to organize regular seminars, and Kevin Higgins edited a bulletin that kept participants informed of progress in the various studies.

Many other people gave of their time and energy to ensure the success of the project. Specific contributions are acknowledged at the end of each chapter. Those persons who formed the core research team, and assisted with fieldwork, data analysis, preparation of the manuscript, and numerous other tasks, deserve special mention. In this regard we thank Adnaan Abrahams, Gregory Bailey, Hekkie Barnard, Lyn Carelse, Lee Cunningham, Anthony Duckworth, Iris Gouws, Ann Green, Kevin Higgins, Gordon Hockey, Clare Jones, Harold Julius, Godfrey Moses, Gerard Mostert, Eric Prinsloo, Reney Robyntjies, Mina September, Adrian Simmers, Elmarie Theron, Dirkie Willemse, Anthony Winter and Steve Worth. Danie van Wyk and Hekkie Barnard designed and constructed the stream-gauging weir in Swartboskloof against a tight deadline, and Eric Prinsloo installed the rain-gauge network.

The following persons kindly reviewed manuscripts: D.T. Bell (University of Western Australia, Perth, Australia), R. Benyon (Melbourne and Metropolitan Board of Works, Australia), W.J.

Bond (University of Cape Town, South Africa), R.A. Bradstock (National Parks and Wildlife Service, Hurstville, NSW, Australia), C.M. Breen (Institute of Natural Resources, Pietermaritzburg, South Africa), R.E. Burgan (Intermountain Research Station, Missoula, MT, USA), L.F. De Bano (Rocky Mountain Forest and Range Experiment Station, Tempe, AZ, USA), M.J. Delaney (University of Bradford, UK), P.J. Dye (CSIR Division of Forest Science and Technology, Sabie, South Africa), A.M. Gill (CSIRO Division of Plant Industry, Canberra, Australia), P.J. Grubb (University of Cambridge, UK), J.D. Helvey (Wenatchee, WA, USA), R.J. Hobbs (CSIRO Division of Wildlife and Ecology, Perth, Australia), J.E. Keeley (Occidental College, Los Angeles, CA, USA), I.J. Linn (University of Exeter, UK), D.T. Mitchell (University College, Dublin, Ireland), D. O'Dowd (Monash University, Victoria, Australia), P. O'Shaughnessy (Melbourne and Metropolitan Board of Works, Australia), B.N. Richards (University of New England, NSW, Australia), M.J. Samways (University of Natal, Pietermaritzburg, South Africa), R.J. Scholes (Botany Department, University of the Witwatersrand, Johannesburg, South Africa), L. Trabaud (Centre d'Études Phytosociologiques et Ecologiques, Montpellier, France), and J. Walker (CSIRO Division of Water Resources, Canberra, Australia).

We thank the editors and publishers of the following journals for permission to reproduce material previously published: *Forest Ecology and Management* (Figs. 5.4, 5.5 and 5.6; Tables 5.5 and 5.6); the *Journal of Ecology* (Figs. 3.4 and 3.6); the *Journal of Vegetation Science* (Figs. 5.2 and 5.8; Tables 5.1, 5.2, 5.3 and 5.4); and the *South African Journal of Botany* (Figs. 5.9, 5.10, 5.11, 5.12, 5.13, 6.1 and 6.2).

# Contents


**1 The Swartboskloof Experimental Site**
*B.W. van Wilgen and D.J. McDonald* (17 Figures) ... 1

1.1 Introduction ... 1
1.2 Location and Physical Site Description ... 2
1.3 Land-Use History ... 7
1.4 The Vegetation of Swartboskloof ... 10
1.4.1 Specialized Hygrophilous Communities ... 11
1.4.2 Short to Tall Proteoid Shrublands ... 12
1.4.3 Ericoid-Restioid Shrublands ... 14
1.4.4 Short Riparian Forest Communities ... 15
1.4.5 Tall Forests ... 15
1.5 Fauna ... 17
1.6 Research Rationale and Design ... 18
1.7 The Fire of March 1987 ... 18
Acknowledgements ... 20

**2 The Climate of Swartboskloof**
*D.B. Versfeld, D.M. Richardson, B.W. van Wilgen, R.A. Chapman, and G.G. Forsyth* (10 Figures) ... 21

2.1 Introduction ... 21
2.2 The Climate of Swartboskloof in a Mediterranean-Climate Context ... 21
2.3 Important Climatic Parameters ... 25
2.3.1 Rainfall ... 25
2.3.2 Temperature and Vapour Pressure Deficit ... 26
2.3.3 Radiation ... 27
2.3.4 Evaporation ... 29
2.3.5 Wind ... 30
2.3.6 Periods of Drought ... 31
2.4 The Influence of Climate on Fire ... 32
2.4.1 Lightning as an Ignition Source ... 32

| | | |
|---|---|---|
| 2.4.2 | Climate and the Probability of Fire | 33 |
| 2.5 | Conclusions | 35 |
| Acknowledgements | | 35 |
| **3** | **Fuel Properties of Vegetation in Swartboskloof**<br>*B.W. van Wilgen and H.J. van Hensbergen*<br>(7 Figures) | 37 |
| 3.1 | Introduction | 37 |
| 3.2 | The Description of Vegetation as Fuel | 37 |
| 3.2.1 | Subdivision of Biomass into Fuel Classes | 38 |
| 3.2.2 | Fuel Arrangement | 38 |
| 3.2.3 | Fuel Energy Contents | 38 |
| 3.2.4 | Fuel Models | 39 |
| 3.2.5 | Fuel Buildup and Decay | 40 |
| 3.3 | The Fuel Properties of Vegetation in Swartboskloof | 40 |
| 3.3.1 | Fuel Loads, Fuel Buildup<br>and the Probability of Fire | 40 |
| 3.3.2 | Fuel Arrangement | 43 |
| 3.3.3 | Seasonal Trends in Fuel Moisture Content | 43 |
| 3.3.4 | Live Plants | 45 |
| 3.3.5 | Dead Fuel Moisture | 45 |
| 3.3.6 | Energy and Crude Fat Contents of Selected Species | 45 |
| 3.4 | Fire in Fynbos and Forest Patches in Swartboskloof | 48 |
| 3.5 | The Effects of Invasion<br>by Alien Trees and Shrubs on Fuel Properties | 50 |
| 3.6 | Fuel in Swartboskloof and Other Ecosystems | 50 |
| 3.7 | Conclusions | 52 |
| **4** | **Regeneration Strategies in Fynbos Plants<br>and Their Influence on the Stability<br>of Community Boundaries After Fire**<br>*B.W. van Wilgen and G.G. Forsyth* (6 Figures) | 54 |
| 4.1 | Introduction | 54 |
| 4.2 | Survival Mechanisms and Vital Attributes of Species | 54 |
| 4.2.1 | Systems for the Classification of Species | 54 |
| 4.2.2 | Selection and Characterization of Species | 56 |
| 4.3 | Results and Discussion | 57 |
| 4.3.1 | Spectra of Fire Response and Vital Attribute Types | 57 |
| 4.3.2 | Age at Maturity | 58 |
| 4.3.3 | Longevity of Individuals and Seed Banks | 58 |
| 4.3.4 | The Timing of Critical Life History Events | 59 |
| 4.3.5 | Differences Between Communities | 63 |
| 4.3.6 | Fire and the Stability of Community Boundaries | 66 |

4.3.7 Comparison to Other Fire-Prone Shrublands ....... 69
4.3.8 Fire, Resilience and Dynamics in Fynbos ........... 70
Acknowledgements ................................. 71

**5 Is Fynbos a Stage in Succession to Forest? Analysis of the Perceived Ecological Distinction Between Two Communities**
*P.T. Manders, D.M. Richardson, and P.H. Masson* (13 Figures) ............................. 81

5.1 Introduction ................................. 81
5.2 Fire and the Stability of Forest/Fynbos Boundaries .. 84
5.2.1 The Effects of Fire on Forest Boundaries in Swartboskloof ............................. 84
5.2.2 Soil Seed Banks ............................. 85
5.3 Colonization of Fynbos by Forest Species .......... 86
5.3.1 Post-Fire Seed Deposition ..................... 86
5.3.2 Germination Requirements ...................... 89
5.3.3 Development of Forest in Fynbos Between Fires .... 92
5.4 Factors Influencing Forest Development ........... 95
5.4.1 The Role of Nutrients in Forest Development ...... 96
5.4.2 Soil Moisture ............................... 98
5.4.3 Historical Factors ............................ 103
5.5 Synthesis ................................... 104
5.5.1 Fynbos Succession ............................ 104
5.5.2 Determinants of the Rate of Succession ........... 106
Acknowledgements ................................. 107

**6 Coexistence of Seeders and Sprouters in a Fire-Prone Environment: the Role of Ecophysiology and Soil Moisture**
*R.E. Smith, B.W. van Wilgen, G.G. Forsyth, and D.M. Richardson* (4 Figures) ............... 108

6.1 Introduction ................................. 108
6.2 Ecophysiological Characteristics of Seeders and Sprouters in Swartboskloof ........ 110
6.3 Comparison of Swartboskloof with Other Regions .. 116
6.3.1 Other Fynbos Areas .......................... 116
6.3.2 Californian Chaparral ......................... 116
6.3.3 Australian Kwongan .......................... 118
6.4 Community Composition Along a Moisture Gradient .................... 119
6.5 Conclusions.................................. 121
Acknowledgements ................................. 122

**7 The Relative Advantages of Seeding and Sprouting in Fire-Prone Environments: a Comparison of Life Histories of *Protea neriifolia* and *Protea nitida***
*D.C. Le Maitre* (4 Figures) ..................... 123

7.1 Introduction ..................... 123
7.1.1 Evolution of Seed Regeneration in Mediterranean Shrublands ..................... 125
7.1.2 Trade-Offs Linked to Seeding and Sprouting ...... 126
7.1.3 Hypotheses ..................... 126
7.2 Population Growth after Fire ..................... 128
7.3 Mortality Rates ..................... 129
7.3.1 *Protea neriifolia* ..................... 129
7.3.2 *Protea nitida* ..................... 130
7.4 Age at Maturity ..................... 135
7.5 Seed Production and Seed Banks ..................... 135
7.5.1 *Protea neriifolia* ..................... 135
7.5.2 *Protea nitida* ..................... 136
7.6 Seed Germination ..................... 138
7.7 Recruitment Between Fires ..................... 139
7.8 Effects of Fire Frequency ..................... 141
7.9 Coexistence of Seeders and Sprouters ..................... 142
7.10 Implications for Management ..................... 143
7.11 Conclusions ..................... 143
Acknowledgements ..................... 144

**8 Life Cycles and Fire-Stimulated Flowering in Geophytes**
*D.C. Le Maitre and P.J. Brown* (5 Figures) ....... 145

8.1 Introduction ..................... 145
8.1.1 Factors that Stimulate Flowering After Fires ....... 145
8.1.2 Factors Favouring the Evolution of Fire-Stimulated Flowering ..................... 148
8.2 The Life Cycle of *Watsonia borbonica* (Iridaceae) .. 149
8.3 The Life Cycle of *Cyrtanthus ventricosus* (Amaryllidaceae) ........ 152
8.4 Fire Season and the Flowering Response of *Watsonia borbonica* ..................... 153
8.5 Proximate Cues for Flowering ..................... 154
8.6 Factors Favouring the Evolution of Fire-Stimulated Flowering ..................... 156
8.7 Geophyte Life Cycles and the Type of Flowering Response ............. 157
8.8 Effect of Frequent Fires on *Watsonia borbonica* .... 158
8.9 Conclusions ..................... 159
Acknowledgements ..................... 160

**9** **Why Is Mountain Fynbos Invasible and Which Species Invade?**
*D.M. Richardson and R.M. Cowling* (5 Figures) ... 161

9.1 Introduction ... 161
9.2 What Makes Terrestrial Plant Communities Invasible? ... 164
9.3 Plant Invasions in Mountain Fynbos: Why and How? ... 165
9.3.1 Invasions by Trees and Shrubs ... 165
9.3.2 Why is Mountain Fynbos Invasible? ... 167
9.3.3 How is Mountain Fynbos Invaded? ... 172
9.4 Which Species Invade? ... 174
9.4.1 Describing a Good Invader ... 174
9.4.2 Those That Have Made It: Pines and Hakeas ... 175
9.5 Conclusions ... 179
Acknowledgements ... 181

**10** **Do Small Mammals Govern Vegetation Recovery After Fire in Fynbos?**
*H.J. van Hensbergen, S.A. Botha, G.G. Forsyth, and D.C. Le Maitre* (8 Figures) ... 182

10.1 Introduction ... 182
10.2 Methods ... 184
10.2.1 Factors Affecting the Distribution and Size of Small Mammal Populations ... 184
10.2.2 Vegetation Factors Affecting Habitat Selection by Small Mammals ... 185
10.2.3 The Effect of Small Mammals on Plant Regeneration ... 186
10.2.3.1 Factors Affecting Plant Establishment in Mature Fynbos ... 186
10.2.3.2 Factors Affecting Plant Establishment After Fire in Fynbos ... 187
10.3 Results ... 189
10.3.1 Post-Fire Development of Small Mammal Communities ... 189
10.3.2 Vegetation Factors Affecting Habitat Selection by Small Mammals ... 189
10.3.2.1 The Relationship Between Floristic Communities and Small Mammal Occurrence ... 189
10.3.2.2 The Relationship Between Structure of the Vegetation and Small Mammal Densities ... 190
10.3.2.3 Survival and Recolonization Strategies of Small Mammals ... 191

10.3.3 The Effect of Small Mammals on Plant Regeneration .......................... 196
10.3.3.1 Factors Affecting Plant Establishment in Mature Fynbos .......................... 196
10.3.3.2 Factors Affecting Plant Establishment in Burnt Fynbos .......................... 197
10.4 Discussion .......................... 199
10.4.1 Post-Fire Development of Small Mammal Communities .......................... 199
10.4.2 Factors Affecting Habitat Selection by Small Mammals .......................... 200
10.4.3 Do Small Mammals Limit *Protea* Regeneration from Seed? .......................... 200
10.5 Conclusions .......................... 202
Acknowledgements .......................... 202

**11 The Influence of Fire, Vegetation Age and Argentine Ants (*Iridomyrmex humilis*) on Ant Communities in Swartboskloof**
*A.E. de Kock, J.H. Giliomee, K.L. Pringle, and J.D. Majer* (4 Figures) .......................... 203

11.1 Introduction .......................... 203
11.2 Methods .......................... 204
11.2.1 Locality of Study Sites .......................... 204
11.2.2 Extent and Rate of Invasion by Argentine Ants in Swartboskloof .......................... 205
11.2.3 Ant Surveys .......................... 205
11.2.4 Data Analysis .......................... 206
11.3 Results and Discussion .......................... 207
11.3.1 Extent and Rate of Invasion by Argentine Ants in Swartboskloof .......................... 207
11.3.2 Ant Species Collected .......................... 208
11.3.3 Similarity of Ant Species Composition .......................... 211
11.3.4 Ant Diversity Patterns .......................... 213
11.3.5 Impact of Fire .......................... 214
Acknowledgements .......................... 215

**12 The Effects of Fire on Soil Water Repellency, Catchment Sediment Yields and Streamflow**
*D.F. Scott and D.B. van Wyk* (6 Figures) .......................... 216

12.1 Introduction .......................... 216
12.2 Methods and Results .......................... 217
12.2.1 Studies at the Whole Catchment Level .......................... 217

12.2.1.1 Effects of Fire on Streamflow .................... 217
12.2.1.2 Effects of Fire on Catchment Sediment Yields ..... 221
12.2.2 Surface Runoff Plots ............................ 222
12.2.3 Water-Repellent Soils ........................... 226
12.3 Discussion ..................................... 228
12.3.1 Studies at the Whole Catchment Level ............ 228
12.3.1.1 Effects of Fire on Streamflow .................... 228
12.3.1.2 Effects of Fire on Catchment Sediment Yields ..... 229
12.3.2 Surface Runoff Plots ............................ 229
12.3.3 Water-Repellent Soils ........................... 230
12.3.4 Integration of Catchment, Plot and Profile Level Studies ................... 232
12.4 Conclusions .................................. 233
Acknowledgements ..................................... 233

**13 Fire and Catchment Chemical Budgets**
*D.B. van Wyk, W. Lesch, and W.D. Stock*
(3 Figures) ..................................... 240

13.1 Introduction ................................... 240
13.2 Approach and Methods ......................... 241
13.2.1 Data Collection and Analysis .................... 241
13.2.2 Statistical Treatment of Data .................... 242
13.3 Atmospheric Deposition ........................ 244
13.4 Stream Water Chemistry ........................ 245
13.5 Mineral Nutrient Budgets ........................ 251
13.6 The Effect of Fire ............................. 251
13.7 Conclusions .................................. 257
Acknowledgements ..................................... 257

**14 Soil Microorganisms and Activities in Relation to Season, Soil Factors and Fire**
*C.A. Van Reenen, G.J. Visser, and M.A. Loos*
(4 Figures) ..................................... 258

14.1 Introduction ................................... 258
14.2 Soil Chemical, Physical and Microbiological Investigations ............... 259
14.2.1 Experimental Soils ............................ 259
14.2.2 Soil pH ........................................ 259
14.2.3 Soil Moisture ................................. 260
14.2.4 Soil Microbial Respiratory Activity ............... 261
14.2.5 Soil Microbial ATP ............................ 262
14.2.6 Fungal, Actinomycete and Non-Actinomycete Bacterial Biomass ............................ 262

14.2.7 Post-Fire Soil Populations of Fungi and Bacteria by Plate Count ... 263
14.2.8 Post-Fire Soil and Rhizosphere Populations of Bacteria and Apparent Nitrogen-Fixing Bacteria by Plate Count ... 264
14.3 Effects of Season ... 265
14.4 Effects of Soil Type and Depth ... 267
14.5 Effects of Fire ... 268
14.6 Rhizosphere Stimulation of Nitrogen-Fixing Bacteria in Post-Fire Soils in Relation to General Rhizosphere Effect ... 268
14.7 Intercontinental Comparisons ... 269
14.7.1 Microbial Activity and Biomass in Relation to Seasonal Factors ... 269
14.7.2 Microbial Activity and Biomass in Relation to Soil Factors ... 270
14.7.3 Effects of Fire on Soil Microorganisms and Their Activities ... 270
14.7.4 Nitrogen-Fixing Bacteria in Mediterranean Regions in Relation to Fire ... 271
14.8 Conclusions ... 272
Acknowledgements ... 272

**15 Ecosystem, Community and Species Response to Fire in Mountain Fynbos: Conclusions from the Swartboskloof Experiment**
*D.M. Richardson and B.W. van Wilgen* (2 Figures) . 273

15.1 Introduction ... 273
15.2 Population-Level Effects of Fire ... 275
15.3 Community-Level Effects of Fire ... 277
15.4 Ecosystem-Level Effects of Fire ... 280
15.5 Managing Fire in Mountain Fynbos ... 283

**References** ... 285

**Subject Index** ... 315

# List of Contributors

*Botha, S.A.*
Forestry Branch, Natal Forest Region, Private Bag X9029, Pietermaritzburg, 3200, South Africa

*Brown, P.J.*
CSIR Division of Forest Science and Technology, P.O. Box 395, Pretoria 0001, South Africa

*Chapman, R.A.*
CSIR Division of Forest Science and Technology, Jonkershoek Forestry Research Centre, Private Bag X5011, Stellenbosch 7600, South Africa

*Cowling, R.M.*
Botany Department, University of Cape Town, Rondebosch 7700, South Africa

*De Kock, A.E.*
CSIR Division of Forest Science and Technology, P.O. Box 395, Pretoria 0001, South Africa

*Forsyth, G.G.*
CSIR Division of Forest Science and Technology, Jonkershoek Forestry Research Centre, Private Bag X5011, Stellenbosch 7600, South Africa

*Giliomee, J.H.*
Department of Entomology and Nematology, University of Stellenbosch, Stellenbosch 7600, South Africa

*Kruger, F.J.*
CSIR Division of Forest Science and Technology, P.O. Box 395, Pretoria 0001, South Africa

*Le Maitre, D.C.*
CSIR Division of Forest Science and Technology, Jonkershoek Forestry Research Centre, Private Bag X5011, Stellenbosch 7600, South Africa

*Lesch, W.*
CSIR Division of Forest Science and Technology, Jonkershoek Forestry Research Centre, Private Bag X5011, Stellenbosch 7600, South Africa

*Loos, M.A.*
Department of Microbiology, University of Stellenbosch, Stellenbosch 7600, South Africa

*Majer, J.D.*
School of Environmental Biology, Curtin University of Technology, P.O. Box U1987, Perth, 6000, Australia

*Manders, P.T.*
CSIR Division of Forest Science and Technology, P.O. Box 395, Pretoria 0001, South Africa

*Masson, P.H.*
Botany Department, University of Cape Town, Rondebosch 7700, South Africa

*McDonald, D.J.*
National Botanical Institute, P.O. Box 471, Stellenbosch 7600, South Africa

*Pringle, K.L.*
Department of Entomology and Nematology, University of Stellenbosch, Stellenbosch 7600, South Africa

*Richardson, D.M.*
CSIR Division of Forest Science and Technology, Jonkershoek Forestry Research Centre, Private Bag X5011, Stellenbosch 7600, South Africa

*Scott, D.F.*
CSIR Division of Forest Science and Technology, Jonkershoek Forestry Research Centre, Private Bag X5011, Stellenbosch 7600, South Africa

*Smith, R.E.*
CSIR Division of Forest Science and Technology, P.O. Box 395, Pretoria 0001, South Africa

*Stock, W.D.*
Botany Department, University of Cape Town, Rondebosch 7700, South Africa

*van Hensbergen, H.J.*
Department of Nature Conservation, Forestry Faculty, University of Stellenbosch, Stellenbosch 7600, South Africa

*van Reenen, C.A.*
Department of Microbiology, University of Stellenbosch, Stellenbosch 7600, South Africa

*van Wilgen, B.W.*
CSIR Division of Forest Science and Technology, Jonkershoek Forestry Research Centre, Private Bag X5011, Stellenbosch 7600, South Africa

*van Wyk, D.B.*
CSIR Division of Forest Science and Technology, Jonkershoek Forestry Research Centre, Private Bag X5011, Stellenbosch 7600, South Africa

*Versfeld, D.B.*
CSIR Division of Forest Science and Technology, Jonkershoek Forestry Research Centre, Private Bag X5011, Stellenbosch 7600, South Africa

*Visser, G.J.*
Department of Microbiology, University of Stellenbosch, Stellenbosch 7600, South Africa

# 1 The Swartboskloof Experimental Site

B.W. van Wilgen and D.J. McDonald

## 1.1 Introduction

Swartboskloof was chosen as one of a network of research sites within the fynbos biome, at which detailed ecological studies were to be undertaken. The research was conducted under the auspices of the Fynbos Biome Project, an enterprise within the National Programme for Environmental Sciences (Jarman 1982; Huntley 1987). The Fynbos Biome Project was initiated in 1979 (Day et al. 1979), and aimed to support interdisciplinary projects that included research designed to meet local needs as well to provide contributions to the international programme of SCOPE (Scientific Committee On Problems of the Environment).

The choice of Swartboskloof as a primary research site was influenced by a number of factors. First, the catchment fell within the Jonkershoek valley, where hydrological and ecological research had been conducted for over 50 years (Bands et al. 1987). This meant that baseline data, particularly pertaining to climate and hydrology, were readily available. In addition, the biota was well documented, and a comprehensive reference herbarium was located on site. The area was to be burnt as part of the normal management of fynbos catchment areas, and this provided the opportunity for research to focus on the effects of the planned fire. Finally, the site was readily accessible to researchers from nearby universities.

In this chapter, we describe the physical and biological features of the ecosystem, to provide the background for the research results presented in subsequent chapters. We summarize the history of land use, and of fire, that would have influenced the current ecosystem. We describe the experimental design of the main studies that were conducted to examine the effects of the planned fire, and give an account of conditions which prevailed during the fire of March 1987.

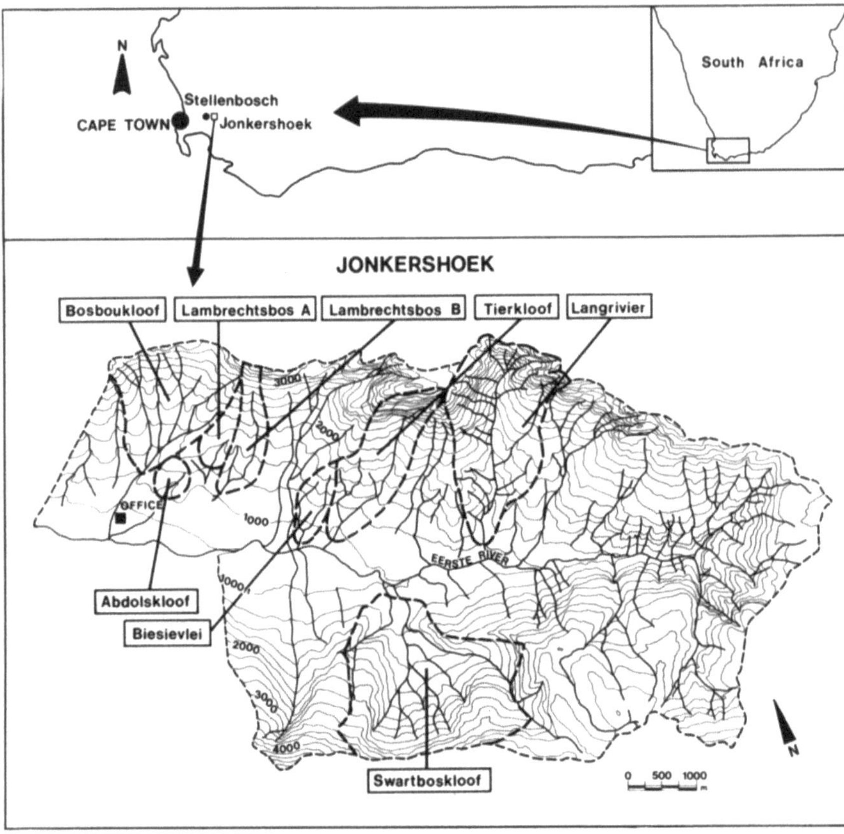


**Fig. 1.1.** The Swartboskloof catchment is situated in the Jonkershoek valley about 15 km east of the town of Stellenbosch. Other research catchments located in the valley, and the Jonkershoek Forestry Research Centre Office are shown on the map. Contour intervals are in feet (1 foot = 0.3 m)

## 1.2 Location and Physical Site Description

Swartboskloof (Swartboschkloof on some maps – both are Afrikaans for "black forest gorge") is situated in the Hottentot's Holland Nature Reserve, about 15 km from the town of Stellenbosch in the Cape Province. The area is centred on 33° 57′ S and 18° 55′ E and is 373 ha in extent (Fig. 1.1). The climate is characterized by cool, wet winters and a warm, dry summer (see Chap. 2 for a detailed description of the climate of the valley).

Swartboskloof is a fan-shaped valley (Fig. 1.2), ranging from 285 to 1200 m a.s.l. and has an equatorial (north-facing) aspect. Slopes range from less than 5° to 45°, with the steep slopes averaging 30°. About 2% of the area consists of almost vertical cliffs. A detailed account of the geology of

**Fig. 1.2.** View of the Swartboskloof catchment from the opposite side of the Jonkershoek valley. Dark vegetation on the *right* and in the *foreground* are plantations of *Pinus radiata*

the area is given by Söhnge (1988). Porphyritic Cape Granite (590 to 525 Ma) underlies the quartzitic sandstones of the Peninsula Formation of the Table Mountain Group (Cape Supergroup, 450 to 430 Ma) which form the rugged mountains at Jonkershoek. The granite forms the undulating floor of the Swartboskloof valley. The sandstone cliffs do not erode readily but debris from these cliffs, together with granite boulders, form a colluvial mixture covering most parts of the valley floor. Apart from the steep cliffs and colluvial slopes, there are also loose sandstone boulder screes. The two main streams in Swartboskloof (Jubilee Creek and the Swartboskloof stream; Fig. 1.3) follow fault lines and join at the lower end of the valley. The large area of scree and talus surface deposits in Swartboskloof aid the rapid infiltration of surface water. In addition, extreme fracturing and faulting allows for the conveyance of large amounts of water (Söhnge 1988).

Soils in Swartboskloof reflect the nature of the parent materials as well as the rugged topography of the site. The soils are acid, with the exchange complex dominated by exchangeable hydrogen and aluminium. The cation exchange capacities are generally low. The soils derived from quartzite tend to be coarser than those derived from granite. Quartzite-derived soils are also lower in organic matter, nitrogen, pH, CEC, and total phosphorus than granite soils. Podzolization has occurred in some areas, but no dark brown illuvial horizons have developed in the lower solum. Orthic, humic and organic surface horizons have been identified; organic and humic surface

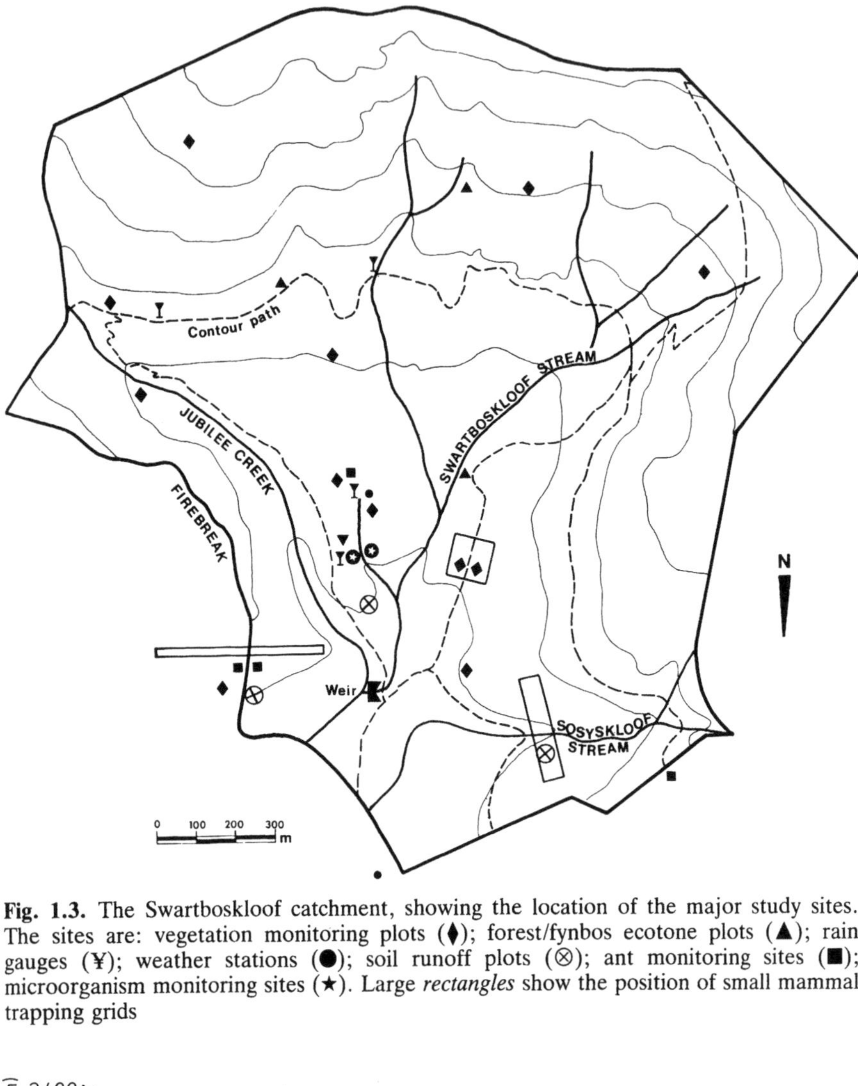


**Fig. 1.3.** The Swartboskloof catchment, showing the location of the major study sites. The sites are: vegetation monitoring plots (⧫); forest/fynbos ecotone plots (▲); rain gauges (¥); weather stations (●); soil runoff plots (⊗); ant monitoring sites (■); microorganism monitoring sites (★). Large *rectangles* show the position of small mammal trapping grids

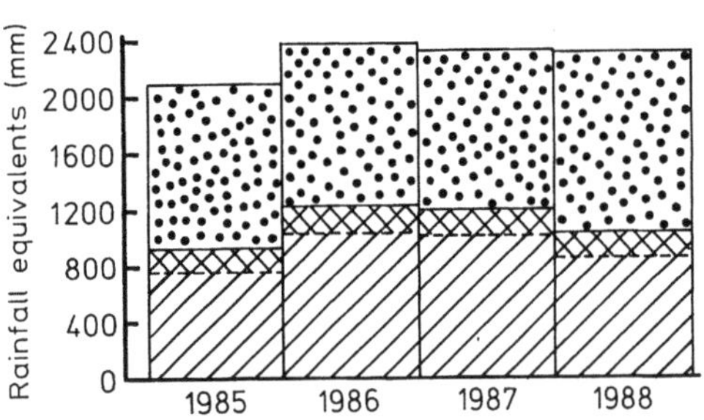


**Fig. 1.4.** The estimated catchment water balance for Swartboskloof from 1985 to 1988. Rainfall is the total of the three shading classes; stormflow (*hatched shading*) is shown as a proportion of total streamflow (*hatched plus striped shading*)

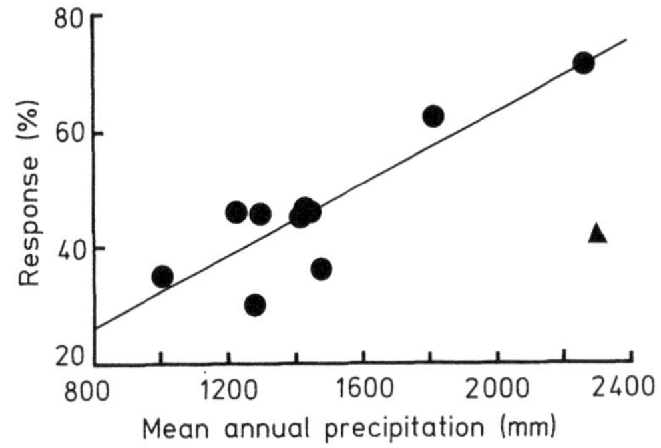


**Fig. 1.5.** Annual catchment response (mean annual runoff/mean annual precipitation) in relation to annual precipitation for 11 catchments in the Jonkershoek valley. Swartboskloof, shown as a *triangle*, is well below the general trend in response

**Table 1.1.** Summary of hydrological features of two gauged catchments in the Jonkershoek valley

| Catchment | Relief (m) | Mean annual precipitation (MAP; mm) | Mean annual runoff (MAR; mm) | Response (MAP/MAR) (%) |
|---|---|---|---|---|
| Swartboskloof | 833 | 2270 | 1060 | 46.7 |
| Tierkloof | 1230 | 1810 | 1128 | 62 |

horizons were associated with finer-textured parent material. Fry (1987) described 11 soil forms and 16 soil series at Swartboskloof, based on the system of MacVicar et al. (1977). The mosaic of soil types in the montane environment is fine, and soils change rapidly from one site to another. This makes the identification of soil-plant community relationships difficult (Fry 1987; but see Chap. 5, Fig. 5.6).

Hydrological research at Jonkershoek has concentrated on the effects of afforestation on water yield and quality. The experimental approach was one of multiple watershed experiments. Details of the experimental design are given by Wicht (1948a, 1967a). The effects of fire on the hydrological characteristics of Swartboskloof are described in Chapters 12 and 13, using the Tierkloof catchment (Fig. 1.1) as a control. Tierkloof, on the opposite side of the Jonkershoek valley, is afforrested with *Pinus radiata*, and has a lower rainfall, and higher runoff to rainfall ratio, than Swartboskloof (Table 1.1). A relatively low proportion (18.2%) of the runoff at Swartboskloof is in the form of quickflow (Fig. 1.4), due to the high infiltration rates, deep soils and low intensity and duration of rainstorms. The relatively low runoff to rainfall ratio at Swartboskloof, compared to other gauged catchments in the Jonkershoek valley (Fig. 1.5) may be due to unsorted colluvial and

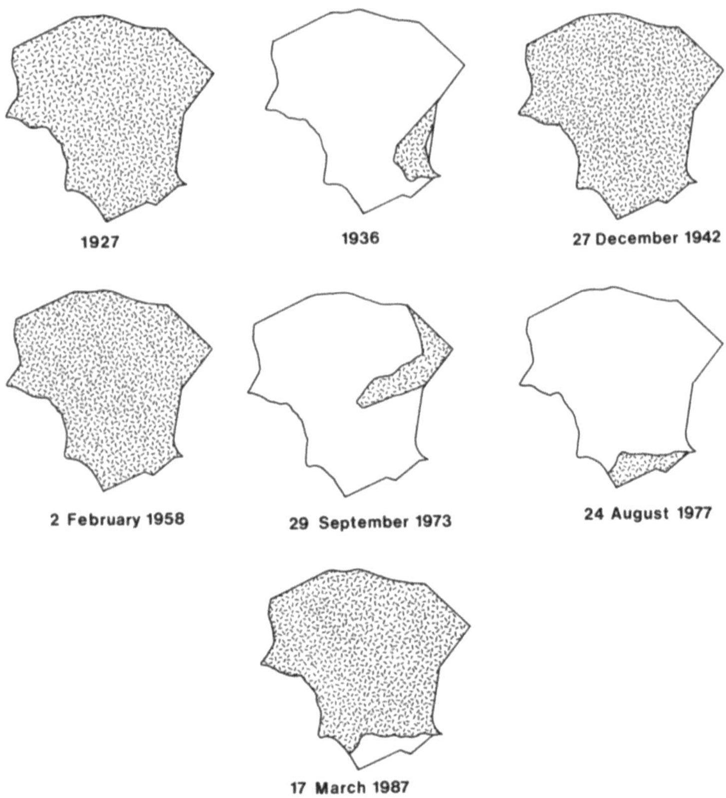


**Fig. 1.6.** Fire history of the Swartboskloof catchment. *Shaded areas* show the extend of fires. The boundaries of the area are as in Fig. 1.3

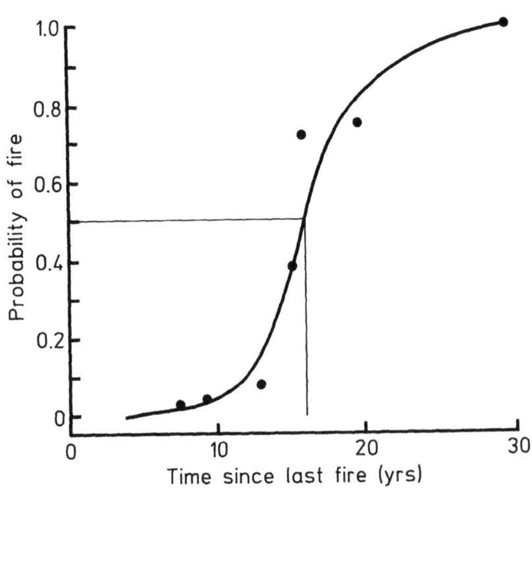


**Fig. 1.7.** The cumulative probability of fire at Swartboskloof, calculated from data shown in Fig. 1.6

alluvial boulder and rock deposits close to the weir. This may cause an unaccounted leakage from the catchment, introducing an error into the calculation of catchment water balance. Secondly, Swartboskloof is on a hotter (equatorial) aspect, which would increase evaporation from the catchment.

Major fires, in which the entire Swartboskloof catchment was burnt, occurred in 1927, 1942 and 1958. Smaller fires burnt portions of the area in 1936, 1973 and 1977 (Fig. 1.6). All of these fires, except the prescribed burn in 1977, were accidental. The 1987 prescribed burn covered the entire area except the portion burnt in 1977. A grid representing $100 \times 100$ m was overlaid on the fire maps (Fig. 1.6), and the interval between fires was assessed at each of the 291 points. The maximum period between fires at any one point was 29 years while the minimum was 7 years; the cumulative probability of fire (Fig. 1.7) was assessed from the data, using the method of Brown et al. (1991). The mean fire return period for the catchment was 16 years (Fig. 1.7).

## 1.3 Land-Use History

Early occupants of the upper Eerste River Valley manufactured hand axes some 70 000 years ago, and probably also used fire (Klein 1974). The earliest occupants of the area were probably San hunter-gatherers. The San would have been superseded by the Khoi, who were predominantly nomadic herdsmen. Swartboskloof may conceivably have been on the cyclical route of nomadic herdsmen for 1000 years or more (Schweitzer and Scott 1973). San hunter-gatherers may have coexisted in the valley with the Khoi, but had certainly been in the area for a much longer period (Klein 1974). A number of stone artifacts, typical of San craft, have been found on the lower mountain slopes and in over-hang shelters in the valley (D.P. Bands, pers. comm.). The San probably utilized the bulbs of various Iridaceae, including those of *Watsonia* (Klein 1974) as food, and may have deliberately burnt the veld in order to stimulate reproduction and spread of Iridaceae over the past several thousand years (Deacon 1972).

After the first European families settled in the vicinity, it is likely that the Khoi were forced to move their traditional grazing lands. The larger wild animals also very soon disappeared (Bigalke 1979a). Within a few years, all suitable grazing land was probably leased out as "vee lecht plaatzen". The farm diagrams drawn by surveyors in the early 19th century still denote virtually the whole of the upper Eerste Rivier valley below the rocky cliffs as "grazing ground". On 15 October 1692 freehold title to three portions of land totalling just over 29 morgen (ca. 25 ha) on the upper reaches of the Eerste Rivier, was granted to Jan Andriessen, also known as Jan De Jonker (Surveyor General's records). The valley up-stream from the property became known as "Jonkers Hoek". Burning and grazing probably continued, although no records exist to show this.

The uppermost portion of the Jonkershoek valley, about 3400 ha, was granted to Christian Ackerman in 1832 (Surveyor General's records). In 1833, about 425 ha of this area (including Swartboskloof) was excised and transferred to Christian Ackerman's son Christiaan (Surveyor General's records). Again, it is probable that this property was regularly burnt for grazing.

Between 1843 and 1858 Jan de Jonker's freehold property and the surrounding land passed to P.G. Neethling, and the Watermeyer family acquired the Ackerman properties in 1844 (Surveyor General's records). The Municipality of Stellenbosch acquired two small portions of land on the river from the Watermeyers in 1911 on which they built an intake weir on the Eerste river, just below the Swartboskloof stream, for a municipal water supply scheme (Surveyor General's records). In 1932 the Watermeyers sold the remainder of their Jonkershoek property to the Municipality of Stellenbosch and took over the adjoining consolidated property from the Neethlings.

Stellenbosch Municipality adopted a policy of protecting the veld from fire, and when the area (including Swartboskloof) was transferred to the Government Forestry Department in March 1934, this policy was retained. Swartboskloof has been managed as a nature reserve since then. The Swartboskloof area was proclaimed a nature reserve in 1935, and was incorporated into the larger Hottentot's-Holland Nature Reserve (24 569 ha) in 1979.

A forest station was established at Jonkershoek in 1934, during a period of economic depression. The Jonkershoek afforestation programme was one of a number of government schemes that provided work for the unemployed. Afforestation (with *Pinus radiata*) commenced in 1935. By 1936 a system of firebreaks had been established around the perimeter with cutoff breaks across the valley.

The issue of forestry and water supplies was thoroughly discussed at the Fourth Empire Forestry Conference hosted by South Africa in 1935. Plans for investigating "the influences of forests on water conservation and allied problems" (1935/36 Annual Report of the Division of Forestry) were endorsed and reinforced by the Conference. Towards the end of 1935 a research station was established at Jonkershoek "to determine how normal afforestation, as carried out in State plantations, would affect climate, water conservation and erosion" (Wicht 1939). This broad objective was subsequently considered to be too ambitious and "research was restricted mainly to the determination of the effects of afforestation on water conservation" (Wicht 1948a). The tasks of establishing the research station, and designing and initiating the research programme, were entrusted to Dr. C.L. Wicht.

Dr. Wicht spoke of forest-influences research. The central idea behind these investigations was to determine how modification of the natural vegetation by different experimental treatments "such as afforestation, veld burning, . . . and complete protection" would affect "the natural circulation of water" (Wicht 1939). The experimental design was based on the paired-catchment principle as used at Emmental in Switzerland and Wagon Wheel Gap in Colorado, USA. However, because no two of the eight catchments identified for experimental work at Jonkershoek were even approximately comparable, a novel approach was required. Dr. Wicht describes this approach as follows: "Each stream is to be studied independently and compared with itself before and after treatment. In each case all factors which might influence streamflow will be observed and correlated. It is hoped that such analysis will disclose general trends common to all catchments. On the basis of these trends it may then be possible to generalise as to the effects of afforestation on streamflow in the winter rainfall region" (Wicht 1939).

A network of rain gauges was installed, soil and vegetation surveys were completed and work commenced on the building of stream-gauging weirs. In the 1936/37 Annual Report of the Division of Forestry, a year after initiation of activity at Jonkershoek, it is reported that readings from 12 rain gauges had been recorded during the year and that the herbarium collection, initiated to facilitate vegetation surveys, already contained 556 specimens. Construction of streamflow-gauging weirs on the Bosboukloof and Biesievlei streams (Fig. 1.1) was completed and recording commenced in 1937.

From the start, ecological studies formed an important part of the work at Jonkershoek. In the 1945/46 Departmental Annual Report, Dr. Wicht writes: "Interesting results at Jonkershoek on the distribution of species indicate that the usual sampling methods applied elsewhere are not adequate for the study of the extremely complex and variable vegetation of the south-western Cape". The first experiment to test the effects of burning season on fynbos was laid out at Jonkershoek in 1945 (Wicht 1948b; Le Maitre 1987a). This experiment also constituted a first attempt to apply statistical design to an investigation of the ecology of the Cape fynbos. Recognizing the need to base catchment management on sound ecological principles, the research programme gained an increased emphasis on ecology from the mid-1960s onwards.

The decision to burn Swartboskloof was made early in 1984. At the same time it was decided to designate the site as one of several at which interdisciplinary ecosystems research would be concentrated (van Wilgen 1986). Although ecological research had been carried out at a low level in the area since the mid-1960s, this was stepped up in 1985. A stream-gauging weir was completed in February 1985, and the site was burnt on 17 March 1987.

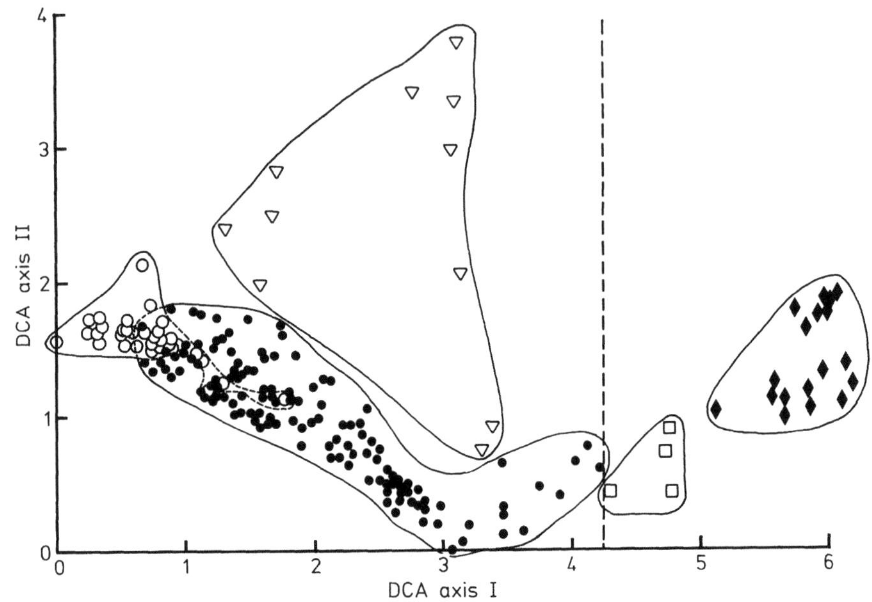


**Fig. 1.8.** Ordination (by Detrended Correspondence Analysis) of 101 vegetation plots in Swartboskloof. The vegetation groups are: (▽), hygrophilous communities; (●), proteoid shrublands; (○), ericoid-restioid shrublands; (□), tall forest; and (⧫), short forest. Data are from McDonald (1987)

## 1.4 The Vegetation of Swartboskloof

The vegetation of Swartboskloof is divided into two main types: mesic mountain fynbos (sensu Moll et al. 1984) and forest communities. The mesic mountain fynbos (one of several subdivisions of fynbos, Moll et al. 1984) is the most extensive type and is found on drier situations than forest. The forests in Swartboskloof are restricted to the riparian zone of the perennial streams and to boulder screes below the cliffs. These forests are Afromontane (sensu White 1978) in floristic composition, but are poor in species. McDonald and Morley (1988) recorded 651 species of flowering plants and ferns in the Swartboskloof valley.

Phytosociological studies, using the Braun-Blanquet vegetation survey method, were conducted in Swartboskloof by Werger et al. (1972) and McDonald (1988). In addition, van der Merwe (1966) analyzed the fire life-forms of all species occurring in the area. Based on McDonald's (1988) community descriptions, three fynbos and two forest community groups are recognized in Swartboskloof (Figs. 1.8 and 1.9). These are described below, and their distribution is shown in Fig. 1.10.

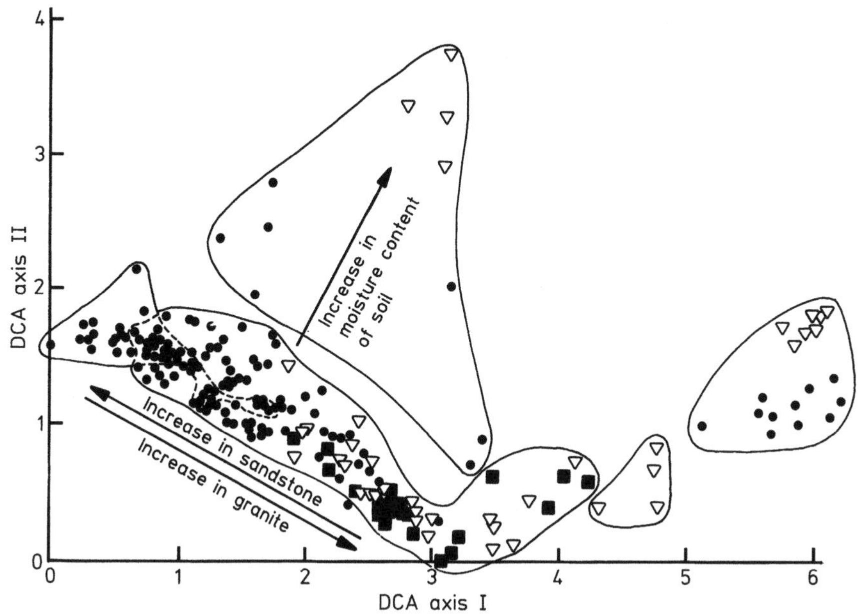


**Fig. 1.9.** Ordination (by Detrended Correspondence Analysis) of 101 vegetation plots in Swartboskloof, with the soil parent material superimposed on the scatter of points. The parent materials are: (●) sandstone-derived soil; (▽) colluvial soil; (■) granite-derived soil. *Arrows* show interpretations of environmental trends. Data are from McDonald (1987)

## 1.4.1 Specialized Hygrophilous Communities (Fig. 1.11)

These communities are found on permanently waterlogged, wet, phreatic sites or seepage zones. Some representative communities of this type are found in the upper reaches of Swartboskloof where there is perennial free water but where forest communities have not developed. In the bottomland situations where they occur, perennial free water overrides the influence of the predominantly sandstone-derived substrate. Young individuals of forest tree species such as *Cunonia capensis* (Cunoniaceae) have been recorded at these sites but no forest development has occurred. The projected canopy cover of these communities exceeds 90%. There are usually two distinct strata: an upper stratum (to 2 m) dominated by the shrub *Berzelia lanuginosa* (Bruniaceae), the reed *Elegia capensis* (Restionaceae) and the swamp daisy *Osmitopsis asteriscoides* (Asteraceae), and a lower stratum (0.5–1 m) dominated by reeds or sedges, notably *Elegia asperiflora* (Restionaceae) and *Tetraria fasciata* (Cyperaceae), or in certain situations overwhelmingly by the grass *Merxmuellera cincta* (Poaceae).

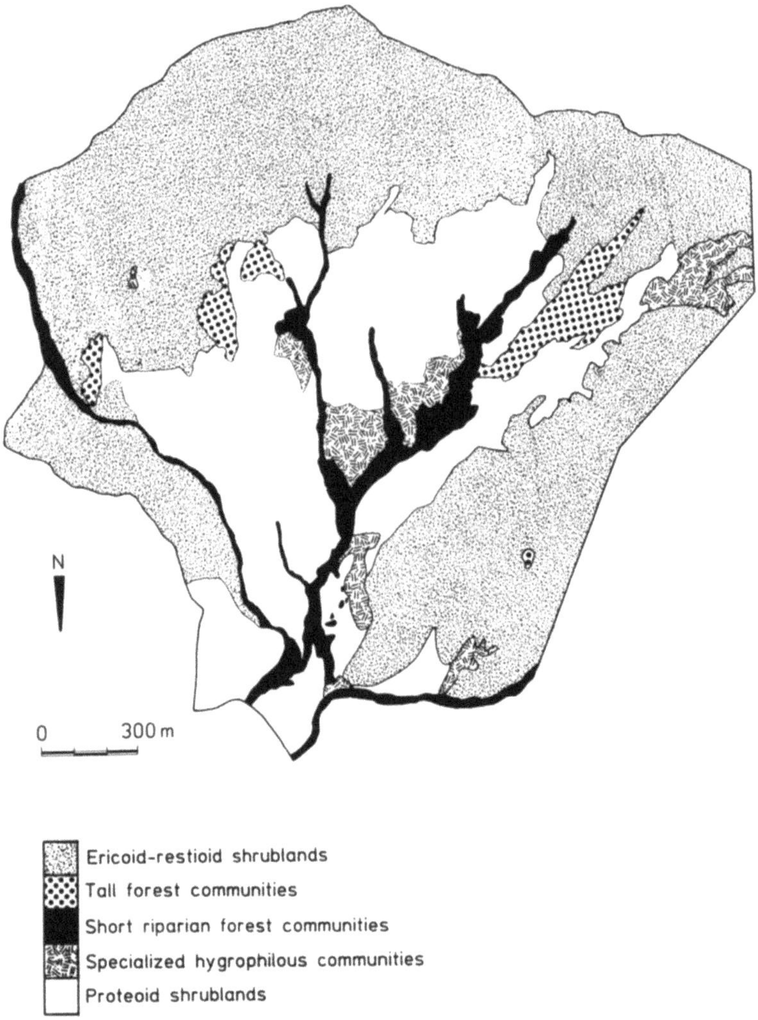


**Fig. 1.10.** Map of Swartboskloof showing the distribution of five major vegetation groupings

## 1.4.2 Short to Tall Proteoid Shrublands (Fig. 1.12)

These shrublands are found mainly in the mid-central part of Swartboskloof (330–700 m), on both sandstone- and granite-derived soils. The common structural-floristic factor linking these shrublands is the presence of *Protea* shrubs dominating the upper strata. *Protea neriifolia* (Proteaceae) is the most prevalent canopy species, but in certain circumstances *P. repens* is co-dominant. Where there are more rocks in the soil profile there is a

**Fig. 1.11.** An example of a specialized hygrophilous community in the central Swartboskloof catchment. The large grass species in the *foreground* is *Merxmuellera cincta* (Photo: P.J. Brown)

**Fig. 1.12.** An example of proteoid shrublands in Swartboskloof. Large shrubs in the *background* are *Protea neriifolia*. The shrub in the *left foreground* is *Brunia nodiflora* (Photo: P.J. Brown)

**Fig. 1.13.** An example of ericoid-restioid shrublands in upper Swartboskloof. The community is dominated by Restionaceae, with several ericoid shrubs co-dominant (Photo: G.G. Forsyth)

shift in dominance to the waboom, *P. nitida*. Other shrubs, such as *Olea europaea* ssp. *africana* (Oleaceae), are sometimes co-dominant on granite-derived soils. Dominant shrubs (ca. 5 m tall when mature) have a canopy cover of 25–70%. A second stratum (1–2 m) is found in most stands, and often has a projected canopy cover exceeding that of the upper stratum. Dominant species include the shrubs *Rhus angustifolia* (Anacardiaceae), *Cliffortia cuneata* and *C. ruscifolia* (Rosaceae), *Brunia nodiflora* (Bruniaceae) and *Erica hispidula* (Ericaceae). The perennial geophyte *Aristea major* (Iridaceae) contributes strongly to the low stratum (<1 m) together with the grasses *Merxmuellera stricta* and *Cymbopogon marginatus* (Poaceae) and the cosmopolitan fern *Pteridium aquilinum* (Dennstaedtiaceae). *Ischyrolepis gaudichaudiana* (Restionaceae) is often important in the low stratum.

### 1.4.3 Ericoid-Restioid Shrublands (Fig. 1.13)

These shrublands are found between 420–960 m on slopes ranging from almost level to steep (38°): The soils are shallow, well-drained lithosols derived exclusively from sandstone. Rock cover varies from low to very high, but averages 44%, about twice the mean for the study area. Two or three strata are found in these communities with the low stratum (<0.5 m) having the highest projected canopy cover (mean of 77%) in all cases. The

**Fig. 1.14.** Short riparian forest along the lower reaches of the Swartboskloof stream. The dominant tree with the light foliage is *Brabejum stellatifolium* (Photo: P.J. Brown)

low stratum is dominated by restios and sedges, the most important of which are *Ischyrolepis sieberi* and *Hypodiscus albo-aristatus* (Restionaceae), and *Tetraria capillacea* and *T. involucrata* (Cyperaceae). The short (0.5–1 m) and tall strata (1–2 m) are dominated by shrubs such as *Brunia nodiflora* and *Nebelia paleacea* (Bruniaceae), *Erica hispidula* and *E. coccinea*, *Cliffortia cuneata* and *C. polygonifolia* and *Protea neriifolia*.

### 1.4.4 Short Riparian Forest Communities (Fig. 1.14)

These communities, which occur along drainage lines and streams in the lower reaches of Swartboskloof, are dominated by tall (3–6 m) *Brabejum stellatifolium* (Proteaceae) trees with a projected canopy cover of 97%. A stratum 1–3 m in height is found with an average projected canopy cover of 35% and dominated by the shrubs *Diospyros glabra* (Ebenaceae), *Halleria elliptica* (Scrophulariaceae) and *Rhus angustifolia*. The field layer (<1 m) is dominated by the fern *Blechnum australe* (Blechnaceae).

### 1.4.5 Tall Forests (Fig. 1.15)

The tall forest communities are found on stabilized boulder screes below the sandstone cliffs and in the riparian zone of the upper reaches of the streams.

**Fig. 1.15.** Tall forest in the upper reaches of the Swartboskloof stream. The community is dominated by *Cunonia capensis* and *Ilex mitis* (Photo: P.J. Brown)

Boulder screes show very little soil development. What soil there is amongst the boulders is humus which results from decayed leaf litter and other organic matter. Soils along the streams amount merely to accumulations of alluvial material. The canopy trees, mainly *Hartogiella schinoides* (Celastraceae), *Olinia ventosa* (Oliniaceae) and *Podocarpus elongatus* (Podocarpaceae) reach 8–16 m in height. The water-loving *Cunonia capensis* and *Ilex mitis* (Aquifoliaceae) are found along the streams but are absent from the boulder screes. The cover of canopy trees is about 80%. The sub-canopy is dominated by *Halleria lucida*, *Maytenus acuminata* (Celastraceae), *Podocarpus elongatus* and *Rapanea melanophloeos* (Myrsinaceae) and has an average canopy cover of 63%. Below the sub-canopy is a field layer (<1 m and covering 10%), with *Myrsiphyllum scandens* (Asparagaceae), *Knowltonia vesicatoria* (Ranunculaceae) and *Zantedeschia aethiopica* (Araceae) most common.

A short forest community (floristic variant of the tall forests) occurs on loose boulder screes and on rocky outcrops. It is found in drier situations than the tall forest communities, where drainage is good. A single tree stratum and a field layer are distinguished. The tree stratum (2–8 m) has a projected cover of 80% and is dominated by *Hartogiella schinoides*, *Heeria argentea* (Anacardiaceae), *Olinia ventosa* and *Maytenus acuminata*. Below this is a short stratum (<1 m) where *Podocarpus elongatus* often spreads in a prostrate fashion on the boulders. Other species in this stratum include *Aloe mitriformis* (Liliaceae), *Knowltonia vesicatoria* and the ferns *Blechnum australe* and *Rumohra adiantiformis* (Aspidiaceae). None of these is dominant and *Aloe mitriformis* is found where the canopy is open. *Secamone alpini* (Asclepiadaceae) is found from ground level into the sub-canopy climbing amongst other plants.

## 1.5 Fauna

The contribution of fauna to fynbos ecosystems, in terms of biomass, is low (Bigalke 1979b; Siegfried 1983; Schlettwein and Giliomee 1987), and endemism is infrequent among faunal groups in the fynbos (Bigalke 1979b). Mammal communities are dominated by smaller species, and include six mice, two molerats, two shrews, a hare, a porcupine, and a hyrax. Small carnivores include four Viverridae and two Mustelidae. Larger herbivores are rare, but include the baboon and three small antelope species. Leopards are rare visitors. Rodents are important granivores (see Chap. 10).

Sixty-seven bird species were recorded in Swartboskloof (Richardson and Fraser in press). The avifauna includes all six species endemic to fynbos: Cape Rockjumper (*Chaetops frenatus*), Victorin's Warbler (*Bradypterus victorini*), Orangebreasted Sunbird (*Nectarinia violacea*), Cape Sugarbird (*Promerops cafer*), Cape Siskin (*Serinus totta*) and Protea Canary (*Serinus leucopterus*). No alien birds were recorded. Only 28% of the species were resident. This is probably a reflection of the transient nature of resources such as insects, fruit and especially nectar. The forests supported a distinct assemblage of bird species, which differed from the shrublands and contained no bird species endemic to the Cape Province. The birds of the *Protea* shrublands are most characteristic of Swartboskloof. The species were mainly nectarivores, insectivores and generalist feeders. Granivores were rare, even immediately post-fire, and no specialized frugivores occurred. The nectar of three *Protea* species was considered very important in determining the composition of the avifauna. Among the herpetofauna, there are 10 frog species, and 35 reptile species (20 snakes and 15 lizards) (A. de Villiers, pers. comm.).

Several studies of the invertebrate fauna of the study site have been published. Ants play an important role in fynbos ecology, particularly in seed dispersal (Bond and Slingsby 1983). Several studies (De Kock 1990b; Donnelly and Giliomee 1985a,b) have revealed 45 ant species at the study site. Dominant species included *Anoplolepis custodiens* and *Pheidole capensis*, as well as the invasive Argentine ant (*Iridomyrmex humilis*). Schlettwein and Giliomee (1987) quantified biomass and community structure with special reference to ants, leafhoppers and grasshoppers at different ages after fire. The effects of insects on certain plant species have also received attention. Coetzee and Giliomee (1985) identified 91 insect species associated with the inflorescence of *Protea repens*. These species include important pollinators and seed predators. MacGillivray (1990) identified 37 hymenopteran insect visitors to the Ericaceae, most of which functioned as potential pollinators. In addition, some information on the effects of fire on the macroinvertebrate fauna in streams of Swartboskloof is available (Britton submitted).

## 1.6 Research Rationale and Design

The studies reported in this volume were mainly undertaken to document the effects of a single early autumn fire on the ecosystem. The research objectives, outlined in the Preface, were: (1) to document the regeneration and recolonization strategies of the biota; (2) to study the stability or otherwise of the boundaries between vegetation communities; and (3) to examine the effects of fire on resources, particularly mineral nutrients and water. Research on the effects of fire in fynbos ecosystems has led to the adoption of a policy of prescribed burning in the late summer – early autumn period, at intervals of 12 to 15 years (see van Wilgen et al. 1990a, 1991 for reviews). Most of the Swartboskloof catchment had been fire-free for a period of 29 years, and senescence was becoming apparent among large *Protea* shrubs. According to policy, the area was scheduled to be burnt in March 1987. Further postponement of the planned fire in order to augment pre-fire research was not considered feasible, in view of the adverse effects of long fire-free intervals (Bond 1980; van Wilgen 1982).

The studies reported in this volume were conducted at various locations within the Swartboskloof catchment (Fig. 1.3). Data from the Swartboskloof catchment were augmented by data from elsewhere in the Jonkershoek valley where appropriate. Gauged catchments elsewhere in the Jonkershoek valley (Fig. 1.1) were used as controls against which the responses to the fire in Swartboskloof were compared in Chapters 11 and 12.

## 1.7 The Fire of March 1987

The Swartboskloof fire was scheduled for the first day in March 1987 which was within the following prescriptions: (1) at least 3 days since last rain; (2) temperature and humidity in the range 15–25°C and 30–60% respectively; and (3) atmospheric stability with a wind speed $<4\,ms^{-1}$. These conditions were met on 17 March following a heavy rainshower of 27 mm on 14 March and relatively cool conditions on the 15th and 16th.

Swartboskloof was ignited at 08:50 on 17 March 1987. Progress with the burn was intentionally slow (Fig. 1.16), with teams working initially along the crest at high altitude, and then from prepared firebreaks and against

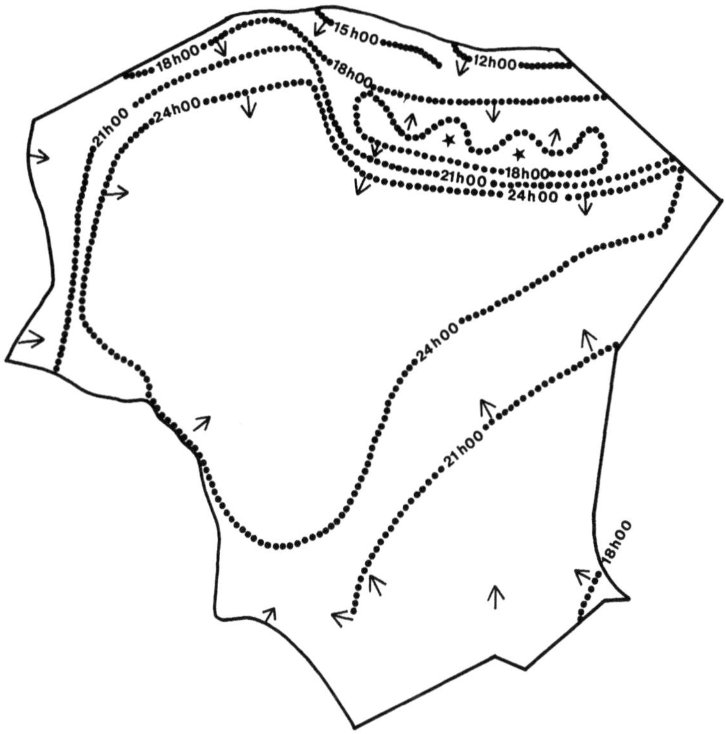


**Fig. 1.16.** Map of Swartboskloof showing the advance of the fire front at 3-h intervals on 17 March 1987. The *central portion* (within the 24:00 line) was burnt on 18 March 1987. *Stars* (★) indicate positions where the fire was ignited using flares

both wind and slope, to ensure control. Once the catchment had been "closed off" the upper reaches were ignited with flares but the burn was nevertheless slow and the central portion of the catchment had still not burnt when a combination of wind, topography, and increasing humidity saw the fire die out after 24:00. The remaining unburnt areas were ignited and burnt successfully between 10:00 and 13:00 on 18 March.

Temperatures on the days of the fire rose to 25°C, while humidities dropped to 36% (Fig. 1.17). Hourly values of temperature and humidity were used to simulate the moisture content of fine dead fuels, using the equilibrium moisture content method (Burgan 1987). The moisture content data, together with hourly wind data and a fynbos fuel model (van Wilgen 1984a) were then used to simulate fire intensity by means of Rothermel's fire model (Rothermel 1972). A slope of 20% with wind blowing across the slope was assumed, and predictions of fire intensity were obtained for a fire burning upslope. Simulated fire intensity peaked at 10 500 $kW\,m^{-1}$ on 17 March, but only at 1800 $kW\,m^{-1}$ on 18 March (Fig. 1.17). Average

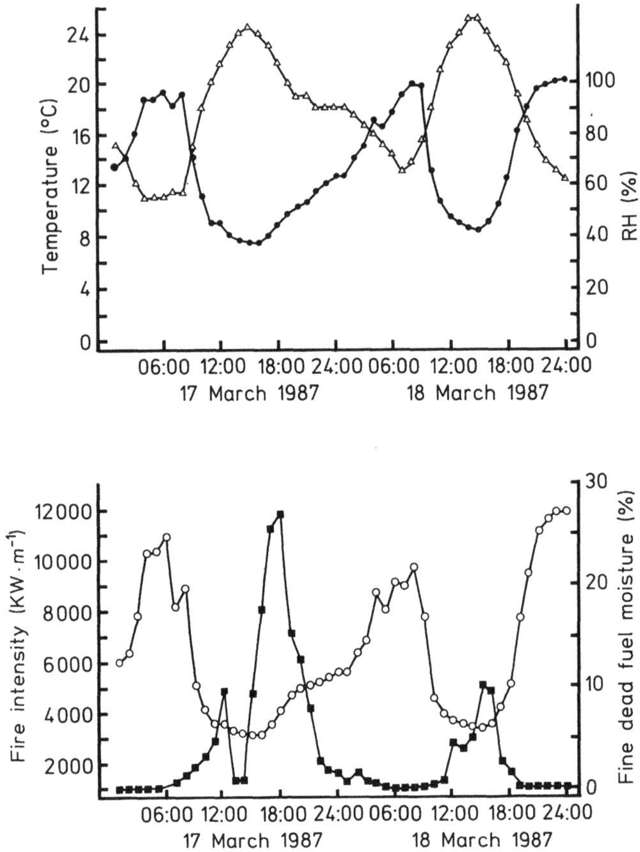


**Fig. 1.17.** Conditions of temperature (△), relative humidity (●), simulated fine dead fuel moisture (○), and simulated fire intensity (■) at Swartboskloof on 17 and 18 March 1987. The fire burnt between 08:50 and 24:00 on 17 March and between 10:00 and 13:00 on 18 March 1987

hourly values of fire intensity for the duration of the fire were 3254 and 950 kW $m^{-1}$ for 17 and 18 March respectively. These values are low compared to other fynbos fires. For example, van Wilgen et al. (1985) measured fire intensities of up to 21 000 kW $m^{-1}$ on experimental plot burns in fynbos. These data, and the fact that the fire burnt itself out against riparian vegetation, attest to the mild nature of the 1987 fire.

*Acknowledgements.* We thank D.P. Bands, A. de Villiers and D.F. Scott for supplying unpublished information.

# 2 The Climate of Swartboskloof

D.B. Versfeld, D.M. Richardson, B.W. van Wilgen,
R.A. Chapman, and G.G. Forsyth

## 2.1 Introduction

The research programme in Swartboskloof evolved with the aim of providing further understanding of the functioning of mountain fynbos ecosystems, through comparison with other fynbos ecosystems, and with mediterranean-type ecosystems on other continents (see Preface). This chapter provides a description of the salient features of the climate of the Swartboskloof study site to facilitate these comparisons. We begin by examining the climate of Swartboskloof in a mediterranean-climate context. We describe features of the climate that influence the biota and hydrology, namely rainfall, temperature, vapour pressure deficit, radiation, evaporation, wind and droughts. We also describe the fire climate of the study site.

## 2.2 The Climate of Swartboskloof in a Mediterranean-Climate Context

Swartboskloof (34° 00′ S; 18° 57′ E) lies within what is known as the mediterranean-climate region of the southwestern Cape Province of South Africa. The climate is characterized by cool, wet winters with low solar irradiance, and hot, dry summers with high solar irradiance (Fig. 2.1; Miller 1983). This bi-seasonality in temperature and precipitation forms the basis of Köppen's (1931) definition of a "mediterranean" region which has been adopted by most geographers. The climate of Swartboskloof may be classified as type Csb (sensu Köppen 1931), that is mesothermal (C), with a warm and dry summer (mean temperature of the hottest summer month ≤22°C) and a relatively wet winter (sb). The mean temperature is 16.2°C and mean annual rainfall is 1523 mm at the Swartboskloof weather station (Fig. 2.1). Mean monthly rainfall exceeds 100 mm month$^{-1}$ between April and September and never drops below 30 mm month$^{-1}$ (14 years data). A period of relatively low rainfall and high temperature extends from

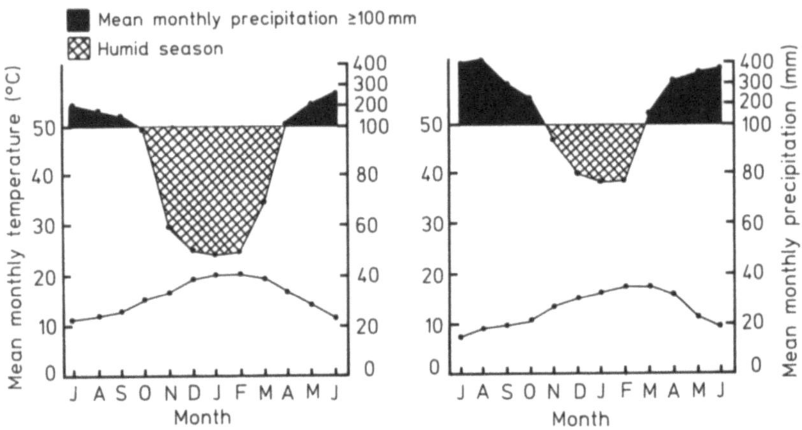


**Fig. 2.1.** Climate diagrams for (*left*) the Swartboskloof weather station (altitude 305 m) and (*right*) a high-altitude site (910 m) in the catchment (temperatures at this site were estimated from lapse rates computed from data from Swartboskloof and Victoria Peak). Mean monthly temperature (°C) is shown by the *lower line* and the *scale at left*. Mean monthly rainfall (mm) is shown by the *upper line* and *scale at right*

November to March. Although drought periods are recorded in some years, the mean rainfall data show that conditions are rarely arid.

Mediterranean-type climates show great diversity, while also sharing some fundamental characteristics (Nahal 1981). Di Castri (1981) defines the mediterranean-climate as transitional between temperate and dry tropical, with rainfall concentrated in winter, a distinct summer drought, highly variable annual precipitation, mild to hot summers, cool to cold winters, and intensive solar radiation especially in summer. No limits are placed on total precipitation. Aschmann (1973) allows 65% of rainfall in the winter half-year as sufficient to define the seasonal difference, but limits the mediterranean-climate to areas with an annual rainfall of betwen 275 and 900 mm.

Winter rainfall is the key characteristic of mediterranean regions but the amount and distribution of rainfall create very different conditions, particularly with regard to the degree of summer dryness (Orshan 1983). The milder mediterranean climate (less severe summer drought) is much more favourable to plant growth than the more typical "mediterranean" climate characterized by a pronounced summer drought. Regions with a mediterranean-type climate differ most markedly in terms of total rainfall. The regions of central Chile and southern California discussed by Miller (1983) have an annual rainfall of 250–350 mm $yr^{-1}$ whereas much of the Swartboskloof catchment receives over 2000 mm $yr^{-1}$ and a *summer* rainfall similar to the California and Chile *annual* total.

In classifying climatic diversity di Castri (1981) adopted the bioclimatic classification of Emberger (1955). Six types of mediterranean climate were defined on the basis of aridity and the severity of the summer drought as expressed by rainfall and temperature. Regions vary from perarid (e.g. Baghdad in Iraq) to perhumid (e.g. Pemberton in Western Australia; see Nahal 1981). A further separation is based on the mean minimum temperature of the coldest month, ranging from extremely cold (<−7°C) to hot (>7°C). A useful index for classifying climates on the basis of aridity is Emberger's (1955) pluviometric quotient (Q):

$$Q = (2000 \times R)/(M^2 - m^2),$$

where R = annual rainfall (mm); M = mean maximum temperature for the hottest month (°C); m = mean minimum temperature for the coldest month (°C).

The pluviometric quotient for Swarboskloof, calculated using data from the weather station at the base of the catchment, is 258 (Fig. 2.2). With a mean minimum temperature in the coldest month of 6.1°C the classification is perhumid temperate, similar to Pemberton in Western Australia (Nahal 1981). Swartboskloof is at the mesic extreme of Emberger's (1955) continuum of mediterranean-type climates. Certainly, when compared with intensive research sites in other "mediterranean-type" climate regions (e.g. Echo Valley and San Dimas in California, Montpellier in France, Fundo Santa Laura in Chile), Swartboskloof is barely "mediterranean". Although fairly typical of the mountain catchments of the southwestern Cape, it is also much more mesic than many other areas within the fynbos biome (Fig. 2.2), for example the Pella research site (cf. Jarman 1988).

Emberger (1955) considered the climate to be mediterranean when the ratio of summer rainfall to average mean maximum temperatures for the summer months was less than 7 (with lower values reflecting more pronounced summer droughts). For the Swartboskloof weather station (altitude 305 m) the ratio is 6.5 for the driest quarter and therefore again close to the limits of the humid range.

Seasonal patterns in the water relations of plants show that the summers at Swartboskloof are mild in a mediterranean-climate context. Deep-rooted adult plants, and even seedlings, of species such as *Protea repens*, *P. nitida* and *Cliffortia cuneata* show little sign of moisture stress during summer (see Chap. 6). Summer soil moisture deficits are of short duration, and are recharged rapidly after relatively light rainfall events (Richardson and Kruger 1990; Smith and Richardson 1990). The lowest leaf xylem pressure potentials recorded for fynbos shrubs in Swartboskloof (−2.75 MPa) are far higher (less negative) than the lowest potentials (−6.0 MPa) quoted for sclerophyllous trees and shrubs in the mediterranean climate of California (see Chap. 6; Sect. 6.3.2).

Swartboskloof is certainly far wetter, and experiences a much less intense summer drought, than the true mediterranean-climate regions. Other parts of the world with equivalent climates, particularly with similarly high rainfall, have a quite different vegetation cover. Miller et al. (1983)

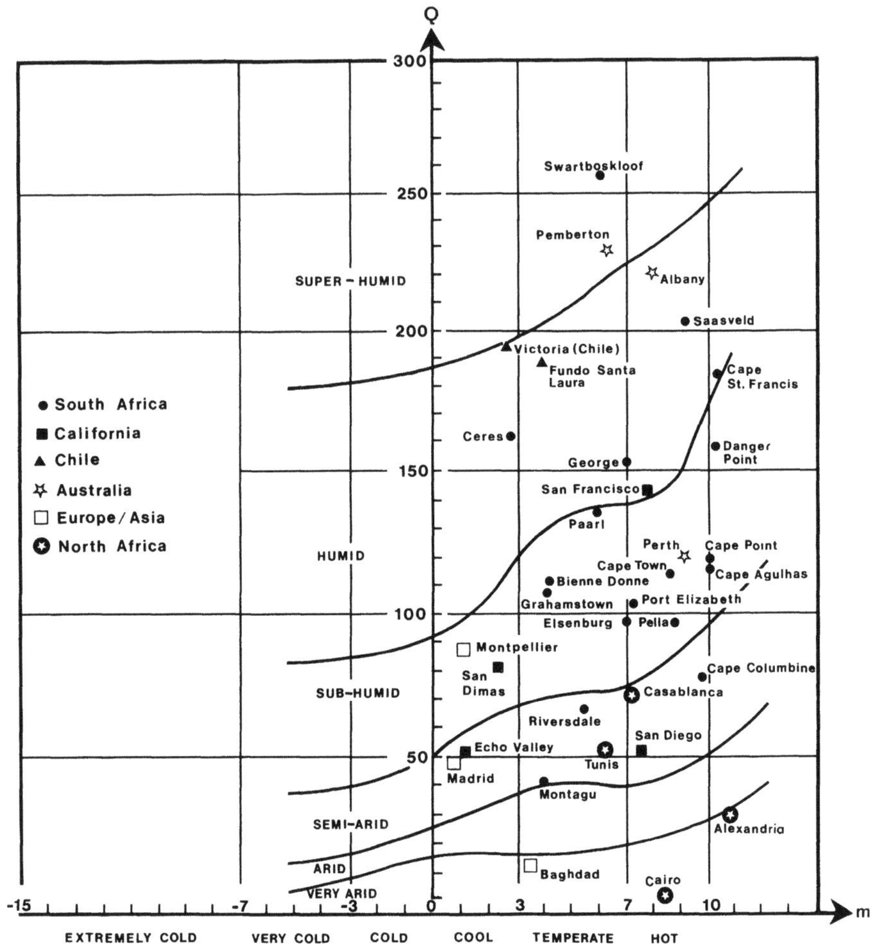


**Fig. 2.2.** Climagram showing Emberger's (1955) pluviometric quotient (Q; see text) and the mean minimum temperature for the coldest month (m) for a range of localities with mediterranean-type climates. Data for South African sites are from Weather Bureau records. Other data are from Miller et al. (1977) and Nahal (1981)

suggested that fynbos with rainfall greater than 650 mm yr$^{-1}$ "should be functionally equivalent to evergreen forests in western North America . . ." and that areas with more than 1250 mm yr$^{-1}$ ". . . may often become equivalent to North American heaths and bogs . . .". One would expect sclerophyllous shrubs to predominate in parts of the western Cape with an annual rainfall of less than 650 mm. However, the predominance of these shrubs throughout the fynbos biome, whether the annual rainfall is 350 mm or more than 1500 mm, remains an enigma (see Chap. 9; Sect. 9.3.2 for further discussion).

## 2.3 Important Climatic Parameters

The Swartboskloof catchment is on the north-facing slopes of the Jonkershoek valley. The climate of the valley has been monitored since 1936, and on a site at the base (altitude 305 m) of the Swartboskloof catchment since 1975. Altitude in Swartboskloof ranges from 285 to 1200 m, and the climate (notably rainfall) at higher altitudes differs markedly from that of the valley floor. A second weather station (at 410 m) and a network of rain gauges were added in 1985. Rainfall in the catchment has been mapped on the strength of these rain gauges. Net radiation and potential evaporation have been modelled using radiation and evaporation budgeting models for mountainous terrain. Additional inferences on temperature and wind in the upper catchment area are based on comparisons between the Swartboskloof weather station and a high altitude automatic weather station on Victoria Peak (1500 m and 8 km to the east). The monitoring network is depicted in Chapter 1 (Fig. 1.3) and is further described in Appendix 2.1.

### 2.3.1 Rainfall

The Jonkershoek valley shows a very strong orographic gradient; annual rainfall along the floor of the valley increases from 786 mm in Stellenbosch (altitude 100 m) to 1110 mm at the offices of the Jonkershoek Forestry Research Centre (244 m), and 1523 mm at the base of Swartboskloof (305 m). The Dwarsberg plateau at the head of the Jonkershoek valley (1220 m) and 6 km from Swartboskloof receives a mean annual rainfall of 3625 mm – the highest recorded in South Africa. Rainfall increases with altitude and with proximity to the head of the Jonkershoek valley (Wicht et al. 1969; van Laar 1984). The increase in rainfall with altitude is equally evident within the Swartboskloof catchment where mean annual rainfall increases from 1523 mm at 305 m to 2815 mm at 910 m.

Mean annual rainfall (RAIN; mm) for any altitude (ALT; m) within Swartboskloof may be approximated using the following function:

$$\mathrm{RAIN} = 3505 - 6.03 \times 10^5 \times 1/\mathrm{ALT} \ldots r^2 = 0.78 \; (n = 8)$$

Mean annual rainfall over the entire catchment has been calculated at 2328 mm using the method of Thiessen polygons.

Rainfall is strongly seasonal with 387 mm (25%) of rain recorded in the summer half-year (October–March) and 1136 mm (75%) in the winter half-year (April–September) (Fig. 2.3). January to March is the driest 3-month period with a total of 174 mm (11.2%) of annual rainfall.

Rainfall intensities are low and storms tend to be intermittent and of long duration. The typical pattern is one of light to steady rain on the arrival of a frontal air mass from the northwest, followed by sharp, intermittent

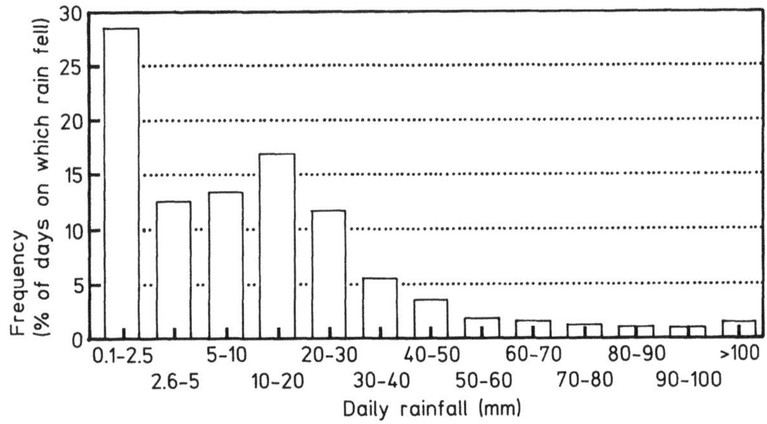


**Fig. 2.3.** The intensity of daily rainfall at the Swartboskloof weather station (305 m) between 1976 and 1989

showers as the wind swings through southwest to southeast (see also Wicht et al. 1969). In an analysis over 12 months on rain gauge R26B, as part of a rainfall interception experiment, Versfeld (1985) recorded 858 h with rainfall ≥0.2 mm, with a total of 1803 mm, and a mean rainfall intensity for all hours with rain of 2.1 mm $h^{-1}$.

The average number of days with measurable rain (>0.1 mm), is 109 per year. Most rain days are in the 0.1–2.5 mm class and very few days with rainfall >50 mm have been recorded (Fig. 2.3)

## 2.3.2 Temperature and Vapour Pressure Deficit

The mean annual temperature for the Swartboskloof weather station is 16.2°C (Table 2.1). Many fynbos taxa have a low tolerance to frost (Levyns 1962; see also Deacon 1983c) but frost is a rare phenomenon in Swartboskloof and the lowest minimum temperature on record is 0.2°C. The highest temperature recorded is 39°C. The vapour pressure deficit (vpd) was calculated from the minimum relative humidity and maximum temperature values for each day, on the assumption that these values would be approximately coincident in time. Vapour pressure deficits are at a minimum in June, while maximum values are recorded in January and February (Table 2.2).

Temperatures in the catchment may be expected to decrease with altitude. The mean temperature lapse rate between the Swartboskloof weather station and Victoria Peak is 0.51°C per 100 m (G.G. Forsyth unpubl. data). With an altitudinal range of 700 m, a temperature depression of 3.8°C can be expected across the Swartboskloof catchment. Frost is

**Table 2.1.** Mean monthly climate parameters for the Swartboskloof weather station (altitude 305 m); January 1976–December 1989

| Month | Rain (mm) | Temperature (°C) | Solar radiation ($MJ\,m^{-2}\,d^{-1}$) | Evaporation (mm) | Wind speed ($m\,s^{-1}$) |
|---|---|---|---|---|---|
| January | 49.7 | 20.6 | 24.6 | 295.4 | 4.0 |
| February | 52.8 | 21.1 | 23.3 | 254.4 | 4.0 |
| March | 71.2 | 19.5 | 19.8 | 213.5 | 3.4 |
| April | 132.2 | 17.2 | 13.0 | 120.8 | 2.5 |
| May | 201.0 | 14.4 | 7.7 | 75.9 | 2.5 |
| June | 258.1 | 12.2 | 7.0 | 53.8 | 2.2 |
| July | 201.5 | 11.4 | 7.0 | 56.7 | 2.3 |
| August | 189.7 | 12.0 | 10.5 | 81.1 | 2.9 |
| September | 153.3 | 13.1 | 13.3 | 110.3 | 2.8 |
| October | 98.9 | 15.5 | 21.1 | 180.5 | 3.5 |
| November | 59.8 | 17.8 | 24.0 | 235.0 | 3.9 |
| December | 55.0 | 19.4 | 24.7 | 271.2 | 3.6 |
| *Annual* | *1523.2* | *16.18* | *16.30* | *1948.6* | *3.13* |

**Table 2.2.** Mean monthly temperature (°C) and vapour pressure deficits (vpd; kPa) for Swartboskloof (1975–1989)

| Month | Mean temperature | Mean monthly maximum temperature | Mean monthly minimum temperature | Mean absolute maximum temperature | Mean absolute minimum temperature | Mean monthly maximum vpd |
|---|---|---|---|---|---|---|
| January | 20.6 | 27.1 | 14.2 | 34.9 | 8.8 | 2.39 |
| February | 21.1 | 27.6 | 14.5 | 35.7 | 8.4 | 2.40 |
| March | 19.5 | 25.8 | 13.2 | 34.4 | 7.9 | 2.18 |
| April | 17.2 | 23.1 | 11.2 | 32.4 | 6.0 | 1.81 |
| May | 14.4 | 19.6 | 9.2 | 28.6 | 4.9 | 1.32 |
| June | 12.2 | 17.2 | 7.1 | 25.6 | 2.4 | 1.12 |
| July | 11.4 | 16.7 | 6.1 | 25.0 | 1.8 | 1.14 |
| August | 12.0 | 17.2 | 6.8 | 27.0 | 2.2 | 1.17 |
| September | 13.1 | 18.5 | 7.8 | 28.3 | 2.9 | 1.31 |
| October | 15.5 | 21.1 | 9.8 | 31.3 | 3.8 | 1.66 |
| November | 17.8 | 23.9 | 11.8 | 32.5 | 5.9 | 1.94 |
| December | 19.4 | 25.4 | 13.3 | 34.6 | 7.3 | 2.13 |
| *Annual* | *16.18* | *27.6* | *6.1* | *35.7* | *1.8* | *1.12* |

likely in pockets at higher altitudes and occasional snowfalls occur in winter above 600 m.

### 2.3.3 Radiation

Solar irradiance (incoming shortwave radiation) is modified substantially by physical conditions within and around Swartboskloof, which is more

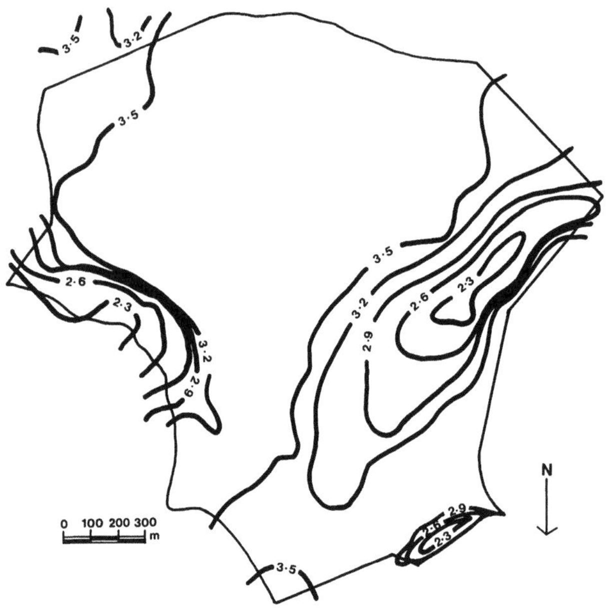


**Fig. 2.4.** Spatial variation of net solar radiation ($\text{MJ m}^{-2}\text{yr}^{-}\text{k}^{1} \times 10^{3}$) in Swartboskloof. Radiation loadings were mapped on a 200-m grid using a digital terrain model which incorporated slope, aspect and topgraphic shading from surrounding terrain

**Table 2.3.** Total incoming solar radiation and potential evapotranspiration ($\text{mm yr}^{-1}$) in Swartboskloof and selected mediterranean-climate study sites in California and Chile (source Miller and Hajek 1981)

| Study Site | Region | Altitude (m) | Incoming solar radiation ($\text{MJ m}^{-2}\text{yr}^{-1}$) | Potential evapotranspiration ($\text{mm yr}^{-1}$) |
|---|---|---|---|---|
| Swartboskloof | South Africa | 305 | 5954 | 1011 |
| Mount Laguna | California | 2030 | 5602 | 1182 |
| Echo Valley | California | 1000 | 6184 | 1129 |
| Camp Pendelton | California | 110 | 5234 | 1117 |
| Fundo Santo Laura | Chile | 1000 | 6125 | 1303 |

mountainous than other intensive research sites in mediterranean-climate regions around the world (see Chap. 1 for site description). The steep slopes and varying aspects have an important influence on the radiation budget. Mean daily levels of incoming and net solar radiation (after accounting for slope and reflectance) vary considerably over time and space within the catchment. Incoming solar radiation (at the weather station) is lowest in June ($7\,\text{MJ m}^{-2}\text{day}^{-1}$) and highest in December and January ($24.7\,\text{MJ m}^{-2}\text{day}^{-1}$). The annual mean is $16.3\,\text{MJ m}^{-2}\text{day}^{-1}$. The total

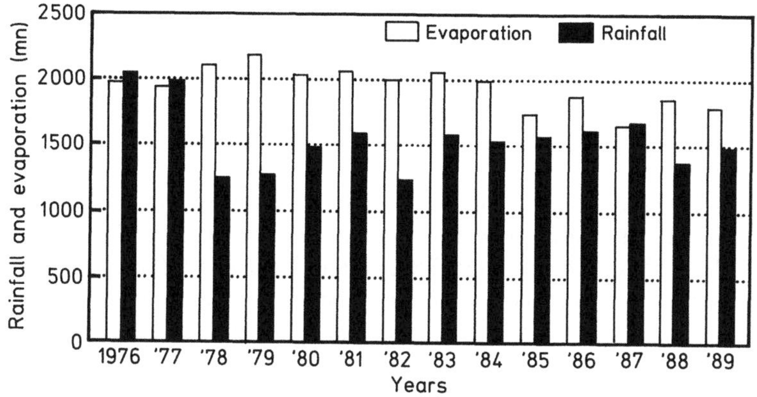


**Fig. 2.5.** Annual evaporation and rainfall at Swartboskloof weather station between 1976 and 1989

solar radiation loading at this location is similar to that of other intensive research sites in the mediterranean climates of California and Chile (Table 2.3).

Albedo in Swartboskloof is approximately 0.13 over mature vegetation and 0.08 under post-fire conditions (D.B. Versfeld unpubl. data). Cloud cover is an important factor, especially in winter, in reducing radiation levels. A substantial component of incoming radiation, however, is reflected onto opposing surfaces from the surrounding terrain, so that diffuse and reflected radiation may constitute up to 30% of total incoming radiation.

Because of the highly variable slopes and aspects in Swartboskloof, spatial variation in net solar radiation (the solar radiation balance) was mapped using a modified version of Schulze and Lambson's (1986) model RADSLOPE. Net solar radiation was calculated on a grid with 200 m intervals for the year 1988, using as core data the net radiation (total incoming shortwave) recorded at hourly intervals at the Swartboskloof weather station. Slopes facing north receive the most radiation (Fig. 2.4), but are also shaded at times. Those slopes receiving the least radiation are often steep and face either southeast or southwest. The nature of the surrounding terrain ensures that most parts of the area are shaded to some extent by nearby topography.

## 2.3.4 Evaporation

Annual pan evaporation (mean = 1948.6 mm) generally exceeds rainfall by 25% with the exception of the "wet" years 1976, 1977, and 1987 when rainfall marginally exceeded pan evaporation (Fig. 2.5). Pan evaporation exceeded 2000 mm $yr^{-1}$ every year between 1978 and 1983. These were years of below-average rainfall and the deficit between rainfall and evaporation averaged 672 mm. The period 1984–1987 was more moderate with the deficit averaging only 225 mm. The mean daily pan evaporation rate is 5.3 mm. The summer months December–February have the highest evaporation rates at 8.5 mm $day^{-1}$ while rates for June and July average 1.8 mm $day^{-1}$.

The spatial variation of net solar radiation across Swartboskloof (Fig. 2.4) also indicates how the amount of energy available for evaporation differs from point to point. Schulze and Lambson (1986) note the commonly observed phenomenon in natural landscapes that certain slopes may be relatively warmer or cooler than others, because of different solar radiation loadings. The soil moisture response to differing energy loadings is also seasonal and elevational, and Nash (1963) has noted longer periods of stress due to soil water deficit on equator-facing slopes than on poleward-facing slopes. Schulze and Lambson (1986) indicate that between 65 and 85% of the variance in potential evapotranspiration may be explained by shortwave radiation alone.

Annual potential evapotranspiration in Swartboskloof is similar to that of intensive research sites in mediterranean-climate California and Chile (Table 2.3). Spatial variation in potential evapotranspiration at Swartboskloof was calculated and mapped (Fig. 2.6), using the Penman equation (Penman 1948; Fig. 2.6). Potential evapotranspiration patterns closely follow the radiation patterns, with north-facing slopes experiencing the most severe soil water deficits. The mapped potential evapotranspiration does not take into account lateral movement of soil water through the shallow soils on steep slopes.

### 2.3.5 Wind

Wind is a major energy factor driving ecosystem processes, particularly with regard to fire weather. The dominant winds are the summer southeast and rain-bearing winter northwest trade winds. Warm northeasterly winds ("bergwinds") are a less important feature in the winter months and are not as important in creating fire hazard conditions as they are in the constant rainfall area of the southern Cape (see van Wilgen 1984b).

Local winds tend to flow up and down the Jonkershoek valley which has an approximately northwest/southeast orientation. The Swartboskloof weather station is exposed to the macroclimate of the Jonkershoek valley, although wind patterns will be different in protected parts of the catchment. Data from Victoria Peak shows that there is far more wind at high altitude. During 1985, the total wind run at the Swartboskloof weather station was 85 115 km and on Victoria Peak 195 037 km (average wind speeds were 2.7 and 6.2 m $s^{-1}$ respectively). Calm conditions (wind speed $\leq 1.5$ m $s^{-1}$) prevailed for 13.5% of 1985 at the Swartboskloof, but only for 0.8% of the

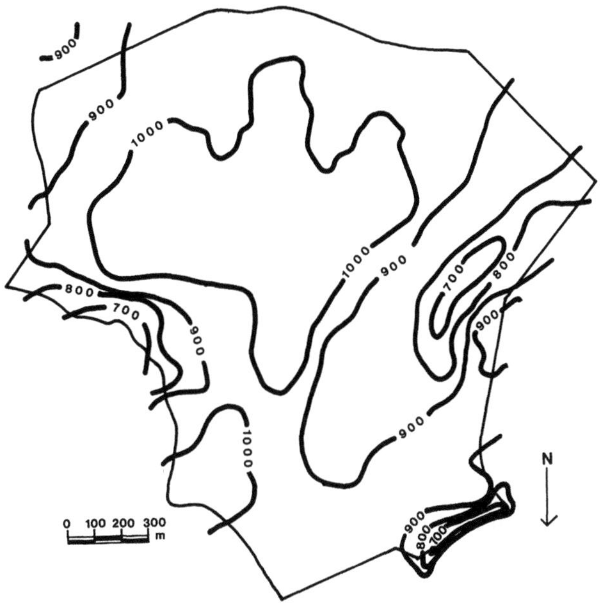


**Fig. 2.6.** Spatial variation of potential evapotranspiration (mm) in Swartboskloof. Potential evapotranspiration depths were calculated using the Penman (1948) method, in conjunction with calculations of net solar radiation (see Fig. 2.4)

year on Victoria Peak. Winds in the upper reaches of Swartboskloof will tend towards the levels recorded on Victoria Peak, and contribute to the stunted nature of the vegetation (see van Wilgen and Kruger 1985). The amount of wind recorded is also highly dependent on the season, with by far the most wind in the summer months of November through February (Table 2.1).

### 2.3.6 Periods of Drought

From ecophysiological studies in the Jonkershoek valley it is apparent that many fynbos species are rarely, if ever, under severe water stress (see Chap. 6). Nevertheless, the summer months, and particularly the months November to March, are characterized by periods of extreme fire hazard. High winds, high radiation loading and low rainfall coincide during these months, although rain-free periods (daily rain $\leq$0.5 mm) seldom persist for more than 3 weeks (Fig. 2.7). Rain-free periods of more than 30 days occur very rarely but do tend to be between December and March, and certainly never in the winter months (Fig. 2.7; Table 2.4). Only three rain-free periods (daily rain $\leq$0.5 mm) of longer than 30 days have been recorded over 14 years and of these only one was longer than 60 days. Richardson

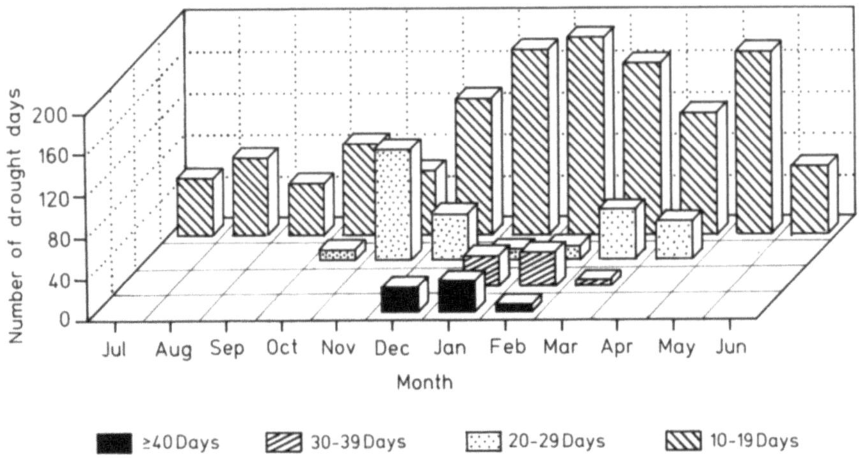


**Fig. 2.7.** Seasonal distribution of droughts (periods during which daily rainfall never exceeded 0.5 mm for 10 days or more) at the Swartboskloof weather station between 1976 and 1989

**Table 2.4.** Frequency of droughts in Swartboskloof between 1976 and 1989. A drought is defined as an unbroken sequence of days without daily rain of 0.5 mm or more. The longest drought on record is 65 days

| Length of drought period (days) | No. of events |
|---|---|
| 10–20 | 124 |
| 20–30 | 18 |
| 30–40 | 3 |
| >40 | 1 |

and Kruger (1990) in an analysis of the complete rainfall record from 1936 to 1986, and using the same definition of drought, found that droughts of 30 days occurred on 12 occasions since 1936 (every 4.2 years), and droughts of 40 days every 8.3 years.

## 2.4 The Influence of Climate on Fire

### 2.4.1 Lightning as an Ignition Source

The role of lightning as a source of ignition in fynbos is discussed by Kruger and Bigalke (1984; see also Chap. 3; Sect. 3.3.1). The entire western Cape region has a lightning flash density of <1 flash $km^{-2} yr^{-1}$ and the value for the Stellenbosch region (including Swartboskloof) is 0.3 (Division of Energy Technology, CSIR). Despite this low density, lightning fires are commonly reported and may be very extensive. For example, many fires were almost simultaneously ignited across the western Cape during a lightning storm in December 1988, and eventually burned about 109 000 ha (Thompson 1990). These lightning fires would almost certainly have included Swartboskloof had the catchment not burnt the previous March.

### 2.4.2 Climate and the Probability of Fire

The United States National Fire Danger Rating System (NFDRS) has been successfully adapted and applied in the western Cape (van Wilgen and Burgan 1984). Climatic inputs include maximum and minimum temperature and relative humidity, hours of sunshine, and total wind run – all on a daily basis. The probability of fire in Swartboskloof was analyzed using this sytem, for the period January 1976 to December 1986. Results are expressed in terms of an index (the burning index, BI) of fire potential for each day. The BI relates to the fire intensity that would be expected under the prevailing conditions.

Four classes of fire danger were defined for the range of fire danger indices in the data set. One third of the days, with the highest values of BI, were designated high fire danger days. Moderate and low fire danger days were represented by the middle and lower third of the BI values. Extreme fire danger days were taken as the upper 5% of all days. This classification gave ranges of 0–39, 40–76, 77–175, and >175 for low, moderate, high and extreme values of BI. While this classification is not strictly valid in terms of the occurrence of fire, the seasonal patterns in terms of high and low values in the data set are illustrated. January is the month with the highest risk of fire, although November to March can be rated as the true fire season months, with the mean BI values in excess of 77 (Fig. 2.8).

In examining the seasonal distribution of each of these categories of fire danger (Fig. 2.9), extreme fire danger days (BI > 175) are never experienced in May through July but may occur in any of the months of August to April, although mean frequency only exceeds 2 days per month from November to March. High fire danger days (BI values > 77) occur in all months of the year, but are infrequent in the winter months (May to August). Thus, whilst fires could occur at any time of the year, the greatest likelihood is during the midsummer months of November to March.

Large fires in the western Cape are associated with prolonged periods of high fire danger. Large wildfires, which all persisted for periods of a week or more, were associated with BI values remaining at greater than 95 on each day of the fire (van Wilgen and Burgan 1984). From the occurrence of periods of seven or more consecutive days with a BI > 95 it is apparent that

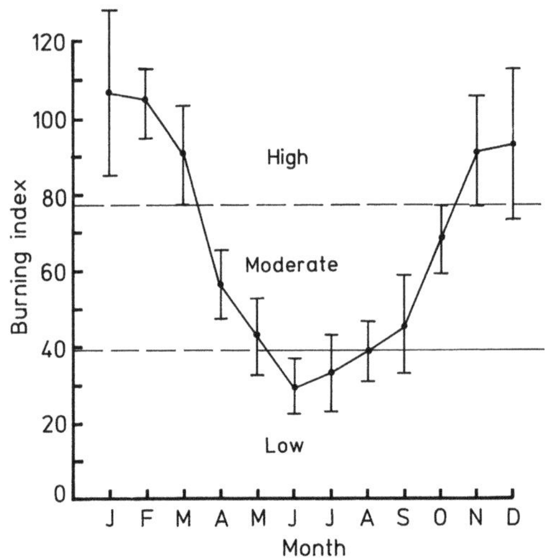


**Fig. 2.8.** Annual cycle of mean monthly burning index at the Swartboskloof weather station between 1976 and 1986. *Bars* represent the 5% confidence intervals of the means. *Horizontal lines* represent cutoff limits for low, moderate and high fire danger

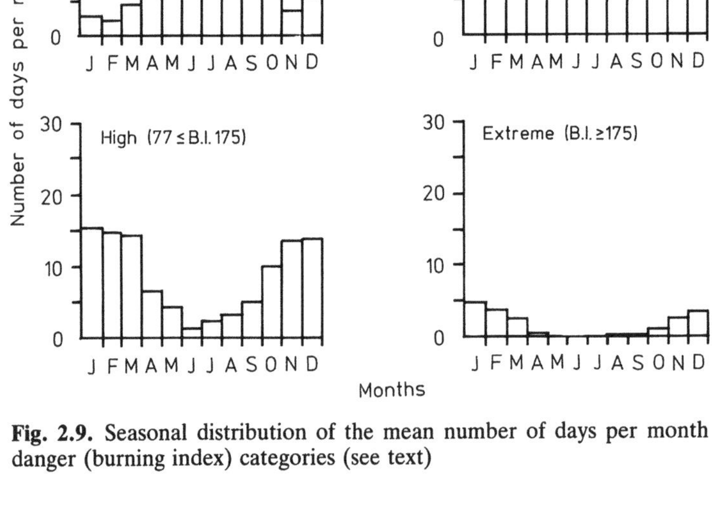


**Fig. 2.9.** Seasonal distribution of the mean number of days per month in different fire danger (burning index) categories (see text)

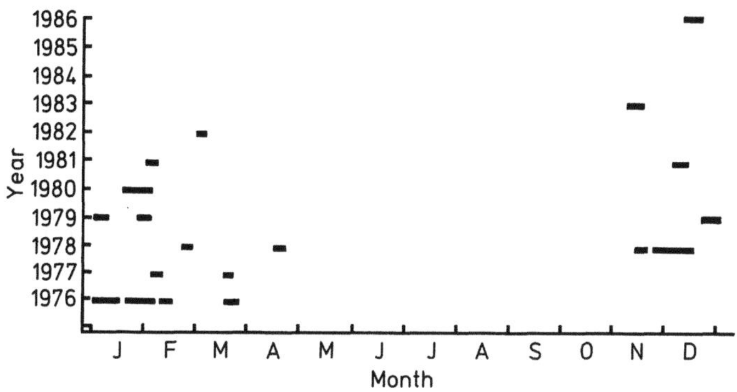


**Fig. 2.10.** The occurrence of extended periods (≥7 day) of high fire danger (≥ daily Burning Index values of 95) at the Swartboskloof weather station from 1976 to 1986

extended periods of high fire danger are strictly limited to the months November through March (with one exception in April; Fig. 2.10). It is thus probable that large fires would have been concentrated in this season for as long as the modern climate has prevailed.

## 2.5 Conclusions

When making comparisons between Swartboskloof and the intensive research sites in other "mediterranean-type" ecosystems, one must bear in mind that, according to most definitions, Swartboskloof is only marginally "mediterranean". The total annual rainfall, even the mean value of 1523 mm at the base of the catchment, is much higher than areas with structurally similar vegetation on other continents. Summer droughts are mild in comparison with those in other mediterranean-type ecosystems; droughts of more than 1 month duration are rare. Climatic conditions do, however, generate high fire danger in summer and early autumn.

*Acknowledgements.* We thank H.J. Barnard, H.P. Julius, F.W. Prinsloo and D. Willemse for data collection, and R.E. Smith for assistance with data analysis.

**Appendix 2.1**
**The Climate Monitoring Network in Swartboskloof**

Data used to describe the climate of Swartboskloof is from a standard weather station sited at the base of the catchment (altitude 305 m) and exposed to the major weather trends of the Jonkershoek valley (see Wicht et al. 1969). This station has been operational since April 1975 and is attended daily at 08:00. Rainfall has been measured on the site since

1936. Parameters monitored include temperature (maximum, minimum, thermograph), humidity (hygrograph), rainfall, wind speed and direction, sunshine hours, and A-pan evaporation. Hourly average values of solar and net radiation were recorded from 1985 onwards.

An automatic weather station was installed in the centre of the catchment (altitude = 410 m; see Fig. 1.3) in 1985 to provide detailed climate and weather data for the period of intensive research before and after the prescribed burn of 1987. Hourly averages of 1-min readings of temperature, humidity, wind speed and direction, solar and net radiation, and rainfall were recorded. A network of Nipher wind-shielded monthly gauges was also installed in 1985 to supplement the daily recording gauges at the standard weather station site and the daily rain gauge at Uitkyk (R20). All instrumentation is marked on the site map in Chapter 1 (Fig. 1.3).

# 3 Fuel Properties of Vegetation in Swartboskloof

B.W. van Wilgen and H.J. van Hensbergen

## 3.1 Introduction

The fuel properties of natural vegetation, and their dynamics, influence the occurrence of fires to a large degree. They also govern the nature of fires, and therefore, to some extent, their effects. Ecologists are interested in fuel properties for a variety of reasons. The most important of these are to reconstruct a plausible fire regime under which the biota evolved and to which they are currently subjected, as well as to provide the necessary inputs for predicting fire behaviour under a variety of weather conditions. Such predictions, combined with an understanding of how fire behaviour determines the response of the biota, are a powerful means of enhancing vegetation management through the use of fire. Fynbos is regarded as a highly flammable vegetation type (Kruger and Bigalke 1984). Despite this, little is known about the fuel characteristics of fynbos and how they affect the spread and intensity of fires.

In this chapter, we describe the framework within which fuel studies were conducted in Swartboskloof, and present data on the fuel properties of the vegetation. These include variations in fuel properties over time (seasonally and during post-fire succession), and the role of fuel properties in determining the nature of fires in the major communities (fynbos, forest and thickets of alien trees and shrubs). Comparisons are made to other fire-prone ecosystems, with a view to quantifying the important differences in fuel properties between such ecosystems, and the effects that they have on the resultant fire regimes.

## 3.2 The Description of Vegetation as Fuel

In this section we describe the framework within which fuel studies were conducted in Swartboskloof as a background to the data presented later. Since fuel studies are rare, and because they vary considerably in approach, it is necessary to review the conventions followed in the description of vegetation as fuel.

### 3.2.1 Subdivision of Biomass into Fuel Classes

The amount of fuel (fuel load or biomass) is usually expressed as mass per unit area. It is important, when describing vegetation as fuel, to distinguish between dead and live plant parts. Because the moisture content of dead plant parts responds relatively rapidly to ambient conditions of temperature and humidity, moisture contents are low – usually between 5 and 15%. Live plant parts have moisture contents between 50 and 250% or greater. Obviously, it is dead fuel rather than live that will actually carry a fire. A fire will not normally be sustained if there is not enough dead fuel in the vegetation, or should the dead fuel be too moist. Combustion of dead fuels will drive the moisture out of living fuels. Live fuels therefore initially act as a heat sink, but once they have dried to a point where they begin to burn, they contribute to energy release in fires. Dead fuels can be partitioned into classes based on rates of moisture loss. Water loss from organic materials tends to follow a logarithmic rate for which the time required to reach an equilibrium moisture content can be divided into periods (Pyne 1984). The time-lag constant of any given fuel particle is the time for it to lose 63% of the difference between its initial moisture content and a new equilibrium moisture content under standard laboratory conditions of 27°C and 20% relative humidity (Byram 1963). The standard moisture time-lag classes used to partition dead fuels based on diameter are: 1 h (0–6 mm), 10 h (6–25 mm) and 100 h (25–75 mm) (Deeming and Brown 1975).

### 3.2.2 Fuel Arrangement

Fuel particle sizes and the spatial arrangement of fuel have important effects on fire behaviour (Rothermel 1972). Several measures of fuel arrangement are used to describe vegetation. Packing ratios are defined as the ratio of bulk density (fuel mass per unit volume occupied by fuel and air) to the specific gravity (fuel mass per unit volume occupied by fuel alone) (Rothermel 1972; Pyne 1984). This is essentially a measure of the fuel: air mix, and is an important input into some fire behaviour prediction models (see below). Surface-area: volume ratios provide an estimate of how finely divided the fuel is, and are used in conjunction with estimates of the fuel bed depth to arrive at packing ratios (Deeming and Brown 1975).

### 3.2.3 Fuel Energy Contents

Energy contents (heat of combustion) of fuels are important as they influence, among other things, the intensity of fires. The heat of combustion refers to the total amount of heat contained per unit mass of fuel. Not all of this energy is released during a fire. The actual amount of heat released during a fire decreases with increasing moisture content of the fuel, and is termed the heat yield. Byram (1959) gives correction factors for determining heat yield from heat of combustion and moisture content values. The most commonly used measure of fire intensity is Byram's (1959) fire line intensity:

$$I = Hwr,$$

where I is the fire line intensity ($kW\,m^{-1}$), H is the heat yield of the fuel ($J\,g^{-1}$), w is the mass of fuel consumed ($g\,m^{-2}$), and r is the rate of spread of the fire ($m\,s^{-1}$).

Oils, fats, waxes and terpenes (hereafter referred to as crude fats) influence the fuel properties of plant material. Crude fats would make plants more flammable as these substances are volatile. They are driven out of plant material at lower temperatures than those required to ignite the plant material itself, and then burn as gases. In addition, oils and waxes have up to twice the energy contents of other plant material (Philpot 1969). Fynbos is often stated to be more flammable due to such properties. Kruger and Bigalke (1984) mention oil contents of buchu (*Agathosma* sp.) of around 1.4–4%, and layers of cuticular wax on restioid plants to support this belief.

### 3.2.4 Fuel Models

Data on the fuel properties of vegetation are essential inputs into fire behaviour prediction models. Descriptions of the fuels of Swartboskloof were designed to meet the requirements for use in formal fire behaviour prediction models (Rothermel 1972; Kessell et al. 1978; Burgan and Rothermel 1984). The Rothermel model was designed to predict fire behaviour in a single continuous stratum of fuel contiguous to the ground. Fire behaviour predictions from the model are based on the physical and chemical makeup of the fuel and the environmental conditions under which it burns. Fuel characteristics are summarized in fuel models (Deeming and Brown 1975). The following data are required to define a fuel model: (1) fuel loads divided into live herbs, live woody fuel (<6 mm), and dead (1, 10 and 100 h) fuel; (2) surface area to volume ratios for each of the above categories; (3) the fuel bed depth; and (4) a mean fuel energy content for the combined fuel mass. Fuel models of this nature would suit structurally homogeneous vegetation, but the assumption of single stratum in the vegetation leads to inaccurate predictions of fire behaviour where vegetation displays vertical discontinuities. Kessell et al. (1978) modified Rothermel's model to allow for multi-layered vegetation fuel models. This was done by accommodating separate fuel models for up to three strata, and calculating the rate of spread for individual strata, which could vary in cover. Fuel inputs for the three strata model include (1) fuel loads divided into the litter layer (first stratum), live grass and herbs with standing dead fuel of <6, 6–25 and >25 mm (second stratum), and shrub foliage and shrub branchwood (<6 mm) (third stratum); (2) a percentage cover value for each stratum; and (3) a packing ratio for each stratum. A stratified fuel modelling approach could allow for the simulation of conditions likely to lead to crown fires in vegetation with marked vertical discontinuities in fuel distribution, such as found in forest patches in Swartboskloof.

### 3.2.5 Fuel Buildup and Decay

The balance between accumulation and decay is used in several important models of fuel dynamics (e.g. Walker 1981; Kessell et al. 1982, 1984). Such models express the amount of dead, fine litter fuel (X) at time t (in years) as:

$$X_t = L/K\ (1 - e^{-Kt}),\ \text{where}\ L = K\ X_{max}$$

L is the litter production rate (tonnes $ha^{-1}\ yr^{-1}$), $X_{max}$ is the maximum litter standing crop at equilibrium, K is the decomposition constant, and e is the base of the natural logarithms. The model assumes that litter is input at a steady rate, and decomposes at a steady rate. Given that post-fire age, and litterfall and decomposition rates are known, it would be possible to calculate litter fuel loads. Equilibrium fuel loads ($X_{max}$) would occur when litterfall was balanced by decomposition.

## 3.3 The Fuel Properties of Vegetation in Swartboskloof

Fuel studies were conducted in representative stands of proteoid shrublands, forest and invaded thickets (see Chap. 1 for vegetation types). Data from these communities are summarized below.

### 3.3.1 Fuel Loads, Fuel Buildup and the Probability of Fire

In the proteoid shrublands in Swartboskloof, live biomass increases with increasing post-fire age until about 25 years after fire (Fig. 3.1). At this point, live biomass declines as the dominant *Protea* shrubs start to senesce and die. Dead mass continues to increase with post-fire age. As litter accumulates with increasing post-fire age of the vegetation, the probability of fire increases. Very few fynbos stands survive periods of longer than 40 years without fire (Kruger and Bigalke 1984; Brown et al. 1991).

Annual litterfall rates below the canopies of *Protea repens* shrubs in Swartboskloof were between 200 and 450 $g\,m^{-2}$, and the decomposition constant K was 0.07 (Mitchell and Coley 1987). This value is similar to those obtained in the only other study done in fynbos (Mitchell et al. 1986),

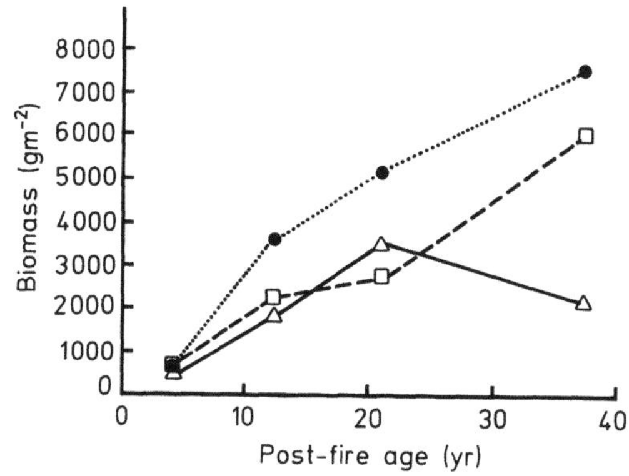


**Fig. 3.1.** The effects of post-fire age on above-ground biomass and fuel mass in Swartboskloof. Data are from van Wilgen (1982) and van Wilgen et al. (1990a). The curves show total (dead plus live) biomass (●), live biomass only (△), and fuel (all dead material, plus live material <6 mm) (□)

and shows that the decomposition of fynbos litter is extremely slow when compared to other vegetation types. Values for K of between 0.09 and 0.75 are quoted for Australian ecosystems (Walker 1981), and between 0.21 and 0.23 for chaparral shrubs (Schlesinger and Hasey 1981). Using the model above, equilibrium litter crops should be around 4300 $g\,m^{-2}$ (below the canopies of *P. repens*), with 50% of the equilibrium mass reached at 10 years after fire, 75% at 20 years, and 95% at 40 years. This figure is an estimate of the long-term fuel load in fynbos in the absence of fire. Most fires occur at intervals of between 10 and 30 years, with a mean around 15 years, however (van Wilgen 1987; Brown et al. 1991). In such cases, a steady state, or equilibrium, litter mass is never attained. In addition, litter input increases dramatically following senescence of the dominant *Protea* shrubs at about 30 years after fire. Steady state or equilibrium models of fuel dynamics could therefore not normally be applied to fynbos ecosystems, and were not considered further at Swartboskloof.

In order for fires to occur, sufficient fuel must be present in conjunction with suitable weather conditions and a source of ignition. The proteoid shrublands in Swartboskloof accumulate enough fuel to burn at 4- to 6-year intervals (van Wilgen 1982). While data for the other major fynbos vegetation groups in Swartboskloof (Chap. 1) are not available, field observations and data from elsewhere (Kruger 1977a) indicate that fuel buildup rates are at least similar for ericoid shrublands, and much faster for the hygrophilous communities. In addition, weather conditions suitable for fires occur annually (see Chap. 2), while those suitable for large conflagrations occur for at least 1 week every 4.2 years on average (van Wilgen 1985; see also Chap. 2). Empirical data (Brown et al. 1991) indicate that fire frequencies of between 11 and 15 years are the norm for fynbos.

Swartboskloof has been subjected to an average fire frequency of around 16 years (Fig. 3.2), with a range of fire frequencies from 7 to 29 years. Minimum fire frequencies in fynbos are therefore obviously not determined by fuel buildup alone. The question of why fynbos does not burn more frequently would probably be explained by a scarcity of ignition sources. Natural fires in fynbos are caused by lightning, and sparks from falling rocks (Edwards 1984). Lightning is rare in the fynbos biome, and flash densities are at least eight times lower than those found in the grassland regions of southern Africa (<1 strike $km^{-2} yr^{-1}$; Edwards 1984). The chances of rockfalls co-occurring with suitable weather conditions are probably also low, but data are not available.

We constructed a simple stochastic (Monte Carlo) model to investigate the distribution of the fire return interval in fynbos. The model considers a single patch of vegetation, and uses two parameters to determine the likelihood of a fire in a given year. These parameters are a base probability (p) of fire in a given year, and an annual growth parameter (g). The risk of a fire in a given year is calculated as:

$$\text{Risk} = \{\Sigma(i = 1 \text{ to } n)\ (g_i)/100\}p,$$

where n is the number of years since the last fire. When $\Sigma\ g_i$ became greater than 100, it was taken to be 100. Fire occurrence in the patch was modelled for 700 consecutive years using different values of p and g, and the distribution of fire return intervals was obtained. This distribution was compared to known fire return intervals for the Cederberg (Brown et al. 1991) and Swartboskloof (see Chap. 1). The model parameters were varied until a well-fitted model for the site was obtained using a simple least squares measure of the discrepancy (Linhart and Zucchini 1986).

The mean discrepancy of ten runs of the model was calculated, and based on this the combination of g and p that gave the lowest discrepancy was selected. The best fitted models were obtained when $g = 15$ and $p = 0.09$ for the Cederberg data, and when $g = 5$ and $p = 0.13$ for Swartboskloof (Fig. 3.2). The results show that it is necessary to take a growth factor for the vegetation into account when modelling the return interval of fires. Models with very high growth rates fitted the observed data very poorly. The growth rates suggest that the likelihood of vegetation carrying a fire increase for longer periods in Swartboskloof than in the Cederberg. In the Cederberg, the vegetation reaches maximum flammability after only 7 years. In all runs of the model for a particular growth rate, the data for the Cederberg fitted best with a higher intrinsic probability of fire than for Swartboskloof. The best fitted combination of growth rate and intrinsic probability gives a lower intrinsic probability of fire for the Cederberg, but it reaches 100% of this after only 7 years of vegetation growth, whereas in Swartboskloof the higher intrinsic probability is only reached after 20 years of growth. The model has some difficulty fitting the short interval fires for Swartboskloof (Fig. 3.2). These are lower than expected by the model and are presumably a result of the fire protection

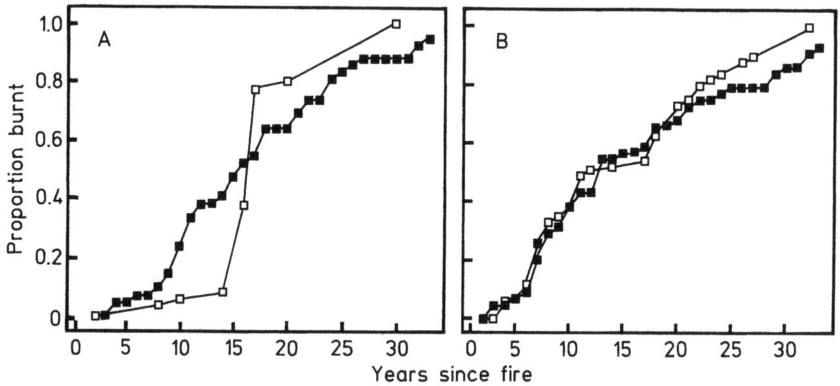


**Fig. 3.2.** The probability of fire (□) for two areas in the fynbos, compared to predictions (■) from a single run of a stochastic model of fire probability (see text). The areas are (**A**) Swartboskloof, and (**B**) Cederberg

of the area, which would have resulted in fewer fires than would normally have occurred in young vegetation.

### 3.3.2 Fuel Arrangement

Fine, medium and coarse fuels have surface area to volume ratios of 8000–10 000, 6500–8000, and 5000–6500 $m^2 m^{-3}$ respectively (Burgan and Rothermel 1984). Fynbos fuels are relatively coarse. Examples from Swartboskloof include the reedlike herb *Restio filiformis* (1820–6880 $m^2 m^{-3}$), the sedge *Tetraria cuspidata* (7000 $m^2 m^{-3}$), and leaves of the shrubs *Protea repens* (4200 $m^2 m^{-3}$) and *Erica plukenetii* (7150 $m^2 m^{-3}$) (van Wilgen 1984a). Most grasses, for example, are fine. Surface area to volume ratios for various grasses can range from 18 900 to 38 000 $m^2 m^{-3}$, while those for the needles of coniferous trees range from 5400 to 9900 $m^2 m^{-3}$ (Rundel 1981b).

Packing ratios (the ratio of fuel volume to fuel bed volume) are useful for comparison of fuel arrangement between vegetation types. For example, the packing ratios of forest species are half those of fynbos species, reflecting the sparse nature of the crowns of forest trees (Table 3.1). The significance of these differences for fires in fynbos and forest patches is discussed below. Values for the packing ratio for the chaparral shrub *Adenostoma fasciculatum* lie closer to those found in forest than in fynbos species in Swartboskloof (Table 3.1).

### 3.3.3 Seasonal Trends in Fuel Moisture Content

An important feature of fynbos fuels is that they are not characterized by seasonal curing. Curing can be defined as the process whereby live plant

**Table 3.1.** Comparative fuel properties for selected fynbos and forest plant species from the Swartboskloof catchment, chaparral shrubs and coniferous trees

| Species | Growth form | Energy contents ($J g^{-1}$) | Packing ratio | Crude fat content (%) | Moisture content (%) | Source[a] |
|---|---|---|---|---|---|---|
| **Fynbos species** | | | | | | |
| *Widdringtonia nodiflora* | Tree | 23 195 | 0.0044 | 1.6–2.2 | 54–167 | 1 |
| *Protea neriifolia* | Tall shrub | 22 695 | 0.0131 | 2.9–5.9 | 118–189 | 1 |
| *Cliffortia cuneata* | Tall shrub | 19 683 | 0.0016 | 2.6–3.3 | 91–178 | 1 |
| *Brunia nodiflora* | Tall shrub | 23 793 | 0.0094 | 8.1–10.3 | 86–120 | 1 |
| *Erica plukenetii* | Low shrub | 20 959 | 0.0077 | 3.6–7.1 | 51–119 | 1 |
| *Ischyrolepis gaudichaudiana* | Aphyllous herb | 20 838 | 0.0078 | 0.4–0.6 | 70–87 | 1 |
| **Forest species** | | | | | | |
| *Cunonia capensis* | Tree | 20 100 | 0.0029 | 2.0–3.5 | 136–243 | 1 |
| *Ilex mitis* | Tree | 20 258 | 0.0016 | 3.3–5.7 | 120–195 | 1 |
| *Kiggelaria africana* | Tree | 19 752 | 0.0016 | 2.9–3.4 | 152–233 | 1 |
| *Maytenus oleoides* | Tree | 20 329 | 0.0092 | 1.6–2.4 | 148–214 | 1 |
| *Brachylaena neriifolia* | Tree | 21 665 | 0.0014 | 4.3–5.2 | 166–319 | 1 |
| *Brabejum stellatifolium* | Tree | 22 116 | 0.0056 | 1.0–1.3 | 141–210 | 1 |
| **Chaparral shrubs** | | | | | | |
| *Adenostoma fasciculatum* | Tall shrub | 19 400 | 0.0007–0.0037 | 9.0–11.0 | 48–115 | 2,3,4 |
| *Arctostaphylos pungens* | Tall shrub | 21 000 | – | 9.0–9.6 | 50–90 | 2,5 |
| **Coniferous trees** | | | | | | |
| *Pseudotsuga menziesii* | Tree | 20 900–22 500 | – | 6.0–9.0 | 7.0–110 | 6 |
| *Pinus ponderosa* | Tree | 21 600–22 500 | – | 8.0–10.0 | 85–115 | 6,7,8 |

[a] Sources: 1, van Wilgen et al. (1990b); 2, Countryman and Philpot (1970); 3 Dell and Philpot (1965); 4, Philpot (1969); 5, Trujillo (1976); 6, Philpot and Mutch (1971); 7, Philpot (1963); 8, Mutch (1970)

material dries out following death, for example due to frost or drought, of above-ground plant parts. Seasonal curring is a common feature of grasslands, for example, and is important in determining whether certain vegetation types will burn or not. The proportion of dead to live fuel remains relatively constant in fynbos. Fires can therefore occur at any time of the year, provided that sufficient dead material has accumulated, and that the weather conditions are suitable (van Wilgen 1984a).

### 3.3.4 Live Plants

Foliar moisture contents of several species, including dominant fynbos shrubs and herbs, and forest trees, were monitored in Swartboskloof monthly for 1 year (van Wilgen et al. 1990b). No strong seasonal trends in the moisture content of live foliage were detected in any of the species (Fig. 3.3). The moisture contents in fynbos plants are equivalent to those found in coniferous trees (*Pinus* and *Pseudotsuga* species), but are generally higher than those recorded for Californian chaparral shrubs (*Arctostaphylos* and *Adenostoma* species, Table 3.1). Unlike fynbos, chaparral shrubs showed strong seasonal trends, with low (50%) moisture contents at the height of the fire season (Dell and Philpot 1965). Although fynbos has much lower moisture contents than co-occurring forest vegetation (Fig. 3.4), most species do not have the same low values found in chaparral shrubs in the fire season.

### 3.3.5 Dead Fuel Moisture

Seasonal trends in the moisture contents of fine dead fuels were simulated using weather data from the Swartboskloof weather station (1975–1986) in the United States National Fire Danger Rating System (Deeming et al. 1978). The system estimates the moisture content of fine dead fuels daily at 14:00, when fuel moisture contents are usually at the lowest. Results are shown in Fig. 3.5. There is a clear annual trend in fine dead fuel moisture, with low fuel moistures occurring from October to March; high fuel moistures prevail from April to September. There is far less of a trend in the minimum fuel moisture, however (Fig. 3.5). The minimum daily value for fuel moisture was never higher than 3% in any month. This indicates that fires could potentially occur in any month of the year.

### 3.3.6 Energy and Crude Fat Contents of Selected Species

The variation in the energy contents of selected species in Swartboskloof is not great (Table 3.1). The maximum and minimum values for fynbos species

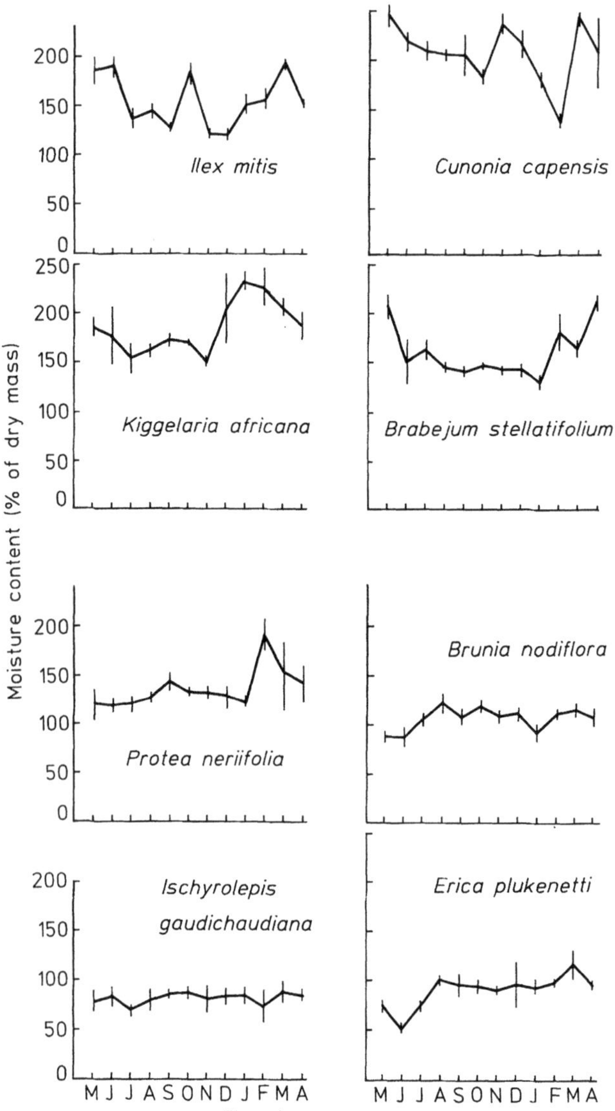


**Fig. 3.3.** Trends in the moisture content of live foliage of selected plant species in Swartboskloof. The species include typical fynbos herbs and shrubs (*Restio*, *Erica*, *Brunia* and *Protea*), and forest patch trees (*Kiggelaria*, *Cunonia*, *Ilex* and *Brabejum*). *Bars* are the 95% confidence limits of the mean (n = 5)

(23 793 J $g^{-1}$ for *Brunia nodiflora* and 19 683 J $g^{-1}$ for *Cliffortia cuneata*) deviate by only 6 and 10% from the mean respectively. Fire line intensity in fynbos fires can vary by several orders of magnitude (van Wilgen et al. 1985). The relatively small variations in the heats of combustion between

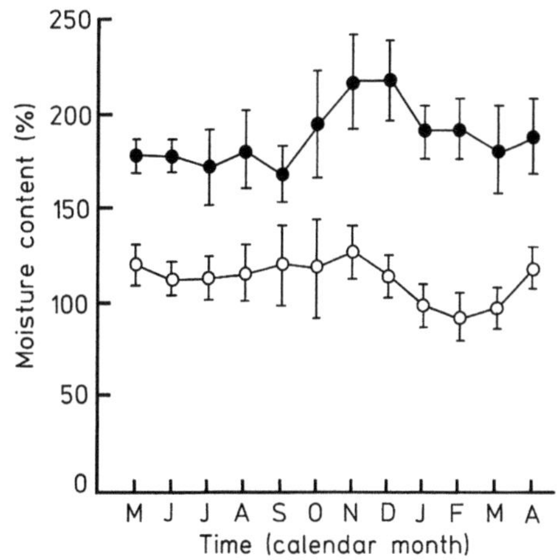


**Fig. 3.4.** Seasonal trends in the mean moisture contents of six fynbos (○) and six forest (●) species. Data are from van Wilgen et al. (1990b). *Bars* are the 99% confidence intervals of the mean (n = 30)

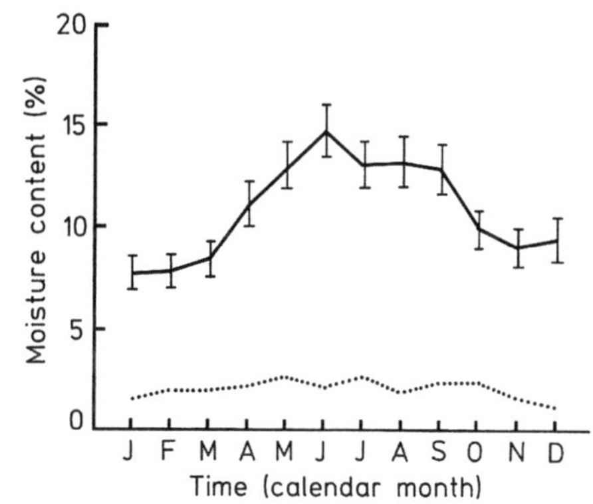


**Fig. 3.5.** Mean monthly trends in simulated fine, dead fuel (1-h fuel) moisture content. *Bars* are the 95% confidence intervals of the mean. The *lower curve* shows minimum values recorded in each month

species will not contribute significantly to the observed differences in fire intensity. Sensitivity analysis by means of the Rothermel model shows that moisture content (which influences the heat yield) and wind (which influences the rate of spread) will be far more important determinants of fire intensity (Burgan and Rothermel 1984; van Wilgen and Wills 1988).

Crude fat contents, expressed as a percentage of the dry mass, varied between 2 and 5% in forest species (Table 3.1). The range of crude fat contents between fynbos species was much greater. Very low fat contents (<1%) were found in *Ischyrolepis gaudichaudiana*. *Brunia nodiflora*, on the other hand, had a fat content predominantly above 10%, while both *Erica plukenetii* and *Protea repens* had high (6–7%) fat contents at times. As these shrubs make up the bulk of live biomass in fynbos, it is apparent that the vegetation can have a higher crude fat content than forest. Californian chaparral shrubs can have fat contents between 9 and 11% (Table 3.1), coniferous trees have fat contents between 8 and 10% (Table 3.1), while

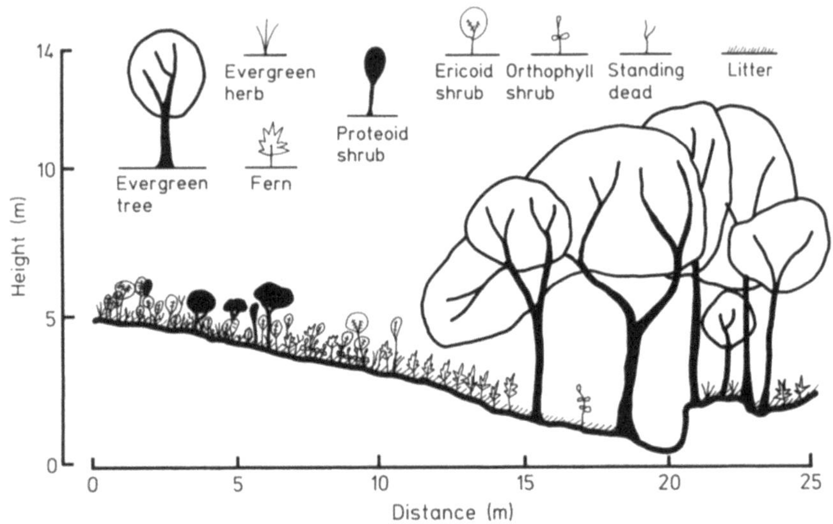


**Fig. 3.6.** Profile diagram from a 1-m-wide transect across the boundary between fynbos and forest communities in Swartboskloof (van Wilgen et al. 1990b)

Australian *Eucalyptus* trees have fat contents as high as 19.1% (Mutch 1970). Thus, although the chemical characteristics of fynbos plants indicate a higher flammability than forest plants, they are not as high as those found in other fire-prone ecosystems.

## 3.4 Fire in Fynbos and Forest Patches in Swartboskloof

The striking structural contrasts between fynbos and forest communities in Swartboskloof are reflected in the divergent nature of the fire regime that they support (see Chap. 5). While fires are common in fynbos, they seldom occur in the forest patches (van Wilgen et al. 1990b). The failure of fires to penetrate forest patches in Swartboskloof can be explained in terms of differences in the fuel properties of the two vegetation types. The moisture content of the live foliage of forest trees is generally higher than that of the fynbos shrubs and herbs (Fig. 3.4), while the crude fat contents are generally lower (Table 3.1). Fuel loads in forests in Swartboskloof can be as high or higher ($3685\,g\,m^{-2}$) than those found in the surrounding fynbos (van Wilgen et al. 1990b), but the spatial distribution of fuel differs markedly. This appears to be more important than other factors such as fuel loads, fuel moisture contents or fuel chemistry in determining the incidence of fire. Fynbos contains graminoid and restioid herbs and ericoid shrubs with fine leaves, and significant proportions of dead material that is held aloft to form a porous fuel bed about 1 m deep. Larger proteoid shrubs (2–3 m) dominate above this fuel bed, and burn readily under warm, dry conditions. Forests, on the other hand, are dominated by large trees (up to 25 m) with canopies of live foliage. Dead material is restricted to the compact litter layer below the tree canopies (see Fig. 3.6).

Fires could burn in the litter layers of forest patches (surface fires), or spread to the canopies of trees (crown fires). Usually crown fires are sustained by surface fires below them, but under severe conditions fires can advance through the crowns of trees leaving the surface fire some distance behind them. Crown fires have only been studied in coniferous woodlands and forests of the northern hemisphere, and to some extent in the *Eucalyptus* woodlands of Australia, and are generally poorly understood (Albini 1984). Van Wagner (1977) proposed a model for predicting conditions that would lead to crown fires in coniferous timber stands. The model assumes that crown fires are dependent on surface fires for active spreading. In order for crown fires to develop, surface fires have to reach a critical intensity (I, $kW m^{-1}$) and rate of spread (R, $m min^{-1}$), which varies according to crown bulk density (CBD, $kg m^{-3}$), foliar moisture content (FMC, percent) in the crowns of trees, and the lower crown branch height above the ground (LCBD in m). The critical intensity necessary is given by:

$$I = [(0.01\ LCBH\ (460 + 26\ FMC)]^{1.5}.$$

The critical rate of spread is given by:

$$R = 3/CBD.$$

Critical intensities and rates of spread needed to sustain a crown fire in Swartboskloof forests would be $5415 kW m^{-1}$ and $3.75 m min^{-1}$, using the above equations (van Wilgen et al. 1990b). Although these levels of fire behaviour are often exceeded in fynbos fires (van Wilgen et al. 1985), van Wagner's model assumes that the surface fire sustaining the crown fire is burning below the trees, and not in adjacent vegetation. In addition, "bridging fuels" are often necessary for the onset of crown fires (van Wagner 1977). Forest patches in fynbos have only dense litter layers to support fires, and lack bridging fuels. This makes the application of van Wagner's model inappropriate for the conditions found in Swartboskloof.

In Canada, only coniferous trees support crown fires, but broad-leaved trees and shrubs in other areas, such as Australian *Eucalyptus* species, and chaparral shrubs, do support crown fires. Although crude fats may play a role in this, van Wagner (1977) argues that foliar moisture content alone will explain the inability of fires to burn in the crowns of broad-leaved trees. For example, the moisture contents in foliage of conifers is around 70 to 130%, while chaparral and *Eucalyptus* species often have foliar moisture contents below 100%. Aspen, birch and maple trees in Canada, on the other hand, have foliar moisture contents of between 140 and 200% (van Wagner 1977). The spatial distribution of fuel, coupled with the high moisture contents of forest trees observed in Swartboskloof (Fig. 3.4) could therefore explain their inability to burn under most conditions.

## 3.5 The Effects of Invasion by Alien Trees and Shrubs on Fuel Properties

Swartboskloof is prone to invasion, and is at considerable risk from surrounding sources of woody weeds (see Chap. 9). Invasion of the habitat by introduced plants affects fire behaviour by changing the structure of the fuel bed. Invasion of mountain fynbos by *Hakea sericea* can increase fuel loads (dead material plus live material <6 mm) by 60%, as well as lower the moisture content of live foliage, thus increasing fire hazard (van Wilgen and Richardson 1985). Invasion of mountain fynbos by pines could increase biomass by up to 300% (Versfeld and van Wilgen 1986). The fuel bed in such stands is effectively confined to the litter layer below the trees, under all but severe conditions of fire hazard. This would cause fires to spread slowly and burn at low intensities under moderate weather conditions. Under extreme weather conditions, however, fires in such invaded stands are known to burn live canopies in thickets of invasive shrubs and coniferous trees. The increased energy released by such fires makes them self-sustaining, difficult to control and potentially more damaging to ecosystems than fires in natural vegetation. The fire regime under such conditions is likely to change. Fires will become less frequent (due to the changes in fuel distribution) and more intense (due to the higher fuel loads and extreme weather conditions under which fires will burn).

## 3.6 Fuel in Swartboskloof and Other Ecosystems

Comprehensive studies that quantify the full range of fuel properties for a given vegetation type are rare. Most studies aimed to provide data for fire behaviour prediction models, and have concentrated on single species, or have been restricted to one or a few descriptors of fuel, such as biomass. Empirical fire behaviour prediction studies have used either fuel mass data (e.g. McArthur 1966, 1967) or data on spatial arrangement (e.g. van Wagner 1977). Models of fuel accumulation and decay (Walker 1981) have been used to support the McArthur models (e.g. Kessell et al. 1982). The development of physical fire spread models (Rothermel 1972; Kessell et al. 1978) led to the collection of fuels and fire behaviour data to support these models (Sneeuwjagt and Frandsen 1977; van Wilgen et al. 1985, 1990a; van Wilgen and Wills 1988; Everson et al. 1988). Fuel data needed for the Rothermel approach were drawn from sometimes extensive studies of the

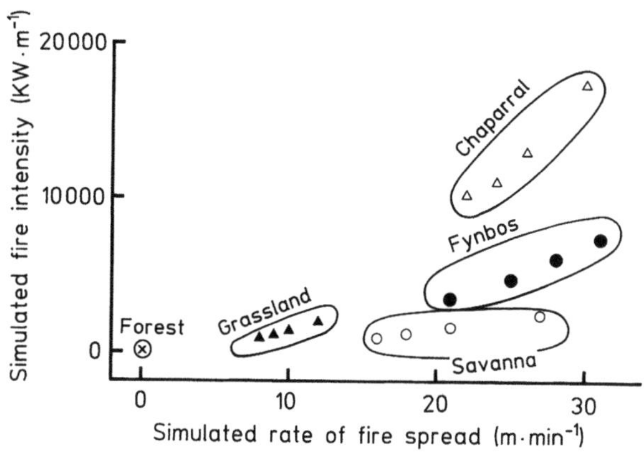


**Fig. 3.7.** Comparisons of predicted fire behaviour from Rothermel's (1972) fire model in five vegetation types under various conditions of fine, dead fuel moisture (2, 5, 8 and 11%). Other conditions used in the simulations were 8 km $h^{-1}$ for wind, zero slope, and 100% fuel moisture content for live foliage. Data for the fuel models are from van Wilgen (1984a) (fynbos); Everson et al. (1988) (grassland); van Wilgen and Wills (1988) (savanna); van Wilgen et al. (1990a) (Afromontane forest); and Andrews (1986) (chaparral)

fuel properties of single species, mainly in North America. The chaparral shrub *Adenostoma fasciculatum* (chamise) is a good example (Dell and Philpot 1965; Countryman and Philpot 1970; Rothermel and Philpot 1973). Studies of the fuel properties of vegetation communities, in order to clarify the fire regimes which characterize them, are rare, but some studies are emerging (Kauffman et al. 1988; van Wilgen et al. 1990b).

Data on fuel loads, fuel moisture or fuel chemistry alone would be insufficient to clarify differences between ecosystems. Full data sets suitable for use as inputs to fire behaviour prediction models would allow for comparisons between vegetation types. Due to the diverse nature of fuel studies, complete comparison of the data collected at Swartboskloof with those collected elsewhere are not always possible. However, examination of predictions from several fuel models, where these are available, can be used to illustrate the effect of fuel differences on the nature of fires. Under similar conditions of wind and fuel moisture, fires in fynbos will burn at a lower intensity than those in chaparral, but at a higher intensity than those in grasslands, savanna of Afromontane forest (Fig. 3.7).

The differences between fynbos and chaparral are of interest in terms of comparisons between mediterranean-type ecosystems. Fire-free intervals in chaparral are generally longer than in fynbos (Kruger 1983) and at least some of this could be explained by differences in the fuel properties of the vegetation. Many chaparral stands are dominated by one or a few shrub species (Hanes 1971), and the herbaceous understorey is almost always absent. Consequently, the shrubs must build up sufficient fuel to support a fire in the shrub crowns; this only occurs at a minimum of 25 years after the previous fire (Green 1981). Fynbos, on the other hand, burns readily at 4 to 6 years after a previous fire, primarily due to the herbaceous and fine leaved shrubby fuels. The nature of fuels therefore dictate, to some degree at least, the fire regime. Chaparral fires are less frequent, and burn at relatively high intensity, as longer intervals between fires mean that greater fuel loads accumulate. Dominant fynbos shrubs that regenerate from seed would not survive such regimes due to early senescence (see Chap. 4).

## 3.7 Conclusions

Comprehensive data on the fuel properties of vegetation communities are rare, but have been well documented at the Swartboskloof site. The data allow for an examination of the factors that influence the occurrence and behaviour of fires. The studies conducted at Swartboskloof have highlighted the importance of vegetation structure in determining the fire regime. Differences in the vertical structure of plants and dead plant material, rather than fuel loads or fuel chemistry, account for the difference in fire behaviour in forests and fynbos, for example. While there is a wealth of data on the fuel properties of Californian chaparral, data for other mediterranean-type ecosystems in Europe, Australia and South America are lacking. For example, although there are data on fuel loads for Australian heaths, no systematic accounts of the oil contents or moisture contents of the vegetation exist (Gill 1981b). Comprehensive data on the full range of fuel properties for the vegetation as a whole would be of considerable interest for examining the nature of fire regimes in other mediterranean-type ecosystems.

Fires burn with ease across community boundaries within the *fynbos* communities in Swartboskloof. The proteoid shrublands, ericoid-restioid shrublands and hygrophilous communities burn with equal ease, indicating similar fuel properties. It is therefore unlikely that the fuel properties of these communities play a role in maintaining their boundaries, even if some differences in fuel accumulation rates and structure do exist between them. Fire is far more likely to play a role in the dynamics of the forest-fynbos boundary, and the relative contribution of fire and other factors in these dynamics are addressed in Chap. 5. Invasion by trees and shrubs (see Chap. 9) will also bring about changes in the fuel properties, and therefore the fire regime, of fynbos. Such changes to the fire regime will potentially eliminate fynbos plant species that are fire-adapted and fire-dependent (van Wilgen 1982).

Finally, comprehensive data on the fuel properties of vegetation are necessary for the prediction of fire behaviour, and indirectly therefore of fire effects. The use of fire behaviour prediction models to optimize the application of fire in vegetation management will become more important as the cost of management increases. Such models have not yet found widespread use in ecosystem management, but deserve more attention from resource managers.

# 4 Regeneration Strategies in Fynbos Plants and Their Influence on the Stability of Community Boundaries After Fire

B.W. van Wilgen and G.G. Forsyth

## 4.1 Introduction

An analysis of the range of regeneration strategies present in fynbos communities was a major aim of the work conducted at Swartboskloof. Information on the mechanisms whereby plants survive fire (e.g. Bell et al. 1984) and the attributes which allow them to persist in vegetation subjected to repeated fire (e.g. Noble and Slatyer 1980) is required to develop a sound understanding of the dynamics of vegetation subjected to fire. Such information is often only available for all of the plants if the communities are species-poor (e.g. Hobbs et al. 1984). In species-rich communities, it is usually the case that detailed information is only available for the dominant plants. A detailed knowledge of the responses of the constituent species in fynbos is restricted to the dominant shrubs of the family Proteaceae (e.g. Bond et al. 1984). A few studies have listed survival mechanisms in other species (van der Merwe 1966; van Wilgen 1981; van Wilgen and Kruger 1981), but data on other vital attributes (juvenile periods, longevity, modes of seed dispersal, and seed bank longevity) are not available for the vast majority of fynbos species. Fynbos has a rich diversity of species, and fynbos communities may differ in the spectrum of responses (and therefore in their vulnerability to fire). In this chapter we quantify the range of regeneration strategies manifested in a sample of fynbos plants subjected to fire, and examine the differences between major vegetation communities in terms of these strategies. We also compare the data to those collected in other fire prone ecosystems.

## 4.2 Survival Mechanisms and Vital Attributes of Species

### 4.2.1 Systems for the Classification of Species

Several systems for the classification of plant species responses to fire have been developed (e.g. van der Merwe 1966; Noble and Slatyer 1980; Gill

1981a; Bell et al. 1984). In this study we examined the responses of plants and communities to fire in terms of the systems described by Bell et al. (1984) and Noble and Slatyer (1980). The systems are outlined briefly below.

Bell et al. (1984) defined five categories of species response to fire. Fire ephemerals (FE) include monocarpic therophytes and annuals, or short-lived polycarpic perennials which exhibit a degree of fire-stimulated germination. They have fast growth rates, early reproductive maturities, high reproductive efforts proportional to biomass, and relatively short (<4 years) life spans. Obligate seeders (OS) are long-lived (>15 years) and are likely to have their growth cycle terminated prematurely by fire, as they do not possess the ability to sprout. Sprouting species are a diverse group, and three categories are recognized. Obligate vegetative reproducing sprouters (OVS) display great powers of multiplication by vegetative (non-sexual) means, but virtually no ability to reproduce by seed. Facultative sprouter-seeders (FSS) sprout weakly, or only after mild fires, but recruitment from seed is highly effective. Autoregenerating long-lived sprouters (ALS) are both successful sprouters, and generate moderately dense stands of seedlings after fire.

Noble and Slatyer (1980) regarded a set of attributes as important in determining the response of a species to disturbance. These included (1) mechanism of persistence during a disturbance, or arrival of propagules afterwards; (2) ability to become established either immediately after a disturbance or later in the post-fire development; (3) the timing of critical life-history events. In regard to the first of these, species are divided into:

D species: arrival of widely dispersed seeds;
S species: seeds with a long viability, stored in the soil;
C species: seeds with short viability surviving disturbance within protective fruits or cones;
V species: persistence by part of the individual surviving the disturbance, and recovery by vegetative growth, with immature tissue;
U species: as for V species, but sprouting mature tissue;
Δ species: a species with both D and U characteristics.

Similarly, three sets of species were recognized in respect of the second group of attributes:

T species: tolerant species, able to establish at any time, tolerating competition with adults of the same or other species;
I species: intolerant species, able to establish only immediately after disturbance, when competition is reduced;
R species: requiring species, unable to establish immediately after a disturbance, but able to establish once mature individuals of the same or other species are present.

Thirdly, with reference to the times taken to reach certain stages in life history, the following were regarded as critical:

1. The time when a disturbance occurs (o) and the time for propagules to become available on a disturbed site (p);
2. The time taken for individuals to reach reproductive maturity (m);
3. The life span of individual plants of the species (l);
4. The time taken for all propagules to be lost from the community (e.g. for the seed store to become depleted) and for the species to become locally extinct (e).

### 4.2.2 Selection and Characterization of Species

We studied pre-fire composition and post-fire succession on ten permanently marked quadrats (5 × 10 m). These were placed in proteoid and ericoid-restioid shrublands prior to the March 1987 fire in Swartboskloof (see Fig. 1.3). Quadrats were surveyed once before the fire, and at intervals of 9, 18 and 30 months after the fire. Post-fire surveys were done in the spring (except the first one), when most species are in flower. At each survey, all the plant species occurring on the quadrats were identified and listed. Two more quadrats were laid out in the specialized hygrophilous communities after the 1987 fire, and these were also surveyed at three intervals after the fire.

The quadrats were subdivided into 50 sub-plots of 1 $m^2$ each. At each survey, the maximum height of the vegetation, and an ocular estimate of the percentage canopy cover was made on each of the 50 sub-plots within each quadrat. In addition, the three dominant species (in terms of visual estimates of live biomass) were listed and ranked on each sub-plot. Species were allocated importance values in proportion to their rank, as follows: rank 1 = 1, rank 2 = ⅔ and rank 3 = ⅓. These importance values were used in comparisons of the relative contribution of fire survival strategies and vital attributes between the major vegetation communities (see below).

We recorded the mode of post-fire regeneration (whether from seed or sprouting) and the date of first flowering after fire for each of the species. Each species was assigned a fire response category, and a description of the mechanism of persistence, based on morphological observations. Non-sprouting species were classified as either C, D, or S based on seed morphology and seed shed characteristics. C species are those where the seeds are retained on the parent plant; D species those with structures (such as wings or pappi) that enable the dispersal of seeds, or fruits that are dispersed by birds. The remaining non-sprouting species were classified as S species. Age to maturity (first flowering), where this was 3 years or less, was obtained from the quadrat data. For species that took longer to mature, data on age to maturity was obtained by searching for mature specimens of

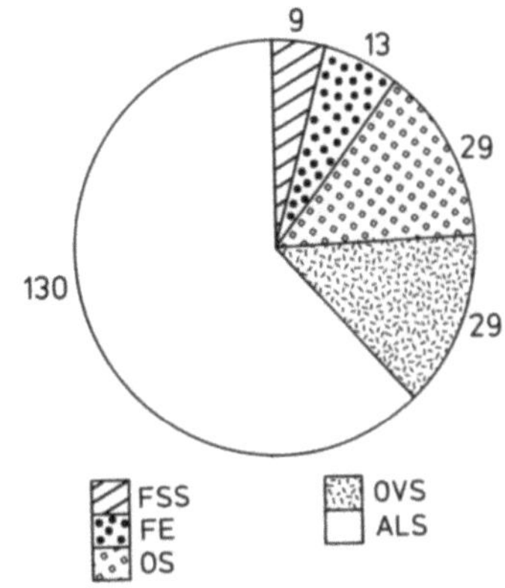


**Fig. 4.1.** Distribution of fire-response types among 210 plant species from Swartboskloof. The fire-response types are: *ALS* autoregenerating long-lived sprouters; *FE* fire ephemerals; *FSS* facultative sprouter-seeders; *OS* obligate seeders; OVS obligate vegetative reproducing sprouters. The number of species in each life-form is indicated on the diagrams

the species in adjacent areas of known post-fire ages. In addition, the comprehensive herbarium collection at Jonkershoek was searched for data on maturity. Data on longevity were obtained in a similar manner, but for species with longevities that exceeded 45 years (the longest that vegetation in the Jonkershoek valley had survived fire), longevity was simply given as >50 years. Fynbos that survives fire for longer than 50 years is almost unknown, and this assumption will have no effect on comparisons of fire frequencies of less than 50 years. For species which rely on soil-stored seed banks to regenerate, seed bank longevity was estimated as greater than the period between longevity and the occurrence of fire, where the given species was known to have regenerated after fire. For other species, seed bank longevity data were not considered essential to the analysis.

## 4.3 Results and Discussion

### 4.3.1 Spectra of Fire Response and Vital Attribute Types

Swartboskloof contains 651 indigenous species of flowering plants and ferns (McDonald and Morley 1988). We encountered 210 species on the 12 quadrats (Appendix 4.1). The most important fire response category (Fig. 4.1) was autoregenerating long-lived sprouters (62% of the species). This category included all of the species where all or most individuals sprouted after fire, and where seedlings of the species appeared after fire. The fate of the seedlings is not known, and many may die before they reach maturity (see Chap. 7). Obligate vegetative reproducing sprouters (14% of the species) included all species which sprouted strongly, but for which we did not find any seedlings. Of the 29 species classified as obligate vegetative reproducing sprouters, 16 were perennial herbs from the family Restionaceae. Species which reproduced by seed were made up by obligate seeders (14%), fire ephemerals (6%) and facultative seeder-sprouters (weak sprouters, 4%).

Of the 51 species that reproduce from seed, most (31) regenerate from soil-stored seed banks (SI and ST, Fig. 4.2). Widely dispersed seeds were found in 18 species, and canopy-stored seeds in only 2 species. Most sprouting species (70%) matured within the first year, and were classified as either U or Δ species. The remainder were classified as V species. Most species (91%) were classified as intolerant (I), while some (7%) were able to establish in mature vegetation and were classified as tolerant (T). Only five species (2%) were classified as requiring (R) a mature overstorey to establish from seed. Obligate vegetative reproducing sprouters were classified as I, but as no seedlings were found, they could be R species. Such a strategy would imply that the species only establish as seedlings if fire-free intervals are very long, as is the case with such species in Californian chaparral (Keeley 1986).

### 4.3.2 Age at Maturity

Of the 210 species, most (86%) flowered in the first year after fire; a further 7% flowered in the second, 5% in the third, 1% in the fourth, and 2% in the fifth year after fire. Almost all of the sprouting species (148 of 159 species) flowered in the first year after fire. Non-sprouting species were somewhat slower to mature, with 30, 10, 6, 2, and 3 out of 51 species flowering in the first, second, third, fourth and fifth years respectively.

### 4.3.3 Longevity of Individuals and Seed Banks

Most species in Swartboskloof are long-lived; 173 of the 210 species occur in vegetation with a post-fire age of >45 years as mature, long-established individuals. Twenty species were short-lived (<10 years), and only one annual species (*Sebaea exacoides*) occurred. There were 18 species of moderate longevity (15–40 years). Seed bank longevities were estimated only for those species that rely totally on a soil-stored seed bank in order to survive fires (see Appendix 4.1). Short-lived species, such as some *Aspalathus*, *Roella*, and *Thesium* species, must have fairly long-lived seed banks as they disappear early in the post-fire succession, and reappear after fire from the soil-stored seed bank. For longer-lived species, such as some *Blaeria*, *Cliffortia* and *Phylica* species, seed bank longevity could be much

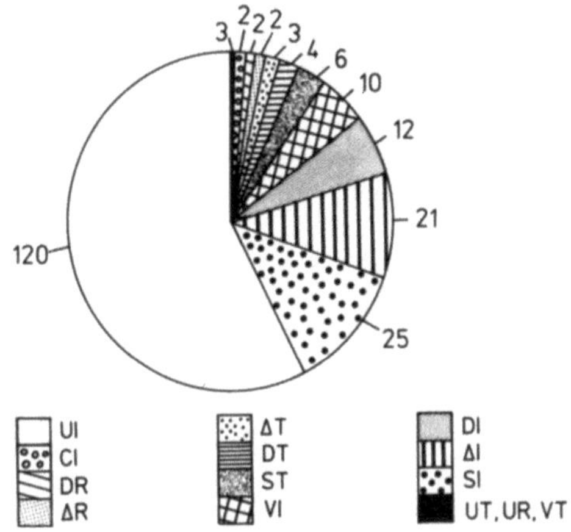


**Fig. 4.2.** Distribution of vital attribute types among 210 plant species from Swartboskloof. Each type indicates both the method of persistence and the requirement for establishment. The methods of persistence are: *C* canopy-stored seeds; *D* dispersed seeds; *S* soil-stored seeds; *U* sprouting mature tissue; *V* sprouting immature tissue; and Δ U and D features combined. The requirements for establishment are: *I* intolerant; *T* tolerant; and *R* requiring. The number of species in each type is indicated on the diagram

shorter. The population sizes of these species, and many *Erica* species, decline markedly with increasing age, but some individuals do remain in old vegetation.

### 4.3.4 The Timing of Critical Life History Events

The time at which plants reach critical stages in their life histories can be summarized with reference to the five recognized phases of post-fire succession in fynbos (Kruger and Bigalke 1984; Table 4.1). We recognized nine fire-response patterns (Table 4.2). A given pattern may apply to a range of different vital attributes or fire-response types, as illustrated in Table 4.2. Each of the patterns is discussed below.

The first pattern was the most common, and applied to 160 of the 210 species in the sample. These species persist on, or return to the site and mature within the youth period (1 year after fire). They have life spans in excess of 50 years, or are able to establish in the interfire period, and are found at all stages of the post-fire succession. The vast majority of these species have the ability to sprout, and are classified as ALS or OVS in terms of fire-response types, and as UI or ΔI in terms of vital attributes. Species classified as UI and ALS (the most common combination) may differ considerably in their ecology, although the survival mechanisms and outcomes of disturbances at different stages of their life cycles remain the same. Consider for example *Maytenus oleoides* (a tree) and *Cyrthanthus ven-*

**Table 4.1.** Post-fire successional phases in fynbos (After Kruger and Bigalke 1984)

| Phase | Period after fire (years) | Characteristics |
|---|---|---|
| Immediate post-fire | 0–1 | Seed germination and vegetative regeneration takes place. Most annuals and some geophytes reproduce only in this phase. |
| Youth | 1–5 | Graminoid herbs and sprouting shrubs dominate. Canopy cover approaches pre-fire levels. |
| Transitional | 5–10 | All plants reach reproductive maturity. Non-sprouting shrubs emerge from the canopy. |
| Mature | 10–30 | Tall shrubs reach maximum height and flowering activity. Low shrubs begin to die. |
| Senescent | 30–60? | Mortality of tall shrubs accelerates. Crowns open and litter builds up. |

*tricosus* (a geophyte). Both sprout after fire, and both are long-lived. *Maytenus*, where it occurs, remains a dominant feature of the above-ground vegetation, and flowers and produces fruit annually. *Cyrthanthus*, on the other hand, flowers only immediately after fires, and appears to survive even long interfire periods as a dormant bulb. The first pattern also applies to a small number of long-lived, non-sprouting species, which are killed by fire.

The second pattern is almost identical to the first, but contains species which mature in the immediate post-fire phase (2–5 years after fire), rather than in the first year after fire. About half of these species possess the ability to sprout. The non-sprouters include plants such as the three *Cliffortia* species, which are long-lived but establish only immediately after fire, and

**Table 4.2.** Fire-response patterns (illustrated by the timing of critical life history events) in 210 species in Swartboskloof, and the range of vital attributes (Noble and Slatyer 1980) and fire-response types (Bell et al. 1984) in each response pattern

| Phase of fynbos succession[a] Immediate post-fire | Youth | Transitional | Mature | Senescent | Vital attributes[b] | | Fire-response types[b] | | Total number of species |
|---|---|---|---|---|---|---|---|---|---|
| opm | | | | le | UI | (122) | ALS | (122) | 160 |
| | | | | | ΔI | (21) | OVS | (26) | |
| | | | | | ST | (5) | OS | (10) | |
| | | | | | DT | (3) | FSS | (2) | |
| | | | | | SI | (4) | | | |
| | | | | | ΔT | (2) | | | |
| | | | | | UR | (1) | | | |
| | | | | | UT | (1) | | | |
| | | | | | ΔR | (1) | | | |
| op | m | | | le | SI | (10) | OS | (10) | 24 |
| | | | | | VI | (8) | ALS | (7) | |
| | | | | | DI | (2) | FSS | (4) | |
| | | | | | ΔT | (1) | OVS | (3) | |
| | | | | | ST | (1) | | | |
| | | | | | DT | (1) | | | |
| | | | | | VT | (1) | | | |
| opm | l | | | e | SI | (6) | FE | (11) | 11 |
| | | | | | DI | (5) | | | |
| opm | | | l | e | DI | (4) | OS | (5) | 6 |
| | | | | | SI | (2) | FSS | (1) | |
| o | | p | m | le | DR | (2) | FSS | (2) | 3 |
| | | | | | ΔR | (1) | ALS | (1) | |
| op | m | | l | e | DI | (1) | OS | (2) | 2 |
| | | | | | SI | (1) | | | |
| op | m | | le | | CI | (2) | OS | (2) | 2 |
| op | m | l | | e | SI | (1) | FE | (1) | 1 |
| opm | | l | | e | SI | (1) | FE | (1) | 1 |

[a] Symbols for critical life history events are: (o), time of disturbance; (p), time at which propagules are available on site; (m), time at which reproductive maturity is reached; (l), local loss of individuals from the community; (e), local extinction from the community.
[b] Abbreviations for vital attributes and fire-response types are explained in the text; number of species in parentheses.

*Chrysanthemoides monilifera*, which is short-lived but can establish between fires. Sprouting species exhibiting this pattern were classified as V in terms of the mechanism of persistence, as opposed to U, which applies to the first pattern.

The third pattern contains most of the plants classified as fire ephemerals. These species all establish from seed after fire. They complete their life cycles within the immediate post-fire phase, but none are annuals. They then either persist as soil-stored seeds, or disperse back to the site from adjacent areas after the next fire. Of the species that are able to disperse widely, two are exotics; the other three are herbaceous Asteraceae species.

The fourth and sixth patterns are similar, and represent non-sprouting plants that either flower within the first year of fire, or shortly thereafter, respectively. Some of these species, such as *Helichrysum cymosum*, *Pelargonium elongatum* and two *Senecio* species, possess the ability to disperse back into the site after fire. Some, such as *Blaeria dumosa*, and *Pentaschistis malouinensis* rely on a soil-stored seed bank. There are no data on the longevity of soil-stored seed banks for any of the species concerned, and this represents a critical gap in the ability to predict the effects of long interfire periods. *Sebaea exacoides*, the only annual species we encountered, fits this pattern as it is able to establish in the interfire period.

The fifth pattern applies to species which require an overstorey of mature vegetation to establish, and includes two trees (*Maytenus acuminata* and *Kiggelaria africana*) and one shrub (*Myrsine africana*). Seeds of these species are dispersed by birds and establish in the transitional phase of fynbos succession. The two tree species need to reach a large size before they can survive fires by sprouting, but it is possible that, given long (>50 years) interfire periods, the species could establish and create foci for further colonization by tree species with bird-dispersed propagules. Such foci would be fire-resistant (van Wilgen et al. 1990b) and could lead to the establishment of clumped forest patches.

The seventh pattern is essentially the same as the third. It represents plants which could best be described as long-lived fire ephemerals. These plants all regenerate from soil-stored seed banks, and complete their life cycles before the mature stage of fynbos succession is reached. Although no data on seed longevity are available, seed banks must persist for up to 40 years in some cases, as abundant regeneration occurs at some sites after long fire-free periods. The ninth pattern applies to only one species (*Aspalathus ciliaris*), which is a long-lived fire ephemeral that matures in the second year after fire.

The eighth pattern describes two shrub species (*Protea repens* and *P. neriifolia*) which are killed by fire, and rely on seed banks stored in the canopies of the shrubs for regeneration. The species have moderately long life spans (up to 35 years), but seed banks do not persist beyond the life span of the adults. The species will thus not survive fires which occur at intervals of longer than 35 years. Secondly, although the species mature within 3 to 4 years after fire, it takes at least 10 years for adequate seed banks to build up in the canopies of plants. Of the 210 species, these two species were considered the most vulnerable to both short and long interfire periods. Should the species be lost from the community, they represent a small proportion of the total diversity. However, they can contribute up to 89% of the above-ground and 53% of the below-ground biomass in some communities in Swartboskloof respectively (van Wilgen 1982; Higgins et al. 1987). The dominance of these species, combined with their relative vulnerability, has led to fire frequencies in fynbos being prescribed to accommodate these species.

### 4.3.5 Differences Between Communities

We examined the relative contribution of various categories of fire survival strategies and vital attributes to each of the major vegetation communities (see Chap. 1). The quadrat data from both pre- and post-fire surveys were divided into proteoid shrublands (five quadrats), ericoid-restioid shrublands (three quadrats), specialized hygrophilous communities (two quadrats), and a frequently burnt firebreak (one quadrat). For each sub-set thus defined, the contribution of each category of fire survival strategy and vital attributes to the community as a whole was examined in two ways. Firstly, the number of species in each category was listed for the combined pre- and post-fire data. Secondly, the relative contribution of the various categories in terms of surrogate measures of biomass was calculated for the pre-fire and last post-fire surveys. To do this, importance values for each species on each sub-plot (see above) were totalled per species for each quadrat. A surrogate measure of live biomass was calculated for each species, by multiplying the total importance value (described above) of each of the species per quadrat by the mean cover and height of the vegetation for that quadrat. Data for the quadrats were then pooled to obtain means, and summarized by vital attributes and fire survival strategies for each of the major communities.

The communities are remarkably similar in terms of the proportions of species in each of the categories (Figs. 4.3 and 4.4). Obligate seeders and fire ephemerals were less prominent in the firebreak and hygrophilous communities, but differences were not great. Similarly, the relative composition in terms of vital attributes was similar in all communities.

A total of 197 species were encountered on the ten quadrats that were surveyed pre-fire (the two quadrats in hygrophilous communities were not enumerated pre-fire). Of these, 134 occurred pre-fire, and 193 post-fire. Changes in the composition of the vegetation after the fire were not great, and similarity of pre- and post-fire vegetation ranged from 72 to 89% (Table 4.3) on the ten quadrats surveyed. Most of this difference is explained by the addition of new species post-fire. While 63 species were added to the list, only 4 species disappeared.

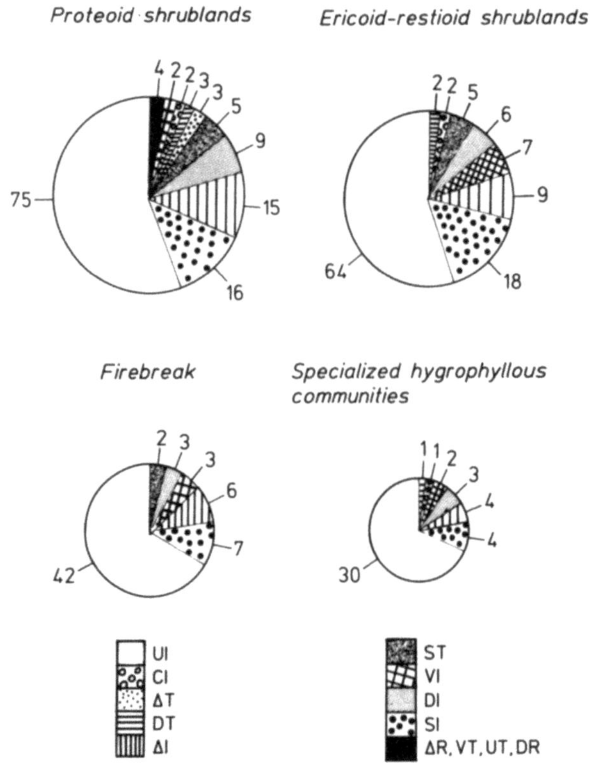


**Fig. 4.3.** The relative contribution of species, in terms of vital attributes, to four vegetation communities in Swartboskloof. The size of the circle reflects the total number of species encountered. The number of species in each category is indicated on the diagrams; abbreviations as in Fig. 4.2

**Table 4.3.** An analysis of the pre- and post-fire species composition on 10 × 10 m quadrats in three vegetation communities subjected to fire in Swartboskloof

| Community | Number of species pre-fire | Number of species post-fire | Species lost after fire | Species gained after fire | Total species | Pre- to post-fire similarity (Sorenson's index) |
|---|---|---|---|---|---|---|
| Proteoid shrublands | 88 | 131 | 3 | 46 | 134 | 77.6 |
| Ericoid shrublands | 67 | 109 | 4 | 46 | 113 | 71.6 |
| Firebreak | 53 | 60 | 3 | 10 | 63 | 88.5 |
| Entire survey | 134 | 193 | 4 | 63 | 197 | 81.7 |

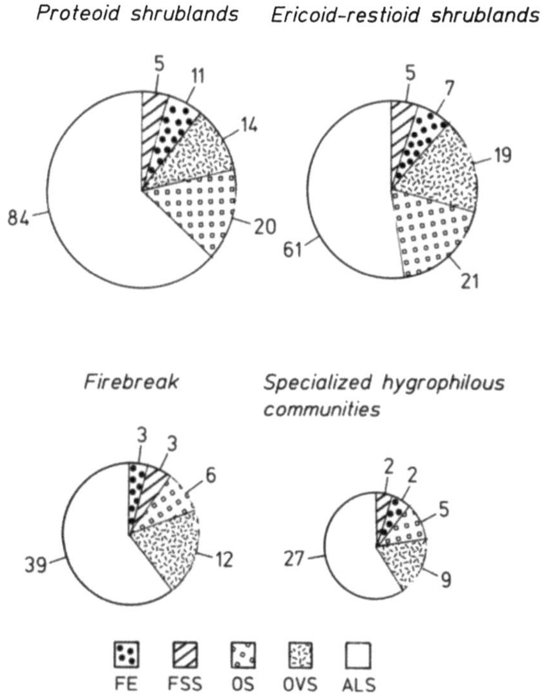


**Fig. 4.4.** The relative contribution of species, in terms of fire-response types, to four vegetation communities in Swartboskloof. The size of the circle reflects the total number of species encountered. The number of species in each category is indicated on the diagrams; abbreviations as in Fig. 4.1

All of the communities examined had more species after fire than before fire. The species that appeared after the fire were either fire ephemerals (regenerating from widely dispersed or soil-stored seed banks), geophytes (which were probably present as dormant bulbs but were undetected pre-fire), species that were present as soil-stored seed pools, or species with the ability to colonize the site by means of soil-stored seed banks (Table 4.4). Examples of the dispersed species include fire ephemerals such as *Ursinia pinnata*, *Solanum tomentosum* and *Senecio pubigerus*. Examples of species regenerating from seed pools includes all species of *Thesium*, *Roella ciliata* and *Aspalathus ciliaris*. Many geophytes, such as *Bulbine tuberosa*, *Cyrtanthus ventricosus*, *Ornithogalum hispidum* and *Urginea dregei*, flowered after the fire, but were either dormant, or remained undetected as they were inconspicuous before the fire. Some species that were added to the list were classified as long-lived sprouters, and should have been present before the fire. These may either have been overlooked in the dense vegetation before the fire, or may have regenerated from seed. Similar numbers of species were added to both the proteoid and ericoid-restioid shrubland communities

**Table 4.4.** An analysis of the number of species gained after fire on 10 × 10 m quadrats in three vegetation communities in Swartboskloof

| Community | Fire ephemerals | Geophytes | Species regenerating from soil seed pools | Widely dispersed species | Long-lived sprouters | Total species gained |
|---|---|---|---|---|---|---|
| Proteoid shrublands | 11 | 20 | 3 | 5 | 7 | 46 |
| Ericoid-restioid shrublands | 7 | 14 | 8 | 9 | 8 | 46 |
| Firebreak | 2 | 5 | 1 | 0 | 2 | 10 |
| Entire survey | 12 | 27 | 5 | 8 | 11 | 63 |

(Table 4.4). In proteoid shrublands, geophytes and fire ephemerals contributed 67% to the list of species gained. This proportion was smaller (46%) in ericoid-restioid shrublands, which had a higher proportion of species than soil-stored or widely dispersed seeds. There was little change in the firebreak flora.

Of the four species that were lost, two were trees (*Pinus radiata*, an exotic, and *Maytenus acuminata*), and two others were only present as isolated individuals before the fire. As the latter two species have been observed to establish well after fire in other areas, their loss was not real.

The contribution of various regeneration strategies to the pre- and post-fire biomass of the various communities is shown in Figs. 4.5 and 4.6. The largest change occurred in proteoid shrublands, where CI species, which make up only 1% of the species (Fig. 4.2), account for almost 30% of the pre-fire biomass. After fire, these plants have all but disappeared, and will only return to dominance after 10 years (Table 4.1). In all communities, fire ephemeral species appeared in the post-fire flora, and their contribution in terms of biomass was greatest in the proteoid shrublands. There was very little change in the firebreak community, as fire-sensitive species no longer occurred there.

### 4.3.6 Fire and the Stability of Community Boundaries

The data were used to examine the theoretical effects of varying the fire frequency on the survival of species (Table 4.5). For the data set as a whole, the highest number of species (195 out of 210) survive fire frequencies of between 10 and 15 years. Only four species are eliminated by frequent (5-yr intervals) fire. These are the two *Protea* species (which would not have time to mature) and the two tree species (*Kiggelaria africana* and *Maytenus*

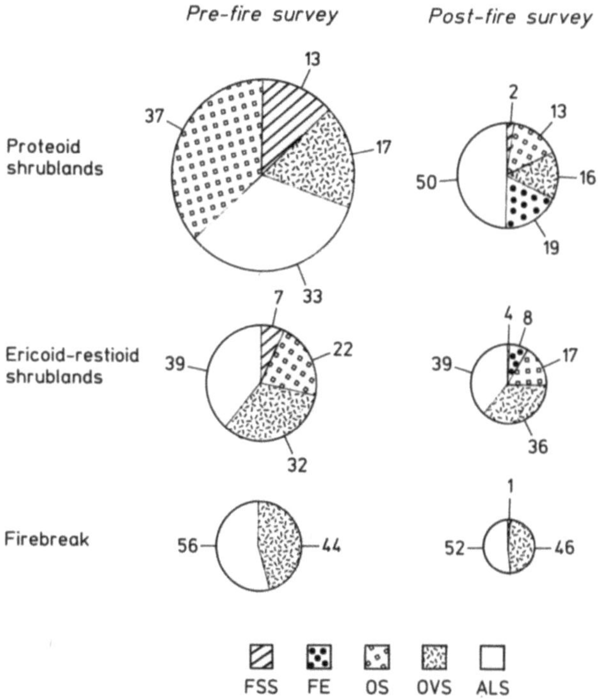


**Fig. 4.5.** The relative contribution of fire-response types, in terms of biomass, to three vegetation communities immediately before and 30 months after fire in Swartboskloof. The size of the circle reflects the relative biomass. The relative contribution (%) of each category is indicated on the diagrams; abbreviations as in Fig. 4.1

*acuminata*) which require a mature overstorey to establish. Only the two *Protea* species would be eliminated by long intervals between fires (due to death of adults and loss of seed pools), but up to 16 species were listed as being potentially eliminated. These included many species in the SI class, which face possible elimination should their seed pools not survive the long interfire periods. This analysis is supported by the results of other studies conducted in the Jonkershoek valley (van Wilgen 1981). Non-sprouting *Protea* species were eliminated from a firebreak burnt at 6-year intervals, as well as being almost eliminated through senescence at 37 years post-fire. Tree species were only found in vegetation with a post-fire age of 37 years. Many of the SI species, particularly in the genera *Erica* and *Thesium*, were not found in older vegetation. These were presumably present as seed pools.

The analysis for separate communities (Table 4.5) shows that, in all cases, fire cycles of 10–15 years maintain the highest number of species. In proteoid shrublands, the three species eliminated by certain fire frequencies are the non-sprouting shrubs *Protea repens* and *P. neriifolia*, and the tree *Maytenus acuminata*. The two *Protea* species are also eliminated in ericoid

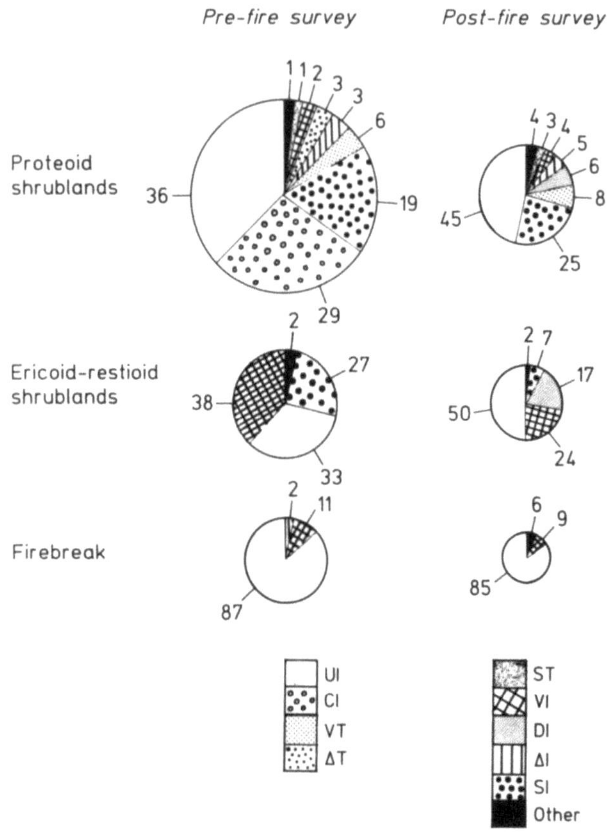


**Fig. 4.6.** The relative contribution of vital attributes, in terms of biomass, to three vegetation communities immediately before and 30 months after fire in Swartboskloof. The size of the circle reflects the relative biomass. The relative contribution (%) of each category is indicated on the diagrams; abbreviations as in Fig. 4.2

shrublands. The single species eliminated in specialized hygrophilous communities is *Protea neriifolia*, which occurred post-fire as isolated seedlings. None of the species occurring in the frequently burnt firebreak are eliminated by any fire frequency, as sensitive species no longer occur there. In all of the communities, some species are listed as potentially eliminated by long interfire periods. These are species with a longevity less than the fire frequency concerned, which rely on soil-stored seed pools for regeneration after fire. In the absence of data on seed pool longevity, these species can only be considered as potential losses from the community.

The hypothesis that a given fire frequency would favour one community over another is not supported by the above data. All communities are maintained by fire frequencies between 10 and 20 years, and no community is favoured over another by a given fire regime, so that other factors must

**Table 4.5.** The fate of species in four vegetation communities subjected to fire at different frequencies. Data show the number of species in each category. The species listed as potentially eliminated represent species for which data on seed bank longevity are not available

| Fire frequency (yr) | 5 | 10 | 15 | 20 | 25 | 30 | 35 | 40 | 45 | 50 |
|---|---|---|---|---|---|---|---|---|---|---|
| *Proteoid shrublands* | | | | | | | | | | |
| Species eliminated | 3 | 1 | 1 | 1 | 0 | 0 | 0 | 2 | 2 | 2 |
| Species potentially eliminated | 0 | 0 | 7 | 8 | 8 | 8 | 8 | 9 | 12 | 12 |
| Species surviving | 131 | 133 | 126 | 125 | 126 | 126 | 126 | 123 | 120 | 120 |
| *Ericoid-restioid shrublands* | | | | | | | | | | |
| Species eliminated | 2 | 0 | 0 | 0 | 0 | 0 | 0 | 3 | 3 | 3 |
| Species potentially eliminated | 0 | 0 | 5 | 5 | 5 | 6 | 6 | 8 | 12 | 12 |
| Species surviving | 111 | 113 | 108 | 108 | 108 | 107 | 107 | 105 | 101 | 101 |
| *Specialized hygrophilous communities* | | | | | | | | | | |
| Species eliminated | 1 | 0 | 0 | 0 | 0 | 0 | 0 | 1 | 1 | 1 |
| Species potentially eliminated | 0 | 0 | 1 | 1 | 1 | 1 | 1 | 2 | 2 | 2 |
| Species surviving | 44 | 45 | 44 | 44 | 44 | 44 | 44 | 43 | 43 | 43 |
| *Firebreak* | | | | | | | | | | |
| Species eliminated | 0 | 0 | 0 | 0 | 0 | 0 | 0 | 0 | 0 | 0 |
| Species potentially eliminated | 0 | 0 | 1 | 2 | 2 | 3 | 3 | 4 | 4 | 4 |
| Species surviving | 63 | 63 | 62 | 61 | 61 | 60 | 60 | 59 | 59 | 59 |
| *Entire survey* | | | | | | | | | | |
| Species eliminated | 4 | 2 | 2 | 2 | 0 | 0 | 0 | 2 | 2 | 2 |
| Species potentially eliminated | 0 | 0 | 9 | 10 | 10 | 11 | 11 | 13 | 17 | 17 |
| Species surviving | 206 | 208 | 199 | 198 | 200 | 199 | 199 | 195 | 193 | 193 |

govern the extent of communities. While vegetation structure can be radically altered by changes in fire frequency through the elimination of non-sprouting *Protea* shrubs, it is only in the case of long intervals between fire that, for proteoid shrublands at least, a transition to forest is possible. This question is addressed in more detail in Chapter 5.

### 4.3.7 Comparison to Other Fire-Prone Shrublands

The flora of the Mediterranean Basin, and the Chilean mattorral, differs from the fynbos in the almost complete absence of non-sprouting plants (Keeley 1986; see also Chap. 7). Californian chaparral differs from fynbos in having an abundant annual flora, a paucity of sprouting herbaceous perennials, and very few non-sprouters that rely on canopy-stored seeds for survival (Keeley 1986). Unlike fynbos, chaparral non-sprouting shrubs are largely resilient to extended fire-free intervals of 100 years or more. Keeley

**Table 4.6.** The number of plant species in five fire-response categories in Swartboskloof, Australian kwongan (Bell et al. 1984) and Scottish heathland (Hobbs et al. 1984)

| Fire-response category | Swartboskloof | Kwongan | Heathland |
|---|---|---|---|
| Fire ephemerals | 13 | 20 | 0 |
| Obligate seeders | 29 | 31 | 2 |
| Facultative sprouter-seeders | 9 | 15 | 2 |
| Autoregenerating long-lived sprouters | 130 | 67 | 20 |
| Obligate vegetative reproducing sprouters | 29 | 18 | 3 |
| Total species | 210 | 151 | 27 |

(1986) also compared the range of post-fire regeneration characteristics for shrublands in mediterranean-climate regions of California, the Mediterranean Basin, Chile, South Africa and Australia. He rated the relative abundance of plants with soil-stored seeds as low, and those with seed storage on the plant as very abundant in fynbos shrublands. The data for Swartboskloof show that, in terms of the number of species present, soil-stored seed is the more important strategy. Fynbos does not necessarily have an abundance of species which store seeds on the plant, but the vegetation is often dominated by a few species which employ this strategy. Keeley also lists seeding herbaceous perennials as absent in fynbos, but the data for Swartboskloof show that some perennial re-seeding grasses (*Pentaschistis* and *Pentameris* species) do occur.

Data sets suitable for direct comparison with the data from Swartboskloof are rare, but two examples from Australian kwongan shrublands and Scottish heathlands are given in Table 4.6. The proportion of plants that sprout is higher in fynbos than in kwongan, and the proportion of non-sprouting plants and ephemerals is correspondingly lower in fynbos. The diversity of fynbos and kwongan, both in terms of the number of species and the range of fire-response types, is high in comparison to the depauperate Scottish heathlands. The heathlands also lack ephemerals, and have a very high proportion (85%) of sprouting plants.

### 4.3.8 Fire, Resilience and Dynamics in Fynbos

The resilience (i.e. the ability to return to the pre-fire equilibrium state) of vegetation is a function of species-specific mechanisms (Keeley 1986). Frequent perturbation often selects for species which specialize, and even become dependent on, the relevant disturbance regime. How resilient is fynbos, and how is this reflected in the spectrum of survival types? Most fynbos species are able to sprout. Sprouters can tolerate large deviations from normal fire frequencies, as they are long-lived and basically unaffected by fire. Fire ephemerals and other non-sprouters that rely on soil-stored seed banks may be vulnerable to long interfire periods that exceed the life span of the seeds. The two non-sprouting *Protea* species are vulnerable to both short and long interfire periods, and are therefore the least resilient component of the vegetation. The genera *Protea* and *Leucadendron* often employ the remarkably tenuous CI method of persistence and establishment after fire, and are frequently dominant in fynbos. This indicates evolution in an environment subjected to a narrow range of fire frequency between 10 and 30 years.

Data on the vital attributes of species can be used to predict the presence or absence of species given different frequencies of disturbance. However, the relative abundance of species within a community cannot be predicted from this knowledge. In addition, variation in the season and intensity of fires can have different effects on plant species under similar fire frequencies. The dynamics of fynbos ecosystems in relation to fire is currently best understood in the case of the Proteaceae, where it has been shown that fire frequency alone cannot explain the variation in recruitment observed after fires in different seasons (Bond et al. 1984), or of different intensities (Bond et al. 1990). In addition, the size of the seed bank varies considerably at different stages within the "mature" period (Kruger and Bigalke 1984), and this will affect the relative size of post-fire populations. Models of the factors controlling species coexistence are currently being developed for the Proteaceae, and these offer a substantial advancement in understanding the dynamics of populations of fynbos plants in relation to fire (Cowling and Gxaba 1990; Yeaton and Bond 1991). However, for many other groups of plants, little or nothing is known of the factors controlling their dynamics in relation to fire. For example, the genus *Erica* has many species which rely on soil-stored seed banks for regeneration, but there are no data on the seed production rates or seed bank dynamics of members of this genus. Data on the range of regeneration strategies, as presented here, represent a starting point from which the dynamics of the vegetation in relation to fire can be examined. However, the data are at best rudimentary for the provision of a sound understanding of vegetation dynamics. Future research should concentrate on selected species within the major fire response patterns (Table 4.2) in order to improve the understanding of fynbos dynamics in relation to fire.

*Acknowledgements.* We thank Adnaan Abrahams, Clare Jones and Gerard Mostert for help with fieldwork and data analysis. Many colleagues assisted in allocating species to species types. We are particularly grateful to Pat Brown, Roger Ellis, Dave Le Maitre, Dave McDonald, Hugh Taylor and Jan Vlok.

**Appendix 4.1.** Life-forms, fire survival types and vital attributes for 210 species in Swartboskloof. Seed longevity is only given for species which rely on soil-stored seed banks for survival

| Species[a] | Life-form[b] | Fire survival type[c] | Vital attributes[c] | Age to maturity (yrs) | Longevity (yrs) | Seed longevity (yrs) |
|---|---|---|---|---|---|---|
| SCHIZAEACEAE | | | | | | |
| *Schizaea pectinata* | H | ALS | ΔI | 1 | >50 | – |
| DENNSTAEDTIACEAE | | | | | | |
| *Pteridium aquilinum* | H | ALS | ΔT | 1 | >50 | – |
| ADIANTACEAE | | | | | | |
| *Pellaea pteroides* | H | ALS | ΔT | 1 | >50 | – |
| CUPRESSACEAE | | | | | | |
| *Widdringtonia nodiflora* | P | ALS | VI | 5 | >50 | – |
| PINACEAE | | | | | | |
| *Pinus radiata* (*) | P | OS | DT | 5 | >50 | – |
| ALLIACEAE | | | | | | |
| *Agapanthus africanus* | EG | ALS | UI | 1 | >50 | – |
| AMARYLLIDACEAE | | | | | | |
| *Cyrtanthus ventricosus* | DG | ALS | UI | 1 | >50 | – |
| *Haemanthus coccineus* | DG | ALS | UI | 1 | >50 | – |
| ASPARAGACEAE | | | | | | |
| *Protasparagus rubicundus* | DG | ALS | UT | 1 | >50 | – |
| ASPHODELACEAE | | | | | | |
| *Bulbine tuberosa* | DG | ALS | UI | 1 | >50 | – |
| *Caesia contorta* | H | ALS | UI | 1 | >50 | – |
| *Trachyandra hirsutiflora* | DG | ALS | UI | 1 | >50 | – |
| *Trachyandra hirsuta* | DG | ALS | UI | 1 | >50 | – |
| *Trachyandra tabularis* | DG | ALS | UI | 1 | >50 | – |
| CYPERACEAE | | | | | | |
| *Ficinia deusta* | H | ALS | UI | 1 | >50 | – |
| *Ficinia filiformis* | H | ALS | UI | 1 | >50 | – |

| | | | | | | |
|---|---|---|---|---|---|---|
| *Ficinia nigrescens* | H | ALS | UI | 1 | >50 | – |
| *Macrochaetum hexandrum* | H | ALS | UI | 1 | >50 | – |
| *Neesenbeckia punctoria* | H | OVS | UI | 1 | >50 | – |
| *Tetraria bromoides* | H | ALS | VI | 2 | >50 | – |
| *Tetraria burmanii* | H | ALS | UI | 1 | >50 | – |
| *Tetraria capillacea* | H | ALS | UI | 1 | >50 | – |
| *Tetraria compar* | H | ALS | UI | 1 | >50 | – |
| *Tetraria crassa* | H | ALS | UI | 1 | >50 | – |
| *Tetraria cuspidata* | H | ALS | UI | 1 | >50 | – |
| *Tetraria fasciata* | H | ALS | UI | 1 | >50 | – |
| *Tetraria fimbriolata* | H | ALS | UI | 1 | >50 | – |
| *Tetraria ustulata* | H | ALS | UI | 1 | >50 | – |
| HAEMODORACEAE | | | | | | |
| *Dilatris pillansii* | EG | ALS | UI | 1 | >50 | – |
| *Wachendorfia paniculata* | DG | ALS | UI | 1 | >50 | – |
| HYACINTHACEAE | | | | | | |
| *Albuca canadensis* | DG | ALS | UI | 1 | >50 | – |
| *Lachenalia orchioides* | DG | ALS | UI | 1 | >50 | – |
| *Ornithogalum hispidum* | DG | ALS | UI | 1 | >50 | – |
| *Urginea dregei* | DG | ALS | UI | 1 | >50 | – |
| HYPOXIDACEAE | | | | | | |
| *Spiloxene capensis* | DG | ALS | UI | 1 | >50 | – |
| IRIDACEAE | | | | | | |
| *Anapalina triticea* | DG | ALS | UI | 1 | >50 | – |
| *Aristea africana* | MiCh | OS | SI | 1 | 15 | >35 |
| *Aristea major* | EG | ALS | UI | 1 | >50 | – |
| *Aristea spiralis* | EG | ALS | UI | 1 | >50 | – |
| *Bobartia gladiata* | EG | ALS | UI | 1 | >50 | – |
| *Bobartia indica* | EG | ALS | UI | 1 | >50 | – |
| *Geissorhiza aspera* | DG | ALS | UI | 1 | >50 | – |
| *Gladiolus blommesteinii* | DG | ALS | ΔI | 1 | >50 | – |
| *Gladiolus carneus* | DG | ALS | ΔI | 1 | >50 | – |
| *Micranthus alopecuroides* | DG | ALS | UI | 1 | >50 | – |
| *Moraea bituminosa* | DG | ALS | UI | 1 | >50 | – |

**Appendix 4.1** (continued)

| Species[a] | Life-form[b] | Fire survival type[c] | Vital attributes[c] | Age to maturity (yrs) | Longevity (yrs) | Seed longevity (yrs) |
|---|---|---|---|---|---|---|
| *Moraea tricuspidata* | DG | ALS | UI | 1 | >50 | – |
| *Thereianthus spicatus* | DG | ALS | UI | 1 | >50 | – |
| *Tritonia crispa* | DG | ALS | UI | 1 | >50 | – |
| *Watsonia borbonica* | DG | ALS | UI | 1 | >50 | – |
| ORCHIDACEAE | | | | | | |
| *Acrolophia capensis* | EG | OVS | UI | 1 | >50 | – |
| *Schizodium obliquum* | DG | ALS | ΔI | 1 | >50 | – |
| POACEAE | | | | | | |
| *Cymbopogon marginatus* | H | ALS | UI | 1 | >50 | – |
| *Ehrharta bulbosa* | EG | ALS | UI | 1 | >50 | – |
| *Ehrharta dura* | H | ALS | UI | 1 | >50 | – |
| *Ehrharta ramosa* | H | FSS | SI | 1 | >50 | – |
| *Festuca scabra* | EG | ALS | UI | 1 | >50 | – |
| *Merxmuellera cincta* | H | ALS | UI | 1 | >50 | – |
| *Merxmuellera rufa* | EG | ALS | UI | 1 | >50 | – |
| *Merxmuellera stricta* | H | ALS | UI | 1 | >50 | – |
| *Pentameris macrocalycina* | H | ALS | UI | 1 | >50 | – |
| *Pentameris thuarii* | H | FSS | SI | 2 | >50 | – |
| *Pentaschistis aristidoides* | EG | ALS | UI | 1 | >50 | – |
| *Pentaschistis colorata* | H | ALS | UI | 1 | >50 | – |
| *Pentaschistis curvifolia* | H | FSS | SI | 1 | >50 | – |
| *Pentaschistis malouinensis* | H | FSS | SI | 1 | 25 | >25 |
| *Pentaschistis pallescens* | H | ALS | UI | 1 | >50 | – |
| *Themeda triandra* | H | ALS | UI | 1 | >50 | – |
| RESTIONACEAE | | | | | | |
| *Calopsis membranaceus* | H | OVS | VI | 2 | >50 | – |
| *Cannomois virgata* | H | OVS | UI | 1 | >50 | – |
| *Elegia asperiflora* | H | OVS | UI | 1 | >50 | – |
| *Elegia juncea* | H | OVS | UI | 1 | >50 | – |
| *Hypodiscus albo-aristatus* | H | OVS | UI | 1 | >50 | – |

| | | | | | | |
|---|---|---|---|---|---|---|
| *Hypodiscus aristatus* | H | OVS | UI | 1 | >50 | – |
| *Hypodiscus willdenowia* | H | OVS | UI | 1 | >50 | – |
| *Ischyrolepis capensis* | H | OVS | UI | 1 | >50 | – |
| *Ischyrolepis gaudichaudiana* | H | OVS | UI | 1 | >50 | – |
| *Ischyrolepis sieberi* | H | OVS | VI | 2 | >50 | – |
| *Platycaulos depauperatus* | H | ALS | UI | 1 | >50 | – |
| *Restio filiformis* | H | OVS | UI | 1 | >50 | – |
| *Restio triticeus* | H | OVS | UI | 1 | >50 | – |
| *Staberoha cernua* | H | OVS | UI | 1 | >50 | – |
| *Thamnochortus fruticosus* | H | OVS | UI | 1 | >50 | – |
| *Thamnochortus gracilis* | H | OVS | UI | 1 | >50 | – |
| *Willdenowia sulcata* | H | OVS | UI | 1 | >50 | – |
| TECOPHILAEACEAE | | | | | | |
| *Cyanella hyacinthoides* | DG | ALS | UI | 1 | >50 | – |
| ANACARDIACEAE | | | | | | |
| *Heeria argentea* | P | OVS | UR | 1 | >50 | – |
| *Rhus angustifolia* | P | ALS | UI | 1 | >50 | – |
| *Rhus rosmarinifolia* | P | ALS | UI | 1 | >50 | – |
| *Rhus tomentosa* | P | ALS | UI | 1 | >50 | – |
| APIACEAE | | | | | | |
| *Annesorhiza inebrians* | H | ALS | UI | 1 | >50 | – |
| *Centella glabrata* | MiCh | ALS | UI | 1 | >50 | – |
| *Lichtensteinia lacera* | DG | ALS | UI | 1 | >50 | – |
| *Peucedanum sieberianum* | EG | ALS | UI | 1 | >50 | – |
| ASTERACEAE | | | | | | |
| *Anaxeton asperum* | NaCh | OS | DI | 2 | 20 | – |
| *Arctotis semipapposa* | MeCh | ALS | UI | 1 | >50 | – |
| *Athrixia heterophylla* | MeCh | OVS | UI | 1 | >50 | – |
| *Berkheya herbacea* | H | ALS | ΔI | 1 | >50 | – |
| *Chrysanthemoides monilifera* | P | OS | ST | 2 | 10 | – |
| *Conyza bonariensis* (*) | MeCh | FE | DI | 1 | 3 | – |
| *Corymbium glabrum* | H | ALS | ΔI | 1 | >50 | – |
| *Corymbium scabrum* | H | ALS | ΔI | 1 | >50 | – |
| *Corymbium villosum* | H | ALS | ΔI | 1 | >50 | – |

**Appendix 4.1** (continued)

| Species[a] | Life-form[b] | Fire survival type[c] | Vital attributes[c] | Age to maturity (yrs) | Longevity (yrs) | Seed longevity (yrs) |
|---|---|---|---|---|---|---|
| *Disparago ericoides* | MeCh | OVS | UI | 1 | >50 | – |
| *Elytropappus glandulosus* | MeCh | ALS | UI | 1 | >50 | – |
| *Gazania serrata* | MiCh | ALS | ΔI | 1 | >50 | – |
| *Gerbera crocea* | H | ALS | ΔI | 1 | >50 | – |
| *Haplocarpha lanata* | H | ALS | UI | 1 | >50 | – |
| *Helichrysum cymosum* | MeCh | OS | DI | 1 | 30 | – |
| *Helichrysum nudifolium* | MeCh | ALS | ΔI | 1 | >50 | – |
| *Helichrysum teretifolium* | MeCh | ALS | ΔI | 1 | >50 | – |
| *Helichrysum zeyheri* | MeCh | ALS | ΔI | 1 | >50 | – |
| *Hypochoeris radicata* (*) | H | FE | DI | 1 | 5 | – |
| *Mairea lasiocarpa* | NaCh | OS | SI | 1 | 35 | >15 |
| *Metalasia muricata* | P | OS | DI | 2 | >50 | – |
| *Osmitopsis afra* | MeCh | OVS | UI | 1 | >50 | – |
| *Osmitopsis asteriscoides* | P | ALS | UI | 1 | >50 | – |
| *Osteospermum junceum* | P | ALS | UI | 1 | >50 | – |
| *Osteospermum tomentosum* | NaCh | OVS | UI | 1 | >50 | – |
| *Senecio grandiflorus* | P | OS | DI | 1 | 25 | – |
| *Senecio paniculatus* | P | OS | DI | 1 | 25 | – |
| *Senecio pinifolius* | MeCh | ALS | UI | 1 | >50 | – |
| *Senecio pubigerus* | P | FE | DI | 1 | 5 | – |
| *Stoebe plumosa* | MeCh | FSS | DI | 2 | >50 | – |
| *Ursinia paleacea* | P | FE | DI | 1 | 5 | – |
| *Ursinia pinnata* | P | FE | DI | 1 | 5 | – |
| BRUNIACEAE | | | | | | |
| *Berzelia lanuginosa* | P | ALS | UI | 1 | >50 | – |
| *Brunia nodiflora* | P | ALS | VI | 3 | >50 | – |
| *Nebelia paleacea* | P | ALS | VI | 3 | >50 | – |
| CAMPANULACEAE | | | | | | |
| *Cyphia bulbosa* | DG | ALS | UI | 1 | >50 | – |
| *Cyphia volubilis* | DG | ALS | UI | 1 | >50 | – |

| | | | | | | |
|---|---|---|---|---|---|---|
| *Lobelia coronopifolia* | MeCh | ALS | UI | 1 | >50 | – |
| *Prismatocarpus diffusus* | MeCh | ALS | UI | 1 | >50 | – |
| *Roella ciliata* | MeCh | FE | SI | 1 | 5 | >45 |
| CELASTRACEAE | | | | | | |
| *Hartogiella schinoides* | P | ALS | ΔR | 1 | >50 | – |
| *Maytenus acuminata* | P | FSS | DR | 5 | >50 | – |
| *Maytenus oleoides* | P | ALS | ΔT | 1 | >50 | – |
| CRASSULACEAE | | | | | | |
| *Crassula capensis* | EG | ALS | UI | 1 | >50 | – |
| *Crassula fascicularis* | MeCh | OS | ST | 1 | 20 | – |
| DIPSACACEAE | | | | | | |
| *Scabiosa columbaria* | MeCh | ALS | ΔI | 1 | >50 | – |
| DROSERACEAE | | | | | | |
| *Drosera aliciae* | EG | ALS | ΔI | 1 | >50 | – |
| *Drosera capensis* | EG | ALS | ΔI | 1 | >50 | – |
| EBENACEAE | | | | | | |
| *Diospyros glabra* | P | ALS | ΔI | 1 | >50 | – |
| ERICACEAE | | | | | | |
| *Blaeria dumosa* | NaCh | OS | SI | 4 | 35 | >15 |
| *Eremia totta* | NaCh | OS | SI | 3 | >50 | – |
| *Erica articularis* | MeCh | ALS | UI | 1 | >50 | – |
| *Erica coccinea* | P | ALS | VI | 3 | >50 | – |
| *Erica hispidula* | P | OS | SI | 3 | >50 | – |
| *Erica longifolia* | P | OS | SI | 3 | >50 | – |
| *Erica sphaeroidea* | MeCh | OS | SI | 2 | >50 | – |
| EUPHORBIACEAE | | | | | | |
| *Clutia alaternoides* | P | ALS | UI | 1 | >50 | – |
| *Clutia rubricaulis* | MeCh | OVS | UI | 1 | >50 | – |
| *Euphorbia genistoides* | NaCh | OVS | UI | 1 | >50 | – |
| FABACEAE | | | | | | |
| *Aspalathus ciliaris* | P | FE | SI | 2 | 6 | >44 |
| *Cyclopia genistoides* | P | ALS | UI | 1 | >50 | – |
| *Indigofera alopecuroides* | NaCh | ALS | UI | 1 | >50 | – |

**Appendix 4.1** (continued)

| Species[a] | Life-form[b] | Fire survival type[c] | Vital attributes[c] | Age to maturity (yrs) | Longevity (yrs) | Seed longevity (yrs) |
|---|---|---|---|---|---|---|
| *Otholobium fruticans* | P | ALS | UI | 1 | >50 | – |
| *Otholobium obliquum* | P | OVS | UI | 1 | >50 | – |
| *Podalyria montana* | P | ALS | UI | 1 | >50 | – |
| *Psoralea aculeata* | P | ALS | UI | 1 | >50 | – |
| *Rafnia capensis* | MeCh | ALS | UI | 1 | >50 | – |
| *Rhynchosia totta* | L | ALS | UI | 1 | >50 | – |
| FLACOURTIACEAE | | | | | | |
| *Kiggelaria africana* | P | FSS | DR | 5 | >50 | – |
| GENTIANACEAE | | | | | | |
| *Chironia baccifera* | MeCh | OS | DT | 1 | 10 | – |
| *Sebaea exacoides* | T | OS | ST | 1 | 1 | – |
| GERANIACEAE | | | | | | |
| *Pelargonium cucullatum* | P | ALS | ΔI | 1 | >50 | – |
| *Pelargonium elongatum* | MeCh | OS | DI | 1 | 35 | – |
| *Pelargonium longifolium* | EG | ALS | ΔI | 1 | >50 | – |
| *Pelargonium myrrhifolium* | MeCh | ALS | ΔI | 1 | >50 | – |
| *Pelargonium tabulare* | MeCh | ALS | ΔI | 1 | >50 | – |
| LAURACEAE | | | | | | |
| *Cassytha ciliolata* | VP | OS | DT | 1 | >50 | – |
| LINACEAE | | | | | | |
| *Linum africanum* | MeCh | ALS | UI | 1 | >50 | – |
| MESEMBRYANTHEMACEAE | | | | | | |
| *Erepsia anceps* | NaCh | FE | SI | 1 | 5 | >45 |
| MONTINIACEAE | | | | | | |
| *Montinia caryophyllacea* | P | ALS | UI | 1 | >50 | – |
| MYRSINACEAE | | | | | | |
| *Myrsine africana* | P | ALS | ΔR | 1 | >50 | – |

| | | | | | | |
|---|---|---|---|---|---|---|
| OXALIDACEAE | | | | | | |
| *Oxalis bifida* | DG | ALS | UI | 1 | >50 | – |
| *Oxalis lanata* | DG | ALS | UI | 1 | >50 | – |
| *Oxalis purpurea* | DG | ALS | UI | 1 | >50 | – |
| *Oxalis strigosa* | DG | ALS | UI | 1 | >50 | – |
| *Oxalis tenuifolia* | DG | ALS | UI | 1 | >50 | – |
| PENAEACEAE | | | | | | |
| *Penaea mucronata* | P | ALS | UI | 1 | >50 | – |
| POLYGALACEAE | | | | | | |
| *Muraltia alopecuroides* | MeCh | ALS | UI | 1 | >50 | – |
| *Muraltia heisteria* | P | ALS | UI | 1 | >50 | – |
| *Polygala bracteolata* | MeCh | FE | SI | 1 | 8 | >42 |
| PROTEACEAE | | | | | | |
| *Aulax pallasia* | P | OVS | VI | 2 | >50 | – |
| *Leucadendron salignum* | P | ALS | UI | 1 | >50 | – |
| *Protea acaulos* | H | OVS | UI | 1 | >50 | – |
| *Protea neriifolia* | P | OS | CI | 4 | 35 | <1 |
| *Protea nitida* | P | ALS | VT | 2 | >50 | – |
| *Protea repens* | P | OS | CI | 3 | 35 | <1 |
| RHAMNACEAE | | | | | | |
| *Phylica imberbis* | MeCh | ALS | UI | 1 | >50 | – |
| *Phylica pubescens* | P | OS | SI | 3 | 40 | >10 |
| ROSACEAE | | | | | | |
| *Cliffortia cuneata* | P | OS | SI | 2 | 40 | >10 |
| *Cliffortia polygonifolia* | P | OS | SI | 3 | 40 | >10 |
| *Cliffortia ruscifolia* | P | FSS | SI | 2 | 40 | >10 |
| RUBIACEAE | | | | | | |
| *Anthospermum aethiopicum* | P | OS | ST | 1 | 8 | – |
| *Anthospermum galioides* | MiCh | OS | ST | 1 | 8 | – |
| *Carpacoce vaginellata* | MeCh | ALS | UI | 1 | >50 | – |
| RUTACEAE | | | | | | |
| *Coleonema juniperinum* | MeCh | FSS | SI | 2 | >50 | – |
| *Diosma hirsuta* | MeCh | ALS | UI | 1 | >50 | – |

**Appendix 4.1** (continued)

| Species[a] | Life-form[b] | Fire survival type[c] | Vital attributes[c] | Age to maturity (yrs) | Longevity (yrs) | Seed longevity (yrs) |
|---|---|---|---|---|---|---|
| SANTALACEAE | | | | | | |
| *Thesium capitatum* | MeCh | FE | SI | 1 | 5 | >45 |
| *Thesium strictum* | P | FE | SI | 1 | 5 | >45 |
| *Thesium virgatum* | MeCh | FE | SI | 1 | 5 | >45 |
| SCROPHULARIACEAE | | | | | | |
| *Agathelpis dubia* | MeCh | OS | ST | 1 | 8 | – |
| *Halleria elliptica* | P | OVS | UI | 1 | >50 | – |
| *Oftia africana* | P | ALS | UI | 1 | >50 | – |
| *Selago spuria* | MeCh | FE | SI | 1 | 5 | >45 |
| SOLANACEAE | | | | | | |
| *Solanum tomentosum* | P | OS | DT | 1 | 5 | – |
| THYMELAEACEAE | | | | | | |
| *Gnidia oppositifolia* | P | ALS | UI | 1 | >50 | – |
| *Struthiola ciliata* | MeCh | OS | SI | 1 | 35 | >15 |

[a] Species marked (*) are exotics.
[b] Life-forms follow the system of Raunkiaer (1934) and modifications thereof (Pate et al. 1984). The abbreviations are: H = hemicryptophytes (renewal buds close to the ground); MiCh = microchamaephytes (renewal buds within 100 mm of the ground); NaCh = nanochamaephytes (renewal buds 100–200 mm from the ground); MeCh = megachamaephytes (renewal buds 200–800 mm from the ground); P = phanerophytes (renewal buds 800 mm or greater from the ground); EG = evergreen geophytes (shoots buried in the ground); DG = deciduous geophytes (shoots buried in the ground); L = lianas (phanerophytes fully supported on other plants); VP = vascular parasites (plants which grow on other living plants); and Th = therophytes (annuals which complete their whole life cycle in 1 year).
[c] Abbreviations are explained in the text.

# 5 Is Fynbos a Stage in Succession to Forest? Analysis of the Perceived Ecological Distinction Between Two Communities

P.T. Manders, D.M. Richardson, and P.H. Masson

## 5.1 Introduction

At first glance, patches of forest form distinct communities within the largely treeless fynbos shrublands. Forests usually occur on granite, shale and sandstone in sheltered ravines, stream banks and patches of rock scree (Fig. 5.1) where the annual precipitation is greater than 650 mm (Werger et al. 1972; McKenzie et al. 1977; Kruger 1979; Campbell 1985). The distribution of forests in the southwestern Cape conforms to the worldwide pattern of vegetation communities along watercourses differing in species composition from adjacent plant communities (Miller and Johnson 1986). The location of forests in moist habitats is often understood to reflect the climatic amelioration in these zones and the preclusion of forest species in the adjacent vegetation by seasonal drought (e.g. White 1978). Although the fynbos is fire-prone, forests seldom burn (van Wilgen et al. 1990b), and an alternative explanation for forest distribution is that the forests are remnants of previously more extensive forests which have been reduced by frequent fires (McKenzie et al. 1977; Campbell et al. 1979; Moll et al. 1980). The treeless nature of the fynbos is puzzling (see Chap. 9). Development of forest species has been noted in fynbos areas, usually in the prolonged absence of fire (Kruger 1984). Succession to forest has been noted after 50 fire-free years in Orange Kloof near Cape Town (mean annual rainfall 1227 mm) (Masson and Moll 1987) and was evident in places in the Swartboskloof valley (mean annual rainfall for the catchment: 2328 mm; see Chap. 2) after 28 fire-free years. However, when forest is destroyed, fynbos will replace it. Such arguments and observations raise questions as to whether the fynbos and forest communities have different edaphic and climatic requirements, or whether they are different successional stages in juxtaposition.

We propose that the relations between forest and fynbos follow the conceptual model of succession in Fig. 5.2. Several of the stages and transitions of the model have been described in other vegetation types. Regeneration of rainforest on clearfelled land, for example, involves the introduction of wind- and bird-dispersed seeds, and amelioration of microclimate (Read and Hill 1983). Woody plant encroachment in grasslands has been described

**Fig. 5.1.** Aerial photograph of Swartboskloof showing the restriction of forests (dark green vegetation) to stream banks and screes. Proteoid shrublands appear grey-green, whereas ericoid-restiod shrublands are lighter

as a process of nucleation (Archer 1989), and invasion of sagebrush by pinyon and juniper has been ascribed to protection from fire (Blackburn and Tueller 1970). The model assumes that disturbance by fire is part of the cycle, and that processes exist whereby the system is reset after the disturbance. In the absence of fire the model predicts that a semi-stable forest community will be established eventually, but that shrublands may predominate under a regime of frequent disturbance.

Such an interpretation of vegetation development has particular theoretical and practical implications for disturbance-driven systems such as those in mediterranean-climate regions. The process of vegetation development may be held at different levels by different disturbance regimes, leading ultimately to different paths of landscape evolution (Naveh 1982). Never-

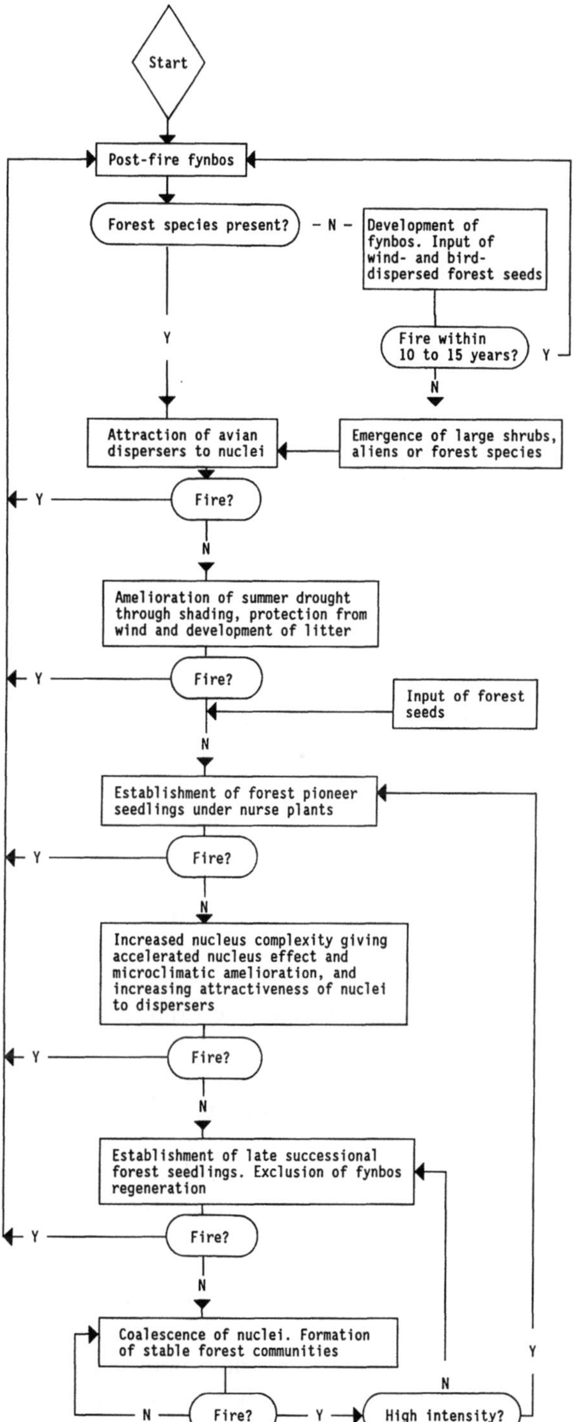


**Fig. 5.2.** A conceptual model of the development of forest species within fynbos communities. The model assumes the availability of forest seeds. Post-fire age increases as the model proceeds from the start situation

theless, apparently stable shrublands may revert to forest if disturbance regimes are altered, as noted in the Mediterranean Basin (Debussche et al. 1980).

To verify the model, we ask if the assumptions regarding the effects of fire are correct, and investigate the processes which reinitiate forest development after disturbance, based on evidence from Swartboskloof. Then we investigate the processes involved in the development of forest in fynbos. Finally, we examine the factors, primarily moisture and nutrient availability and historical events, which may influence these processes.

## 5.2 Fire and the Stability of Forest/Fynbos Boundaries

### 5.2.1 The Effects of Fire on Forest Boundaries in Swartboskloof

The effects of the fire on the location of forest boundaries were studied at three sites in Swartboskloof to determine if the fire diminished the forest areas and how forest communities and individuals recover from fire (Manders 1990a). The main study site (site A, Fig. 2.3) was classified as short riparian forest (Chap. 1), dominated by *Cunonia capensis* and *Ilex mitis* trees. There was a transitional zone between the forest and the fynbos, with *Cliffortia cuneata*, *Rhus angustifolia* and *Halleria elliptica* shrubs predominant. The forest at site B was not well developed, comprising several large *Cunonia capensis* trees, with no transitional zone. Site C was situated on the edge of a scree forest (a short forest variant of tall forest; see Chap. 1) dominated by *Olinea ventosa*, *Maytenus acuminata*, *Olea europaea* ssp. *africana* and *Podocarpus elongatus*.

The intensity of the prescribed burn on 17 March 1987 varied between sites. In the vicinity of site A the fire proceeded downhill and was of low intensity, at site B the fire was more intense, and at site C the fire ran uphill and was very intense. The forest did not burn, apart from scorching of the canopy, in any of the sites.

Studies were based on 2-m-wide transects (five at site A, three at site B and two at site C) from within the forest to within the fynbos, and perpendicular to the boundary. The transects were divided into permanently marked 1 × 1 m sub-plots. In February 1987 and 1 year after the fire each sub-plot was classified as forest, transitional or fynbos according to the dominant species on the sub-plot.

At site A where the transitional zone was well defined, there were changes in post-fire distribution of sub-plots in each vegetation type. Forest and fynbos gained prominence in what had been transitional zone sub-plots (Table 5.1). The apparent movement of forest species resulted from those species which resprouted after fire and contributed relatively more to the post-fire cover in the transitional zone than before the fire. Seedlings of

**Table 5.1.** The median positions of plots, classified according to vegetation type, along transects across the forest/fynbos boundary before and after fire in Swartboskloof; position 1 is in the forest

| Site | Vegetation type | Median plot position (m)[a] | | |
|---|---|---|---|---|
| | | Pre-fire | Post-fire[b] | |
| A | Forest | 5.0 (80) | 8.0 (95) | ** |
| | Transitional | 9.0 (15) | 13.0 (50) | |
| | Fynbos | 15.0 (105) | 13.0 (55) | ** |
| B | Forest | 4.0 (46) | 5.0 (44) | |
| | Transitional | 9.0 (3) | 9.0 (7) | |
| | Fynbos | 10.0 (19) | 9.0 (17) | |
| C | Forest | 4.5 (32) | 5.5 (38) | |
| | Transitional | 13.5 (6) | 8.0 (2) | |
| | Fynbos | 15.0 (42) | 15.0 (40) | |

[a] Numbers of plots in each observation are given in parentheses (Manders 1990a).
[b] ** $P < 0.01$, median test (Zar 1974).

fynbos species in the burnt transitional zone accounted for the changed median position of sub-plots classified as fynbos. The greater number of transitional sub-plots reflects greater mixing of forest and fynbos species after the fire than before the fire.

Population studies were carried out on the transects to determine the relationship between the size of forest plants and their ability to survive fire. Such a relationship, combined with growth rates of forest species, would provide a framework for determining the minimum intervals between fires needed to promote forest development. *Cunonia capensis* was selected for this study as it was the most numerous species among forest saplings and seedlings beyond the forest boundary. Amongst those which had been scorched (some or all leaves killed in the fire), there was an increase in survival with increased plant size. Fire intensity, however, proved to be the overriding factor and no *Cunonia capensis* which had been burnt (all leaves burnt off in the fire) resprouted after the fire.

### 5.2.2 Soil Seed Banks

Post-fire development of fynbos is by resprouting or germination of canopy- or soil-stored seedbanks (Kruger and Bigalke 1984), whereas almost all species in the forests usually resprout after fire (van der Merwe 1966). Trees in forest patches are, however, sometimes killed in high intensity fires. Since no forest species store seeds in the canopy for post-fire release, post-fire regeneration after destruction of the canopy must come from soil-stored seed banks or rely on colonization from distant sources.

Soil seed banks were sampled at site A (Manders 1990b) to determine their role in post-fire regeneration. Each sample was taken from a 0.3 × 0.3 m plot, with litter and soil from 0 to 50 mm, 50 to 100 mm depths taken as separate samples from each plot. Pre-fire soil samples were collected in February and post-fire samples were taken in April 1987. Each sample was halved and seeds were searched for in one half and germinated out of the other. Estimates of seeds per sample were based on the greater of the estimates from the two evaluation techniques, in order to provide the most information on the array of species present (Manders 1990c). These figures are not accurate estimates of viable seed densities as there was no viability testing of seeds found by searching, and not all germination cues were necessarily provided in the germination procedure. The estimates do, however, provide a basis for comparison between vegetation types.

Similar numbers of species occurred in each vegetation type, but seed densities were higher in the transitional and fynbos zones than in the forest (Table 5.2). This follows predictions that vegetation types with low frequencies of disturbance are unlikely to have extensive soil seed banks (Thompson 1978), particularly where the vegetation is not likely to be destroyed in the lifetime of its seeds (Young et al. 1987). There was no evidence of any effect of the fire on the numbers of species and seeds present.

The seed banks in each vegetation type, especially in the forest, contained many seeds from the other vegetation types, suggesting that a transitional or fynbos community will develop after destruction of the forest canopy. Many of the non-forest seeds in the forest were hard seeds of typical transitional zone species of the Fabaceae, stored deep in the soil, which would contribute to the community after a very intense fire.

There were some differences in processes related to seed dispersal and soil storage between the fynbos and the forest. Seeds dispersed by ants are not well represented in the forest, whereas bird dispersal is far more important in the forest than in the fynbos. Although such differences may contribute to the maintenance of distinct communities, no species were excluded from the seed banks of any of the vegetation types. Therefore factors such as germination and establishment requirements, and not the limits to seed dispersal, are important in determining community boundaries.

## 5.3 Colonization of Fynbos by Forest Species

### 5.3.1 Post-Fire Seed Deposition

The movement of seeds between vegetation types was suggested in the study of soil seed banks. The seeds of 40% of forest species in the southwestern Cape are dispersed by vertebrates, mostly birds rather than mammals, in

**Table 5.2.** Numbers of (a) species and (b) seeds per soil sample in each vegetation type before and after fire at site A. Differences between pre- and post-fire observations were assessed with the Mann-Whitney $U$ test, and between vegetation types with the Kruskal-Wallis $H$ test

*a) Species per soil sample*

| Vegetation | Sampling time[a] | | | | |
|---|---|---|---|---|---|
| | Pre-fire | | Post-fire | | |
| | Mean | $n$ | Mean | $n$ | |
| Forest | 7.87 | 15 | 8.80 | 10 | NS |
| Transitional | 8.93 | 15 | 11.40 | 10 | NS |
| Fynbos | 9.13 | 15 | 8.80 | 10 | NS |
| | NS | | NS | | |

*b) Seeds per soil sample*

| Vegetation | Sampling time[a] | | | | |
|---|---|---|---|---|---|
| | Pre-fire | | Post-fire | | |
| | Mean | $n$ | Mean | $n$ | |
| Forest | 13.9 | 15 | 19.6 | 10 | NS |
| Transitional | 35.1 | 15 | 47.7 | 10 | NS |
| Fynbos | 34.7 | 15 | 36.7 | 10 | NS |
| | ** | | * | | |

[a] NS = $P > 0.05$, * = $P < 0.05$, ** = $P < 0.01$ (Manders 1990b).

contrast with less than 10% of fynbos species which are vertebrate dispersed. Wind dispersal is the next most prevalent mode, with 32% of forest species' seeds dispersed in this manner (Knight 1988). The avifauna of Swartbskloof contains nine species that regularly eat and disperse seeds of forest tree species (Richardson and Fraser in press). Most of these species forage in both forest and fynbos, and move frequently between vegetation types.

The seed rain in and adjacent to the forest at the Swartboskloof study site A was monitored for 1 year after the fire to determine whether there was an input of forest seeds, which are not usually part of the soil seed bank, into the fynbos (Manders 1990b). Thirty-five seed traps (plastic pots with an internal upper diameter of 196 mm) were set out in each vegetation type at the beginning of April 1987. Seeds were removed from the traps, identified and counted monthly. Seeds of the dominant forest species, *Cunonia capensis* and *Ilex mitis* were abundant in the transitional zone and did occur in the fynbos (Fig. 5.3).

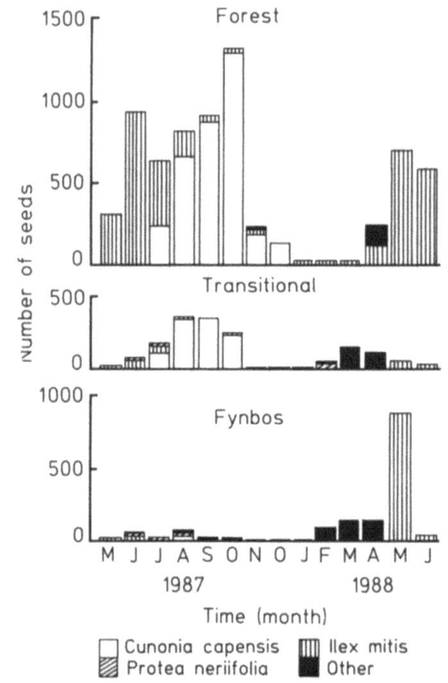


**Fig. 5.3.** The dispersal of forest seeds into transitional zone and fynbos vegetation, shown by the numbers of seeds of tree and shrub species collected monthly in seed traps in each vegetation type at site A in Swartboskloof after fire in March 1987 (Manders 1990b)

Dispersal of forest seeds by birds into fynbos areas is considered to be dependent on the development of perches such as fully-developed shrubs or fruit-bearing shrubs in the fynbos (Masson and Moll 1987; Knight 1988). The degree of seed deposition below a perch is governed by its features such as height, size and complexity and the fruit it bears (McClanahan and Wolfe 1987). In this study, the occurrence of seeds in the traps in the transitional and fynbos zones was related to the type of perch above the trap. Surprisingly, however, branches of dead shrubs appear to act as effective perches (Table 5.3). In May 1988, 675 *I. mitis* seeds were collected from one trap in the fynbos, situated under a single dead branch of *Protea neriifolia*. Effective perches, therefore, need not be a fruiting species or a large complex tree, especially in areas fairly close to the seed source. More distinct perches are likely to attract avian dispersers further from the forest. This study showed that dispersal of forest species into fynbos occurs immediately after fire for both wind- and bird-dispersed species. Therefore colonization of fynbos by forest species soon after fire is not limited by seed dispersal.

**Table 5.3.** Contingency tables of the numbers of seeds of species and berries of *Ilex mitis* recovered from seed traps in the transitional zone and fynbos from May 1987 to June 1988, indicating the relationship between the species of seeds deposited and the type of perch. $\chi^2$ values and levels of significance are given in square brackets below each table

*a) Seed deposition under perches in the transitional zone*

| Perch type | Seed species[a] | | | | |
|---|---|---|---|---|---|
| | *Protea neriifolia* | *Cunonia capensis* | *Ilex mitis* | *Maytenus oleoides* | Totals |
| Forest precursor | 92 (2) + | 628 (0) – | 112 (0) – | 40 (2) – | 872 |
| Dead branch | 35 (2) – | 314 (4) – | 89 (10) + | 50 (16) + | 488 |
| No perch | 11 (1) – | 164 (11) – | 0 (23) – | 0 (10) – | 175 |
| Totals | 138 | 1106 | 201 | 90 | 1535 |

[$\chi^2 = 82.4$, $P < 0.001$]

*b) Seed deposition under perches in the fynbos*

| Perch type | Seed species[a] | | | |
|---|---|---|---|---|
| | *Protea neriifolia* | *Cunonia capensis* | *Ilex mitis* | Totals |
| Forest precursor | 26 (59) + | 1 (0) – | 33 (2) – | 60 |
| Dead branch | 70 (10) – | 15 (7) – | 860 (3) + | 945 |
| No perch | 17 (31) + | 17 (172) + | 11 (20) – | 45 |
| Totals | 113 | 33 | 904 | 1050 |

[$\chi^a = 308$, $P < 0.001$]

[a] Numbers in parentheses show cell contributions to $\chi^2$. Signs indicate whether a cell is over- (+) or under-represented (–) (Manders 1990b).

## 5.3.2 Germination Requirements

Seeds of all communities at site A in Swartboskloof occur in the seed banks of the other communities, and seed dispersal crosses community boundaries. Yet almost all post-fire regeneration on the transects occurred in burnt sub-plots. Most regeneration comprised fynbos and transitional zone species. The small amount of forest regeneration which occurred in the first year after the fire was almost exclusively in the unburnt sub-plots in the forest (Table 5.4). When the species were grouped according to their vegetation type, the association between type and whether the plot was burnt or not was significant for sites A and B ($\chi^2$ = 1305 and 152 respectively, $P <$ 0.001), and for site C ($P < 0.001$ in a Fisher's exact test; Manders 1990a).

The differences in regeneration patterns may be related to different germination requirements of the species in each vegetation type. Regeneration of fynbos and transitional zone species is closely coupled to distrubance (sensu Grubb 1988). Germination and establishment requirements may be

**Table 5.4.** Effects of fire on recruitment of forest, transitional zone and fynbos species, shown by numbers of seedlings on burnt and unburnt 1 m$^2$ plots after fire on transects through the forest/fynbos boundary (Manders 1990a)

| Site | Species | Vegetation | Dispersal[a] | Storage[b] | Number of seedlings | |
|---|---|---|---|---|---|---|
| | | | | | Burnt plots | Unburnt plots |
| A | *Cunonia capensis* | Forest | W | N | 0 | 12 |
| | *Kiggelaria africana* | Forest | B | N | 8 | 36 |
| | *Aspalathus spicata* | Trans. | G | S | 631 | 2 |
| | *Podalyria calyptrata* | Trans. | G | S | 148 | 0 |
| | *Cliffortia cuneata* | Fynbos | G | S | 360 | 4 |
| | *Cliffortia ruscifolia* | Fynbos | A | S | 41 | 0 |
| | *Phylica pubescens* | Fynbos | A | S | 462 | 2 |
| | *Protea neriifolia* | Fynbos | W | C | 90 | 0 |
| B | *Cunonia capensis* | Forest | W | N | 0 | 52 |
| | *Kiggelaria africana* | Forest | B | N | 6 | 39 |
| | *Podalyria calyptrata* | Trans. | G | S | 24 | 3 |
| | *Phylica pubescens* | Fynbos | A | S | 78 | 6 |
| | *Cliffortia cuneata* | Fynbos | G | S | 26 | 9 |
| | *Protea neriifolia* | Fynbos | W | C | 17 | 10 |
| C | *Olea europaea* | Forest | B | S | 4 | 8 |
| | *Cliffortia cuneata* | Fynbos | G | S | 19 | 0 |
| | *Protea neriifolia* | Fynbos | W | C | 13 | 0 |

[a] Dispersal categories: W = wind, B = bird, A = ant, G = gravity.
[b] Storage categories: N = no storage phase, S = soil-stored, C = canopy-stored (released on death of the plant, usually in fire).

related directly to fire (in provision of germination cues, seed release cues or decreased competition) or indirectly via the habitat (soil moisture, cover or litter accumulation). Germination cues for soil-stored seeds may include increased diurnal fluctuations in soil temperature as a result of the removal of the canopy (Thompson et al. 1977).

Reciprocal sowings (Fig. 5.4) showed, however, that germination of the species studied is more or less equally successful in all vegetation types (Manders and Richardson 1992). The fynbos shrubs *Protea neriifolia*, *P. repens* and *P. nitida* and the forest species *Kiggelaria africana* were sown in forest, transitional and fynbos sites near site A immediately after the fire. Germination was similar in all sites, but fynbos seedlings did not survive in the forest and *Kiggelaria africana* did not survive in the fynbos. All species established successfully in the transitional zone, but at slightly lower levels than in their respective vegetation types.

Similar evidence was provided when forest and fynbos species were sown in seedling boxes in fynbos soil, forest soil and forest soil with litter, in the open and under a tree canopy, and with and without additional watering (Manders and Richardson 1992). There was germination of most species

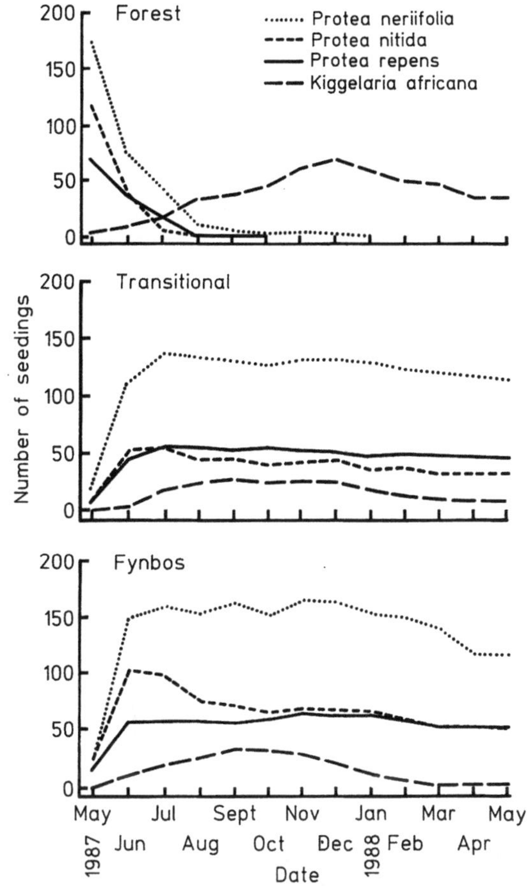


**Fig. 5.4.** Forest species are unable to establish in newly burnt fynbos and fynbos species cannot establish in forest. The data are numbers of seedlings of fynbos species (*Protea neriifolia, P. repens* and *P. nitida*) and a forest species (*Kiggelaria africana*) after sowing in forest, transitional and fynbos areas near site A in Swartboskloof after fire in March 1987 (Manders and Richardson 1992)

under most conditions, but survival of forest and fynbos species followed very different patterns (Fig. 5.5). Fynbos species were able to establish in exposed (post-fire) situations whereas forest species established under closed canopy conditions. The mortality of the fynbos species in the unwatered boxes in the open exceeds that in the field trial and is ascribed to prevention of rapid development of deep roots in these species (Manders 1991) in the boxes. Establishment of forest species is not dependent on high soil moisture availability alone. Conditions of forest soil (i.e. of higher nutrient status than the fynbos soil), litter and added water in the open were not sufficient to achieve survival as high as non-watered seedling boxes with

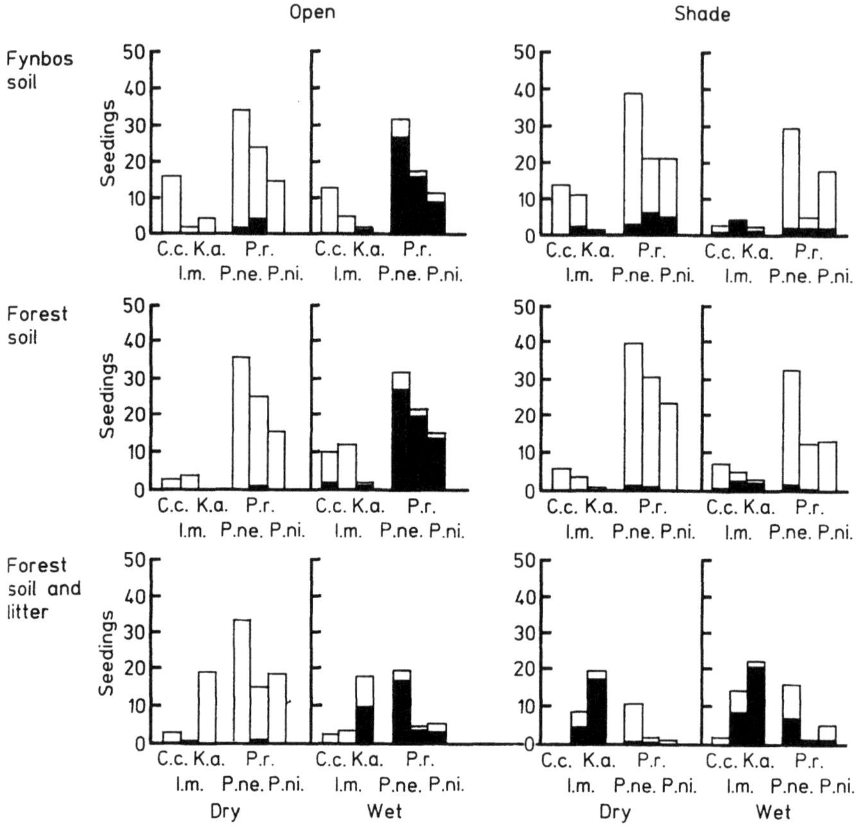


**Fig. 5.5.** The effects of habitat on establishment of forest and fynbos species shown by numbers of seeds germinating (*open bars*) and surviving after a summer (*solid bars*) in the open and in the shade with different soil types and moisture regimes. The data are means of germination and survival in five seedling boxes per species per treatment, each sown with 50 seeds. Forest species: **C.c** *Cunonia capensis*; **I.m.** *Ilex mitis* and **K.a.** *Kiggelaria africana*. Fynbos species: **P.ne.** *Protea neriifolia*; **P.r.** *Protea repens* and **P.ni.** *Protea nitida* (Manders and Richardson 1992)

forest soil and litter under the canopy of a tree. Similarly, forest seedlings under the tree required forest soil and litter as well as added water to achieve maximum survival.

### 5.3.3 Development of Forest in Fynbos Between Fires

If the development of forest in fynbos between fires is not limited by seed dispersal or germination requirements, but by differential abilities of seedlings to survive in the open, there must be some factors which permit such development in some places on some occasions. Two studies were carried out in Assegaaiboskloof to determine these factors (Manders and Richardson 1992). Assegaaiboskloof is on the same slope of the Stellenbosch mountains as Swartboskloof and approximately 4 km to the northwest. The post-fire age of the vegetation was approximately 50 years at the time of the study in 1989.

The first study sought to identify features characterizing the habitat of forest seedlings which have developed in fynbos, and whether these are related to the development of the fynbos vegetation. Habitat features may be related to amelioration of drought stress, including the mulching effects of litter, nurse plants, protection from radiation and wind, and reduced competition for water owing to reduced herbaceous cover. Alternative explanations would be that forest seedlings occurred randomly throughout the area or that seedling occurrence is simply a function of distance from the forest edge.

The characterization of forest seedling habitat was based on observations in four 4-m-wide transects extending from the forest edge to approximately the centre of the fynbos area. The habitat of all forest seedlings and random points along the transects was described. The hypothesis that forest seedling occurrence is related to development of fynbos vegetation was supported. Forest seedlings were associated with tall ground layer vegetation, projected plant cover in excess of 50% and well-developed litter layer (Table 5.5). There was no relationship between the occurrence of seedlings and distance from the forest edge.

The importance of perches in forest development was demonstrated by the paucity of seedlings in the open relative to numbers under perches. Perches were not necessarily fruit-bearing species, and many forest seedlings were found under fynbos shrubs. Perch height was important, and fruiting species less than 1.5 m tall, such as *Rhus tomentosa*, *R. angustifolia* and *Diospyros glabra*, supported relatively few seedlings (Table 5.6).

Groups of forest species are often present in fynbos. We propose in the model in Fig. 5.2 that these act as nuclei of forest development (sensu Yarranton and Morrison 1974). Forest regeneration is enhanced in such nuclei and the development of forest communities involves their coalescence (Janzen 1988). The second part of the Assegaaiboskloof study was to determine if nucleation is involved in the development of forest.

Eighty-one nuclei or plants considered to be potential nuclei were surveyed in Assegaaiboskloof. Nuclei were measured, and the number and size of forest plants were noted. Nuclei were classified according to whether they were dominated by species with fleshy fruit, likely to attract avian seed dispersers, or by species without fruit, such as fynbos shrubs or alien, non-fruiting trees.

Big nuclei had more species and more seedlings than small nuclei. Nuclei dominated by fruiting species tended to be bigger, with more species and regeneration than nuclei dominated by non-fruiting species (Fig. 5.6).

**Table 5.5.** The relationship between (a) height of herb layer, (b) vegetation cover and (c) litter depth and the occurrence of seedlings of forest species in Assegaaiboskloof, compared with the habitat of points without seedlings. $\chi^2$ values and levels of significance are given in square brackets below each table[a]

*a) Herb layer*

| Herb layer height (cm) | Seedlings | Points without seedlings |
|---|---|---|
| 0 | 0 (1.6) − | 7 (0.5) + |
| 1–10 | 2 (0.1) + | 5 (0.0) − |
| 11–50 | 11 (2.4) − | 64 (0.7) + |
| 51–100 | 39 (1.0) − | 156 (0.3) + |
| >100 | 62 (4.4) + | 141 (1.3) − |

[$\chi^2 = 12.5$, $P < 0.05$]

*b) Vegetation cover*

| Vegetation cover (%) | Seedlings | Points without seedlings |
|---|---|---|
| 0 | 0 (1.6) − | 7 (0.5) + |
| 1–50 | 39 (3.1) − | 182 (1.0) + |
| 51–100 | 75 (3.4) + | 184 (1.0) − |

[$\chi^2 = 10.7$, $P < 0.005$]

*c) Litter depth*

| Litter depth (cm) | Seedlings | Points without seedlings |
|---|---|---|
| 0 | 4 (9.1) − | 65 (2.8) + |
| 0–1 | 16 (3.8) − | 95 (1.2) + |
| 1–2 | 39 (0.2) + | 115 (0.1) − |
| 2–3 | 34 (2.5) + | 77 (0.8) − |
| >3 | 21 (12.7) + | 21 (3.9) − |

[$\chi^2 = 37.1$, $P < 0.0001$]

[a] Numbers in parentheses show cell contributions to $\chi^2$. Signs indicate whether a cell is over- (+) or under-represented (−) (Manders and Richardson 1992).

Non-fruiting species, however, do act as nuclei, although the rate of development is slower. Evidence of the development of forests in Assegaaiboskloof is seen in the reverse "J"-shaped curve of the populations of forest species in nuclei. Some species are comparatively rare and exist as juveniles only (e.g. *Hartogiella schinoides*) or as apparently stable populations (e.g. *Maytenus oleoides*, *M. acuminata* and *Olinea ventosa*). *Olea europaea* ssp. *africana* is the major species invading the fynbos. *Rapanea melanophloeos* and *Kiggelaria africana*, and the alien *Pittosporum undulatum* appear to be more recent invaders.

**Table 5.6.** The relationship between (a) distance to the nearest perch and (b) type of perch and the occurrence of seedlings of forest species and points without seedlings in Assegaaiboskloof. $\chi^2$ values and levels of significance are given in square brackets below each table[a]

*a) Perch distance*

| Perch distance (m) | Seedlings | Points without seedlings |
|---|---|---|
| 0 | 83 (22.0) + | 130 (6.7) − |
| 0–1 | 17 (13.6) − | 156 (4.2) + |
| 1–2 | 8 (4.5) − | 63 (1.4) + |
| 2–3 | 5 (0.0) − | 17 (0.0) + |
| >3 | 1 (0.4) − | 7 (0.1) + |

[$\chi^2 = 52.9$, $P < 0.0001$]

*b) Perch type*

| Perch type | Seedlings within 1 m of a perch | Points without seedlings |
|---|---|---|
| Open | 14 (3.9) − | 87 (1.2) + |
| Non-fruiting alien | 4 (2.1) − | 31 (0.6) + |
| Fynbos shrub | 21 (4.0) + | 37 (1.2) − |
| Low fruiting perch | 18 (0.8) − | 77 (0.2) + |
| Tall fruiting perch | 57 (2.4) + | 141 (0.7) − |

[$\chi^2 = 17.5$, $P < 0.01$]

[a] Numbers in parentheses show cell contributions to $\chi^2$. Signs indicate whether a cell is over- (+) or under-represented (−) (Manders and Richardson 1992).

There was no relation between distance from the forest edge and nucleus complexity. This could imply that the distances involved were too short to have any effect, or that the pattern was obscured by other nuclei acting as sources.

## 5.4 Factors Influencing Forest Development

Having established the processes involved in the development of forest, we examine those factors which may influence the operation of the processes. In many other vegetation types herbivory is a determinant factor, interacting with other factors in the dynamics of community boundries (e.g. Blackburn and Tueller 1970; Davis and Mooney 1985; Richardson and Bond 1991). There is no evidence, however, of any significant influence of herbivores on the dynamic balance of forest and fynbos vegetation.

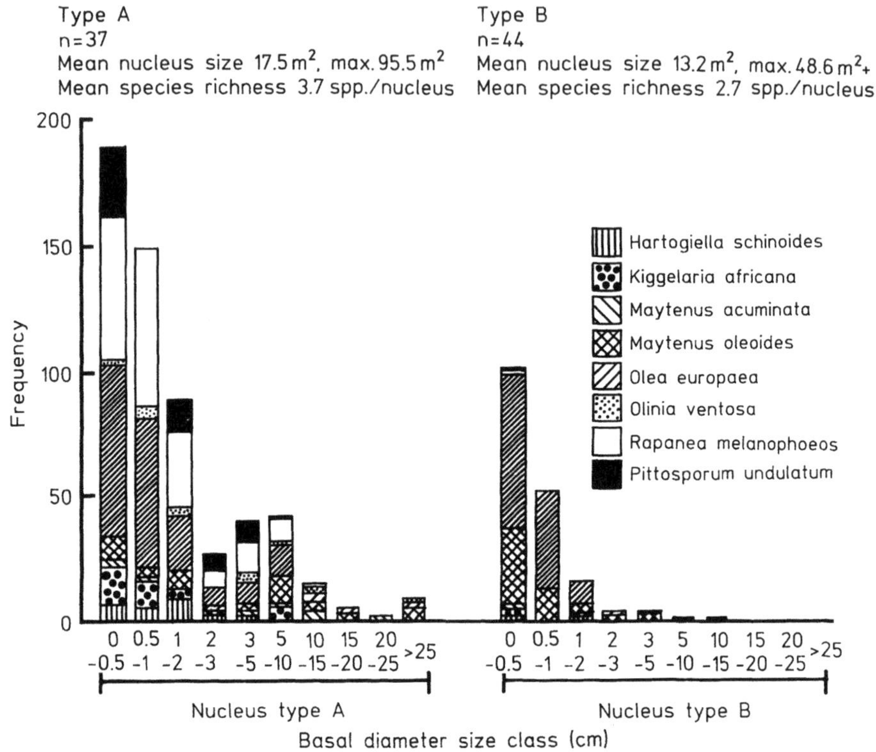


**Fig. 5.6.** Numbers of forest species in 81 forest nuclei in Assegaaiboschkloof in basal diameter size classes, for nuclei dominated by species with fleshy fruits (*type A*) and those dominated by species without fleshy fruits (*type B*) (Manders and Richardson 1992)

## 5.4.1 The Role of Nutrients in Forest Development

Campbell (1985) and Campbell and Werger (1988) predict that high tree cover in the fynbos biome is associated with relatively nutrient-rich soils with high moisture availability. The prediction that forests occur on soils richer in nutrients than nearby fynbos has been supported by several studies, but the magnitude and trend of differences between the soils of forest and fynbos vegetation are very site-specific (Cowling 1984; van Daalen 1984; Lambrechts 1986; Masson and Moll 1987).

P.H. Masson and E.J. Moll (unpubl. data) conducted canonical correspondence analyses on the soil attributes and community composition on plots in forests, forest margins and adjacent fynbos in Swartboskloof (Fig. 5.7). There is separation between forest at the top of axis 1 and transitional zone and fynbos species below, reflecting declining phosphorus levels. Species associated with mature forest are concentrated on the left of axis 2, with high nutrient levels and low aluminium exchangeable acidity.

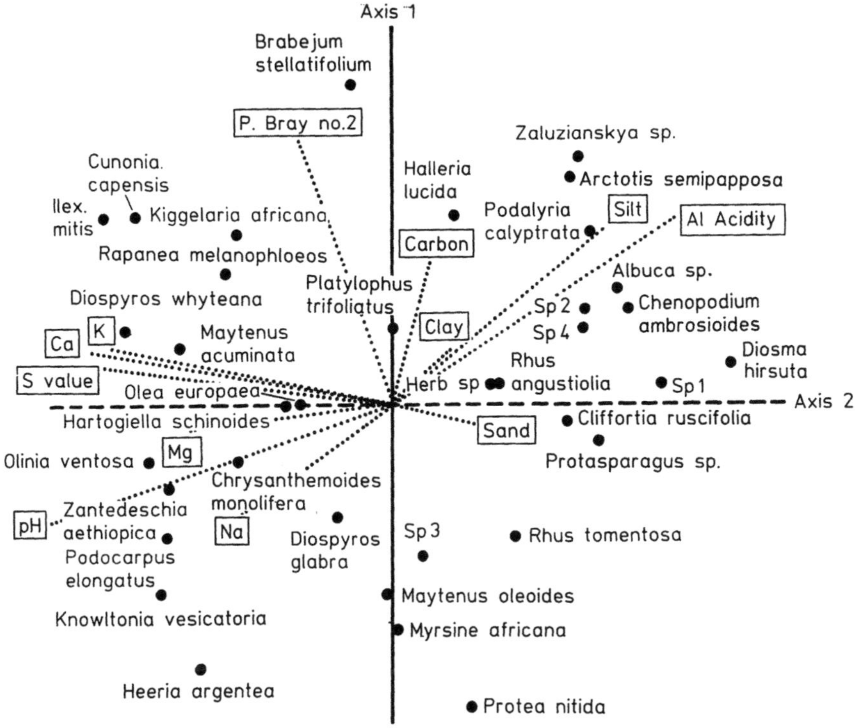


**Fig. 5.7.** The relationship between soil characteristics and the occurrence of forest and fynbos species, shown by an ordination by canonical correspondence analysis. The analysis is based on relative densities of species and soil characteristics of plots in forest, transitional zone and fynbos vegetation in Swartboskloof (P.H. Masson, E.J. Moll unpubl. data)

Soil samples were taken from the A horizon adjacent to the transects at the site A before the fire (Manders 1990a). The soils in the forest were of a finer texture, less acidic, and were consistently richer in nutrients than the other vegetation types. Aluminium levels were lower than in the fynbos, and very much lower than in the transitional zone. Similar differences, with higher levels of available phosphorus and lower levels of aluminium in the forest than in the fynbos (although inorganic levels of phosphorus were similar) have been reported from the southern Cape (Lambrechts 1986).

Our study differed from earlier studies in that the forest and fynbos samples were taken from within a few metres of each other. This indicates strongly that the differences are not related to differences in parent material, but are a result of different nutrient cycling processes operating within each community. It has been suggested that "under similar soil, terrain and climate conditions forest communities will maintain a higher nutrient status in the soils, especially in the surface horizons, compared with fynbos communities" (Lambrechts 1986). Therefore nucleation may involve facilitation through enhanced nutrient status under groups of forest species. The influence of changing species composition during succession to forest on soil properties has been noted elsewhere (Ellis and Graley 1987). Once different soil chemistry has been brought about by vegetation changes, such differences may affect the establishment abilities of species, thus maintaining the identity of the developing community.

### 5.4.2 Soil Moisture

Forests in the southwestern Cape are considered to comprise mesophytic species, confined to sheltered or riparian habitats by the water stress experienced by these species in adjacent fynbos areas (White 1978). In particular, the restriction is considered to reflect intolerance of summer moisture deficits associated with the summer drought of the mediterranean-type climate. Soil water potentials were monitored at 300, 450 and 900 mm on a hillslope and at a riparian site in Swartboskloof from August 1986 to February 1987. Despite the mediterranean rainfall regime, Swartboskloof receives 12.4% (200 mm) of its annual rainfall during the driest months of December through March (Wicht et al. 1969). Droughts of 30 days or more occur only once every 4.2 years, and the longest drought during the monitoring period was 17 days. Consequently, even on hillslopes, soils remained fairly moist and fell to below −0.1 MPa at times only at the 300-mm soil depth (Richardson and Kruger 1990).

The seasonal availability of water along the transects at sites A and B was determined gravimetrically, sampling the top 50 mm of soil adjacent to each transect (Manders 1990a). Surface soil moisture was consistently higher in the forest at both sites, and dried out more slowly in the hot, dry period of December of March (Fig. 5.8).

In general, fynbos areas with precipitation greater than 650 mm year cannot be regarded as a drought-stressed vegetation (Miller et al. 1983). However, soil moisture deficits in fynbos soil surfaces would be heightened after fire, due to increased radiation and wind activity resulting from lost vegetation cover. Therefore the facilitative effects of vegetation cover noted in Assegaaiboskloof may involve amelioration of drought stress.

Adults of mesophytic species in riparian habitats and xerophytic species in hillside habitats exhibit very similar stomatal behaviour, net photosynthesis and water-use efficiencies, and do not reflect any selective influence of contrasting soil moisture regimes (Richardson and Kruger 1990). The absence of differences in water stress between adult riparian and hillslope plants, and that only surface soils in fynbos dry out more than forest soils, suggest that discrimination between forest and fynbos species in non-riparian areas is limited to the establishment phase. Vigour and density of forest precursor seedlings decline with distance from stream bank forest

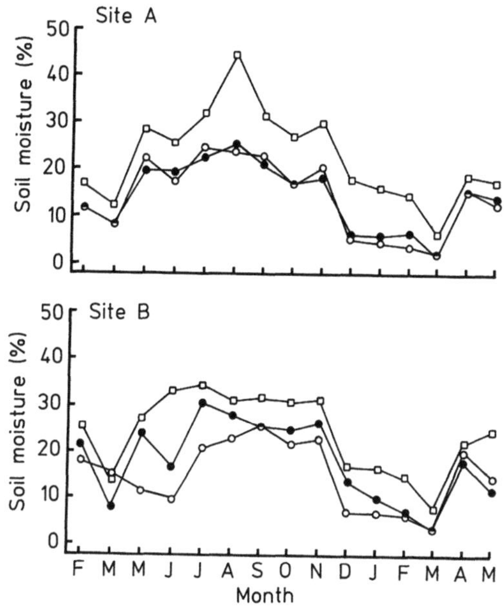


**Fig. 5.8.** Pre-fire and monthly post-fire surface soil moisture in forest, transitional and fynbos vegetation at *sites A* and *B*. ○, Fynbos; ●, transitional zone; □, forest (Manders 1990a)

(Kruger 1984). However, it is very difficult to test in the field the ability of forest species to establish in dry habitats as established individuals of forest species growing in such habitats have survived the period of greatest vulnerability to drought. It is also very difficult to determine what soil moisture is available to plants in the field, some of which may have very deep roots.

To determine the different responses of forest and fynbos plants to levels of soil moisture during and after the establishment phase, while avoiding the problems with field studies, individual plants were cultivated in separate containers, under dry, medium and wet watering regimes (Manders 1991). This excluded competitive effects and did not allow plants to evade moisture stress by developing deep root systems. *Protea neriifolia*, *P. nitida* and *P. repens* were used as representatives of sclerophyllous fynbos shrub species and *Kiggelaria africana* and *Cunonia capensis* were used as representatives of evergreen forest tree species.

No plants died in the wet and medium watering regimes except for 40% of the *C. capensis* in the medium treatment. Greatest mortality in the dry treatment occurred in *C. capensis* (60%). *K. africana* showed an ability to persist in the dry treatment with no deaths, and there were very few deaths in the fynbos species. Growth of surviving forest species in the dry regime was similar to that of the fynbos species (Fig. 5.9). All species had lower root:shoot ratios in the wet watering regime. There was no clear distinction between the conductance rates of forest and fynbos species (Fig. 5.10). The

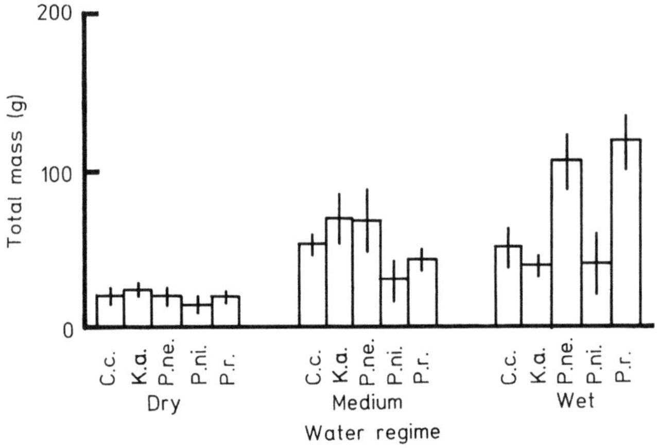


**Fig. 5.9.** The effects of soil moisture availability on the growth of forest and fynbos species shown by the total mass of forest and fynbos plants cultivated from seedlings in July 1987 to December 1988 under three watering regimes. 95% confidence limits about the means are represented by *vertical lines*. See Fig. 5.5 for explanations of species abbreviations (Manders 1991)

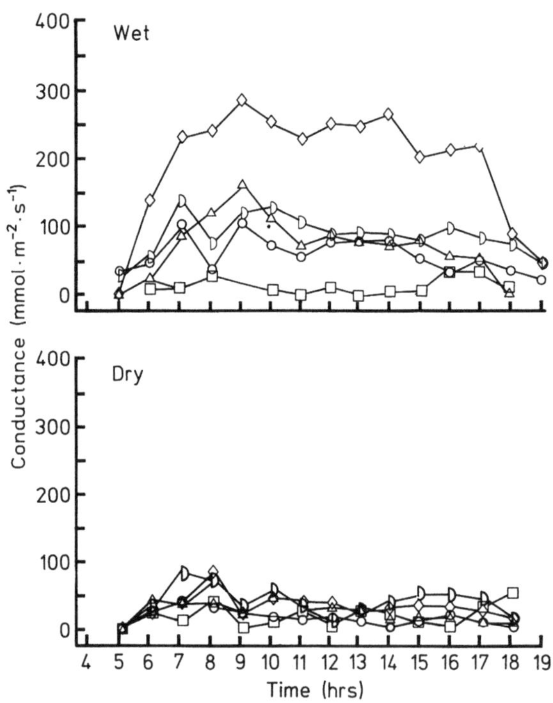


**Fig. 5.10.** Diurnal conductance rates of forest and fynbos species individually cultivated under dry and wet watering regimes. Values represent the median of four leaves per species. ○, *Cunonia capensis*, □, *Kiggelaria africana*, △, *Protea neriifolia*, ◇, *Protea nitida* and D *Protea repens* (Manders 1991)

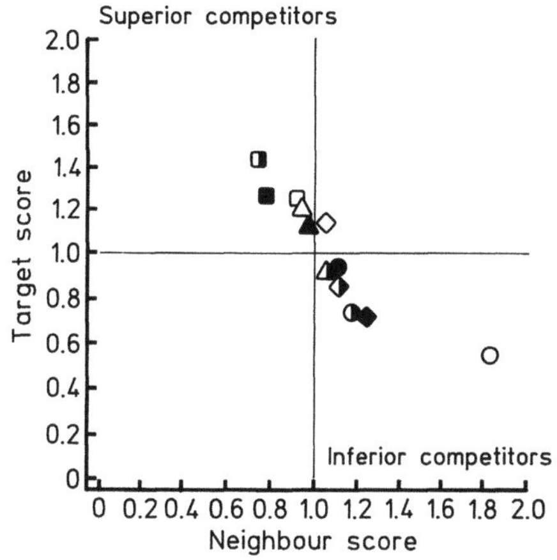


**Fig. 5.11.** Competitive abilities of forest and fynbos species cultivated in a diallel experiment under three watering regimes. ○, *Cunonia capensis*, □, *Kiggelaria africana*, △, *Protea neriifolia* and ◇, *Protea nitida*. *Open*, *half-open* and *closed symbols* represent dry, medium and wet treatments respectively (Manders 1991)

differences in the abilities of forest and fynbos species to survive appear to be related to greater vulnerability to drought of the xylem vessels in forest species (E. February, P.T. Manders, A. Scholtz unpubl. data).

A diallel experiment (sensu Harper 1977) was used to determine competitive abilities (Manders 1991). Two individuals of *P. neriifolia*, *P. nitida*, *K. africana* and *C. capensis* were planted in each container, to produce all possible pairwise species combinations. Analyses were based on the relative increase in plant mass (RIP) of a species when grown with another species compared with that when grown with a conspecific plant. The mean of all RIP values of a species when grown with all species of neighbours provides a target score, and the mean RIP of all neighbour species provides a neighbour score (Wilson and Keddy 1986). Competitive superiority is indicated by high target scores and low neighbour scores.

Significant differences between target scores and between neighbour scores occurred between species within each treatment. *K. africana* was a superior competitor at all levels of water availability, whereas *C. capensis* was a consistently inferior competitor, but tended to greater competitive ability with increased water availability. The competitive abilities of *P. neriifolia* were little influenced by water availability. *P. nitida* was never a strong competitor, and lost competitive ability as the water levels increased (Fig. 5.11). The competitive superiority of *P. nitida* compared with *P. neriifolia* under dry conditions and the loss of this ability under wet conditions are consistent with these species' distribution ranges (Rourke 1980). This consistency provides support for the interpretation that *K. africana*, a forest pioneer species, is able to compete with fynbos species in a dry environment, and is more successful than *C. capensis* which is restricted to wet environments.

Differences between the abilities of forest and fynbos species to survive soil moisture deficits may result from differing efficacies in conservation of water after uptake by the roots, or differing patterns of root development affecting access to soil water. A further study sought to complement the individual cultivation and diallel studies by cultivation of the same species in a situation where root development and competition could play a role, even though the effects of water availability per se may be obscured (Manders 1991). Forest (*Cunonia capensis* and *Kiggelaria africana*) and fynbos (*Protea neriifolia* and *P. nitida*) species were cultivated on an artificially established gradient of depth to water table in troughs similar to the design of Mueller-Dombois (1964). The effects of such gradients will be affected by the capacity of the soil to draw up or retain water, and the nutrient status of the soil. Therefore the experiment was conducted simultaneously in nutrient-poor, sandy fynbos soil and richer forest soil with a greater organic content. Ten evenly spaced rows of eight plants (two per species) were planted across each trough.

The forest soil at the deep end of the trough was drier near the surface than the fynbos soil in the corresponding situation. At depths of 0.75 m and deeper, however, the forest soil held considerably more water than fynbos soil. Plant mass did not vary with position in either trough, but differed between species (Fig. 5.12). The experiment showed the ability of fynbos species to grow under dry, nutrient-poor conditions, but they were not able to respond as well as forest species to improved conditions. Low plasticity of fynbos species in comparison with forest species is evident in the response of xylem vessel anatomy (E. February, P.T. Manders, A. Scholtz unpubl. data). Forest species persisted to some extent in the top (dry) end of the fynbos trough, but the mass of these plants was negligible and further mortality, especially in *Cunonia capensis*, seemed likely at the time of harvesting. In the forest trough *Kiggelaria africana* had no mortality, and mortality of *Cunonia capensis* was very much lower, even at the top end of the trough, than in the fynbos trough.

In the trough of forest soil, *Kiggelaria africana* had considerable lateral root extension and the roots had grown down to the water table. *Cunonia capensis* also had many secondary roots, but these were shorter and finer, and those in the upper rows did not extend to the water table. The *Protea* species in both troughs had relatively very little lateral root development, but had a single main root which usually reached the water table (Fig. 5.13). Roots of the forest species in the fynbos trough were very poorly developed.

There was no evidence of water stress for any of the species in the forest trough. Although the forest species in the deeper soils had slightly higher (more negative) pre-dawn xylem pressure potentials than the fynbos species, no potentials exceeded −1.2 MPa. Water potentials at the bottom of the

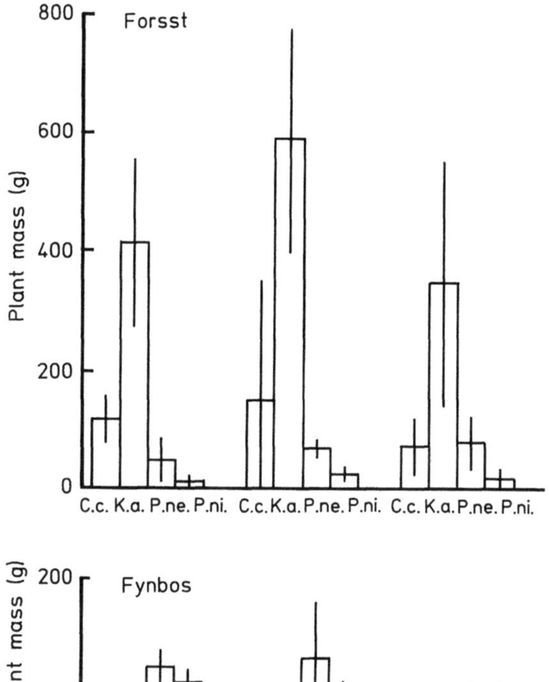


**Fig. 5.12.** Total plant mass of forest and fynbos plants cultivated from seedlings in July 1987 to January 1989 in two troughs, one with forest soil and one with fynbos soil, with gradients of depth to water table. See Fig. 5.5 for explanations of species abbreviations. *Position 1* = rows 1–4 (nearest the water table), *position 2* = rows 5–6, and *position 3* = rows 7–10 (furthest from the water table). 95% confidence limits about the means are represented by *vertical lines* (Manders 1991)

fynbos trough were low (above −0.6 MPa) with no differences between the fynbos and forest species. Forest species which survived in the upper rows of the fynbos trough had higher, although not excessive, water potentials (up to −1.35 MPa). The lack of moisture stress appears to be related to the differing patterns of root development.

### 5.4.3 Historical Factors

The current interactions between forest and fynbos may not be those with which the system evolved, but may be products of changes associated with the alteration of landscapes by human settlement. Before European settlement in 1652, the Khoisan, although not copious users of wood, had certainly influenced the fire regime of the vegetation in the Cape for at least

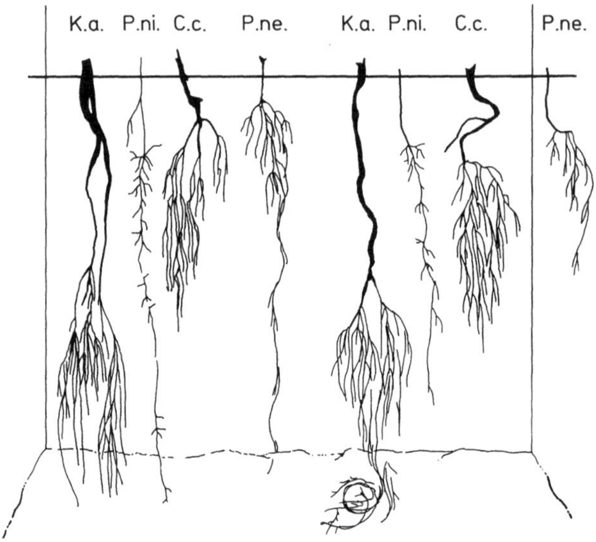


**Fig. 5.13.** Root development of forest and fynbos species situated furthest from the water table in a trough of forest soil. Plant stems were connected to a wire at the original soil level to facilitate excavation. The ground level is the level of the permanent water. See Fig. 5.5 for explanations of species abbreviations (Manders 1991)

100000 years (Deacon 1983a,b; see also Chap. 1, Sect. 1.3). Although there are no reliable records of the extent of southwestern Cape forests prior to 1652, a marked decline may be inferred from historical records of utilization and destruction, and from the laws on forest usage promulgated very soon after settlement (Sim 1907; Wicht 1945; Grut 1977). However, there is no evidence that the southwestern Cape has ever had extensive areas of high forest in historical time (Grut 1977; Campbell 1985; Scholtz 1986). Although forests do not burn readily (van Wilgen et al. 1990b), fires at short intervals may erode the edges of forest patches, and there are occasional reports of very extensive and destructive fires in forest (Sim 1907; Wicht 1945). Exploitation may have allowed fires to penetrate forest patches (Sim 1907; McKenzie 1978). Succession of fynbos to forest is likely to have been more evident before the decline of forests associated with settlement.

## 5.5 Synthesis

### 5.5.1 Fynbos Succession

We can explain the distribution of forest and fynbos, the development of forest species in fynbos and the occasional conversion of fynbos to forest on the basis of successional processes in the model in Fig. 5.2. The lack of forest development in young fynbos (Kruger 1984) is not determined by seed availability as seed banks and dispersal cross community boundaries. Forest seed dispersal by wind is not influenced by fire in the adjacent fynbos, and bird-dispersed seeds are introduced soon after fire. Germination of forest species may occur under post-fire conditions, but establishment is unlikely until habitat amelioration through the development of the fynbos or nuclei of forest species which may have survived the fire.

Gill and Groves (1981) considered fynbos to follow the initial floristic composition (IFC) model of succession (sensu Egler 1954; Connell and Slatyer 1977), but this model fails if forest species are introduced later. The studies reviewed in this chapter support the application of a facilitation model (Connell and Slatyer 1977) in which changes in the community are autogenic (i.e. driven by forces within the community). Evidence for this is provided by the enhanced forest seedling establishment in sheltered areas – under either fynbos vegetation or forest nuclei.

Recently, a reductionist approach has been applied to forest succession (see Finegan 1984) in which the facilitation hypothesis is rejected and autogenic forces are seen as neutral or inhibiting factors rather than the driving force in succession. It could be argued that the alleviation of the constraints on the establishment of forest species, through the provision of shading or litter, represents a release from inhibitory factors and supports the reductionist approach. Furthermore, the reductionist concept of tolerance (the ability of a plant to survive in the presence of high densities of other species) applies to forest species colonizing mature fynbos. Succession of fynbos to forest involves interdependent mechanisms, best understood by a synthetic approach of general and reductionist models (sensu Finegan 1984).

The studies outlined in this chapter also argue strongly for the concept of succession being the consequence of the different life cycles of the species involved (Gómez-Pompa and Vázquez-Yanes 1981). This explanation allows for the predominance of fynbos shrublands under a disturbance regime, although forest may be predicted in the absence of disturbance.

Forest canopies have higher moisture contents, sparser crowns, and are more separated from the litter layer than fynbos canopies, and are therefore less likely to burn than shrublands (van Wilgen et al. 1990b). Therefore, given a standard fire intensity, the size of a forest nucleus or stream bank forest will determine the effect of the fire. As forests develop and nuclei coalesce, higher intensity fires are required to remove the forest, and there will be a greater probability of the forest remaining intact or regressing partially only. Invasions of the fynbos by alien trees and shrubs are usually in response to a disturbance, and usually establish in the disturbed area in a discrete episode (see Chap. 9, Sect. 9.3.3). In contrast, invasions of fynbos by indigenous trees are related to low frequencies of disturbance. This may allow for some manipulation of vegetation through the fire regime. High frequencies of fire aimed at the control of biomass and fuel accumulation may promote invasions by aliens, whereas low fire frequencies may result in a less flammable indigenous community with a greater forest component.

The model in Fig. 5.2 identifies processes which are susceptible to disruption by aliens. Alien trees in fynbos create new perches (Richardson and van Wilgen 1986a) and may initiate nucleation of forest elements. However, utilization of alien fruits by avian dispersers may diminish dispersal of indigenous seeds (Knight 1988) and alien birds may modify dispersal patterns. Therefore the impact of alien invasions may enhance or inhibit forest development, depending on which process is limiting at a site.

The studies described in this chapter have refuted the concept of forest and fynbos as two distinct and mutually exclusive vegetation types. Forest species are a late successional (or tolerant) component of the dynamic vegetation of the fynbos biome, and may potentially predominate some areas currently under fynbos communities. The perceived restriction of forest communities to sheltered environments is principally a function of past and present fire regimes. The distinctness of the forest communities results from the exclusion of fynbos species from forests because of the inferior competitive abilities of fynbos species in these environments.

### 5.5.2 Determinants of the Rate of Succession

It is clear that most of the fynbos biome, and certainly all of Swartboskloof, is bioclimatically suitable for forest. Richardson and Cowling (Chap. 9) have described the factors that mediate the invasion of fynbos by alien trees. They suggest that the fundamental determinants of invasibility are the transient fire-induced openings that allow establishment, and the apparent surplus of available resources in fynbos. Indigenous forest species require different invasion windows (sensu Johnstone 1986), which are created later in the succession.

The assertion by Campbell (1985) that development of forests in the fynbos biome will be enhanced by high nutrient levels and water availability is supported by the studies reviewed in this chapter. At the same time, the studies demonstrated that forest species, particularly those regarded as forest precursors, are able to persist under very xeric conditions. Persistence (rather than competitive ability) is important in the establishment of forest pioneer seedlings as described in Fig. 5.2. Plants that manage to persist through the first few summer droughts may be able to take advantage of any habitat amelioration in later stages of succession. *Kiggelaria africana*, for example, shows greater ability to persist and is usually better able to colonize relatively dry fynbos areas than *Cunonia capensis*. Differences in ability to persist appear related to root growth patterns, and studies in this direction should prove rewarding.

Soil moisture and nutrients per se do not limit forest distribution, but affect the rate of forest development. It is the frequency of fire, in relation to the rate of forest development, which determines the extent of forest development.

*Acknowledgements.* We thank R.M. Cowling, S.M. Cowling, F.J. Kruger, D.C. Le Maitre and B.W. van Wilgen for their helpful advice at various stages of this work. A.C. Duckworth provided valuable advice on data analysis. G. Bailey, C. Jones, K.B. Higgins, G. Mostert, and especially A. Abrahams and A.D.A. Simmers provided invaluable assistance with fieldwork and other tasks.

# 6 Coexistence of Seeders and Sprouters in a Fire-Prone Environment: the Role of Ecophysiology and Soil Moisture

R.E. Smith, B.W. van Wilgen, G.G. Forsyth, and D.M. Richardson

## 6.1 Introduction

The wide variety of flowering plant species that occur in fynbos communities can be divided into two broad categories on the basis of their mode of regeneration after fire: sprouters and non-sprouters (or seeders; see Chap. 4). The loss of the ability to sprout is considered to be a specialized developmental trait in woody dicotyledons, with sprouting being considered conservative in evolutionary terms (Wells 1969; see Chap. 7). Sprouters tend to be more resilient under a range of fire regimes than seeders, often showing remarkable powers of vegetative recovery. Keeley (1977) and Specht (1981b) suggested that the relative dominance of sprouters in fire-prone shrubland communities is a consequence of the competition between adult sprouters and seedlings of non-sprouters. Generally, sprouters have a competitive advantage over seeders after fires because they grow more rapidly, reach reproductive maturity sooner, and are better competitors for light, moisture and space. During long fire-free intervals, sprouter adults become self-thinning, and mortality is driven by density-dependent competition and senescence (Parker 1984). This results in the formation of gaps in the vegetation canopy after fire due to the lack of recovery of senescent sprouters. These gaps can be exploited by seedlings. After long fire-free intervals, the buildup of fuel is greater than over shorter intervals, possibly leading to more intense fires (see Chap. 3) and hence a higher mortality rate of adult sprouters (see Chap. 7), once again opening up gaps or safe sites for seedling establishment after fire (Keeley 1986). Seedlings of non-sprouters must compete with sprouter seedlings to exploit these gaps, and may achieve success either through numerical or physiological superiority. Data do not always support the hypothesis that non-sprouter seedlings are more abundant than sprouter seedlings (see Chap. 7). Therefore something else, perhaps physiological superiority, could explain their success (Parker 1984). Physiological characteristics affecting growth (especially water use and carbon gain) should differ between the seedlings of the two reproductive modes if the non-sprouter seedlings are to gain an advantage (Parker 1984).

Miller (1981b) and Keeley (1986) suggested that there might be certain ecophysiological factors which may separate (and therefore explain the spatial distribution and relative success of) sprouters and seeders in Californian chaparral. The species in chaparral are divided into three major groups (Keeley 1986): the non-sprouters (only expand their populations after fires through seed), facultative sprouters (seeds are fire-resistant and sprout after fires) and obligate sprouters (produce fire-labile seeds which are usually bird-dispersed, and rarely survive the high temperatures of fires, thus relying on sprouting for persistence after fire). Obligate sprouters in chaparral often have more extensive root systems (Hellmers et al. 1955; Kummerow et al. 1977) and are seldom exposed to water stress. Sprouters usually have poorer stomatal control; water stress stops carbon uptake at an earlier stage than seeders. Seeders, on the other hand, are found to have higher stomatal conductances, higher drought tolerance, lower shade tolerance and higher growth rates than sprouters. Higher transpiration rates are usually well correlated with higher photosynthesis rates. Seeders have been found to have photosynthesis rates two to three times greater than those of sprouters (Parker 1984). These higher photosynthesis rates suggest that seeders may have potentially faster growth rates. These characteristics, and the fact that seeders are less demanding than sprouters in terms of site, enable seeders to establish after fire in the gaps between sprouters. These competitive advantages during seedling establishment may, however, disappear in the adult stage (Parker 1984).

The abundance of seeders in an Australian shrubland was inversely proportional to precipitation, because the density of competing sprouters was related to rainfall (Specht 1981b). In more humid environments, sprouters predominate after fire; seedlings of both sprouters and seeders are generally excluded early in the post-fire period by heavy competition from dense stands of resprouters. On xeric, nutrient-deficient sites, the vegetation is more open, and seeders can often survive alongside sprouters (Specht 1981b). In the chaparral, sprouter seedlings are also more drought-sensitive than those of seeders (Keeley 1986). Higher moisture levels and more favourable growing conditions on moist, nutrient-rich sites allow sprouters to out-compete seeders. Accordingly, vegetation on xeric sites should have a high proportion of seeders. The coexistence of sprouters and seeders in mediterranean-climate communities appears to be enhanced by spatial heterogeneity of suitable sites as well as by a temporal heterogeneity due to variation in fire frequency and intensity (see Chap. 7).

In this chapter, we examine the above-mentioned paradigms by collating evidence from several studies in the fire-prone vegetation of Swartboskloof. Specifically, we compare the ecophysiological characteristics of sprouters and seeders to determine whether there are indeed fundamental morphological and physiological differences between these regeneration modes. We also compare the species composition on a number of sites in Swartboskloof, to ascertain whether the soil moisture regime plays a role in determining the relative abundance of sprouters and seeders. The results from Swartboskloof are compared to those from other fynbos sites, and other mediterranean-climate regions.

## 6.2 Ecophysiological Characteristics of Seeders and Sprouters in Swartboskloof

We collated data from several studies on the water relations of selected sprouting and non-sprouting plant species in Swartboskloof before and after fire (P.C. Miller et al. 1983; J.M. Miller et al. 1984; Kruger et al. 1988; Richardson and Kruger 1990; Smith and Richardson 1990). Stomatal conductances, transpiration rates (post-fire only), and leaf xylem pressure potentials (as indicators of water stress) of selected fynbos plant species were measured. Data on the root systems of some species were available from studies by Higgins et al. (1987), Smith and Higgins (1990) and P.T. Manders and R.E. Smith (unpubl. data). We compared fynbos sprouters and seeders to determine whether seeders, facultative sprouters and obligate sprouters differ in their ecophysiological characteristics, and to ascertain whether they follow Parker's (1984) chaparral paradigm.

Xylem pressure potentials could not be used to clearly distinguish between groups of adult seeders, obligate sprouters and facultative sprouters prior to the 1987 fire in Swartboskloof (P.C. Miller et al. 1983; J.M. Miller et al. 1984; Kruger et al. 1988; Richardson and Kruger 1990). Xylem pressure potentials of seeders (e.g. *Protea repens*, *P. neriifolia*), facultative sprouters (e.g. *P. nitida*, *Leucadendron salignum*) and obligate sprouters (e.g. *Cunonia capensis*) were generally moderate (greater than −3 MPa; Table 6.1). This was not entirely unexpected, as the root systems of several of the dominant non-sprouting and sprouting shrubs in the fynbos are of similar structure. For example, the non-sprouting *P. repens* and the sprouting *L. salignum* both have deep roots (≥3 m), and the non-sprouting *Erica plukenetii* and the sprouting *Otholobium fruticans* both have relatively shallow roots (≤1.5 m; Table 6.1; Higgins et al. 1987; Smith and Higgins 1990). Differences in potentials between seedlings of both sprouters and seeders, and resprouts immediately after the fire were only evident during the relatively dry summer. Seedlings of sprouters (*P. nitida*) and seeders (*P. repens*) displayed lower (more negative) potentials (ca. −4 MPa) than post-fire resprouts (>−2 MPa) during summer due to their comparatively shallow roots (Table 6.1). Xylem potentials could also not be used to differentiate between the seedlings of sprouters and seeders (Smith and Richardson 1990).

Pre-fire conductances were variable within and between sprouting and non-sprouting growth forms, but not consistently so. In some months, cer-

**Table 6.1.** A summary of the water use characteristics of mature plants and seedlings of dominant fynbos and chaparral shrubs

| Species | Regeneration form[a] | Maximum root depth (m) | Maximum conductance (mmol $m^{-2} s^{-1}$) | Minimum xylem pressure potential (MPa) | Source[b] |
|---|---|---|---|---|---|
| | | | Mature plants in Swartboskloof | | |
| *Protea repens* | NS | 3+ | 350 | −2.7 | 7,8 |
| *Protea neriifolia* | NS | 3+ | 160 | −1.6 | 7,9 |
| *Erica plukenetii* | NS | 0.5 | 480 | −3.0 | 7,9 |
| *Protea nitida* | FS | 1.5+ | 450 | −2.8 | 10,11 |
| *Leucadendron salignum* | FS | 3.5+ | 500 | −2.5 | 7,11 |
| *Otholobium fruticans* | FS | 1.5 | – | −4.5 | 7,12 |
| *Cunonia capensis* | OS | – | 140 | −1.5 | 8 |
| *Brachylaena neriifolia* | OS | – | 230 | −2.0 | 8 |
| | | | Seedlings in Swartboskloof | | |
| *Protea repens* | NS | – | 650 | −3.6 | 11 |
| *Cliffortia cuneata* | NS | – | 500 | −3.5 | 11 |
| *Protea nitida* | FS | 0.6 (2 yrs) | 1000 | −4.0 | 13 |
| *Kiggelaria africana* | OS | 2.2 (2 yrs) | – | −1.0 | 14 |
| *Cunonia capensis* | OS | 1.0 (2 yrs) | – | −1.0 | 14 |
| | | | Mature chaparral plants | | |
| *Arctostaphylos glauca* | NS | 2.6 | 200 | ≪ −6.5 | 1,2 |
| *Ceanothus greggii* | NS | 1.4 | 225 | ≪ −6.5 | 1,2 |
| *Ceanothus megacarpus* | NS | 2.4 | – | < −8.0 | 3 |
| *Ceanothus spinosus* | FS | 3.1 | – | < −5.0 | 3 |
| *Adenostoma fasciculatum* | FS | 7.6 | 180 | −6.0 | 1,4 |
| *Rhus laurina* | FS | 5.4 | 100 | −2.5 | 3 |
| *Heteromeles arbutifolia* | OS | – | 200 | −3.5 | 5 |
| *Quercus dumosa* | OS | 8.5 | 135 | −5.0 | 1,2 |
| *Rhus ovata* | OS | – | 110 | −2.5 | 2 |

**Table 6.1** (continued)

| Species | Regeneration form[a] | Maximum root depth (m) | Maximum conductance (mmol $m^{-2} s^{-1}$) | Minimum xylem pressure potential (MPa) | Source[b] |
|---|---|---|---|---|---|
| | | | Chaparral seedlings | | |
| *Ceanothus greggii* | NS | – | 70 | −4.3 | 6 |
| *Ceanothus megacarpus* | NS | 2 (2 yrs) | 55 | < −8.0 | 3 |
| *Ceanothus spinosus* | FS | 3 (2 yrs) | 65 | < −5 | 3 |
| *Adenostoma fasciculatum* | FS | – | 25 | −4.0 | 6 |
| *Rhus laurina* | FS | 6 (2 yrs) | 50 | < −4.5 | 3 |
| *Quercus dumosa* | OS | – | 10 | −4.1 | 6 |

[a] NS = non-sprouters (seeders); FS = facultative sprouters; OS = obligate sprouters; (–) indicates no data.

[b] Sources: 1, Kummerow (1981); 2, Poole et al. (1981); 3, Thomas and Davis (1989); 4, Hart and Radosevich (1987); 5, Poole and Miller (1975); 6, Hastings et al. (1989); 7, Higgins et al. (1987); 8, Kruger et al. (1988); 9, Miller et al. (1983); 10, Smith and Higgins (1990); 11, Smith and Richardson (1990); 12, R.E. Smith (unpubl. data); 13, Smith (1990); 14, P.T. Manders and R.E. Smith (unpubl. data).

tain seeders displayed higher conductances than some of the sprouters, while in other months the situation was reversed (P.C. Miller et al. 1983; J.M. Miller et al. 1984; Kruger et al. 1988; Richardson and Kruger 1990). The seasonal range of mean and peak conductances after the fire also varied from species to species rather than from one life-form to the next. Differences in the mean and peak conductances of seedlings (of both sprouters, e.g. *P. nitida*, and seeders, *P. repens*) and sprouters were not clearly evident after the fire. At times (e.g. April, immediately after the first winter rains, Fig. 6.1), conductances were higher in seedlings (of sprouters and seeders) than for resprouts (Fig. 6.1). Maximum conductances also appear to be slightly higher for seedlings of both regeneration modes (500–1000 $mmol\,m^{-2}s^{-1}$) than for resprouts (140–500 $mmol\,m^{-2}s^{-1}$; Table 6.1). Both regeneration forms showed low conductances ($\leqslant$100 $mmol\,m^{-2}s^{-1}$) during the dry summer months after the fire; the effect was more marked in the seedlings (both regeneration modes) with their shallow root systems (Smith and Richardson 1990).

Little is known of the transpiration rates of fynbos species. No estimates were made of transpiration rates of species in Swartboskloof prior to the fire, but data are available for the period after the fire. Mean and peak transpiration rates of resprouts were more consistent than those of the seedlings (of both sprouters and seeders) throughout the year (Fig. 6.2). Non-sprouter seedlings (e.g. *Protea repens* and *Cliffortia cuneata*) showed more fluctuation in transpiration rates through the year than the resprouts of *P. nitida* and *L. salignum*. Seedlings of sprouters (e.g. *P. nitida*) occupied an intermediate position between non-sprouter seedlings and resprouts. If anything, the seedlings probably had slightly higher transpiration rates per unit leaf area than the resprouts, and non-sprouter seedlings generally had rates fractionally higher than sprouter seedlings (Fig. 6.2; Smith and Richardson 1990).

The limited data on root systems in Swartboskloof did not reveal consistent patterns regarding rooting depth of sprouters and seeders. Root systems of dominant seeders (e.g. *Protea repens* and *Protea neriifolia*) were as deep (>3 m) as those of some of the dominant sprouters (e.g. *Leucadendron salignum*, >3.5 m; *P. nitida*, >1.5 m) (Table 6.1). Other sprouters (like *Otholobium fruticans*) had relatively shallow root systems ($\leqslant$1.5 m), similar to seeders like *Erica plukenetii* (Higgins et al. 1987; Smith and Higgins 1990; Table 6.1). In a greenhouse experiment, roots of 2-year-old seedlings of a seeder (*P. neriifolia*; 1 m) were found to be deeper than those of a facultative sprouter (*P. nitida*; 0.5–0.7 m) and two obligate sprouters (*Cunonia capensis*, *Kiggelaria africana*; ca. 0.25 and 0.6 m) when grown in nutrient-deficient sandy soils. On nutrient-rich soils the pattern was different. The roots of one of the obligate sprouters (*K. africana*; 1.7–2.2 m) were longer than those of the seeder (0.7–1.0 m) or facultative sprouter seedlings (0.5–0.9 m). The other obligate sprouter (*C. capensis*; 0.8–1.0 m) occupied an intermediate position (P.T. Manders and R.E. Smith, unpubl. data).

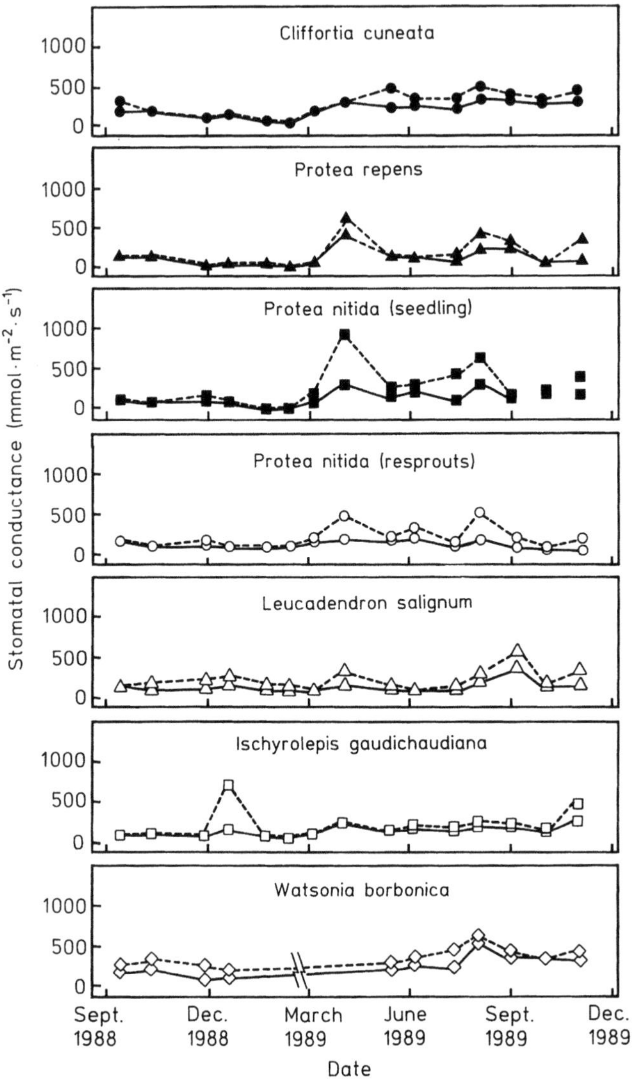


**Fig. 6.1.** The median (*solid line*) and peak (*dotted line*) stomatal conductances for each measurement day (September 1988–December 1989), for selected seeding (*closed symbols*) and sprouting (*open symbols*) species

There have been no studies on the comparative survival rates and relative shade tolerances of seeder and sprouter seedlings in Swartboskloof. The only data available comes from a trial where seedlings of Swartboskloof plants were planted under shade cloth, and out in the open, under two moisture regimes and different soils (see Chap. 5). Germination was slightly higher for non-sprouter seedlings than for sprouter seedlings in the shade

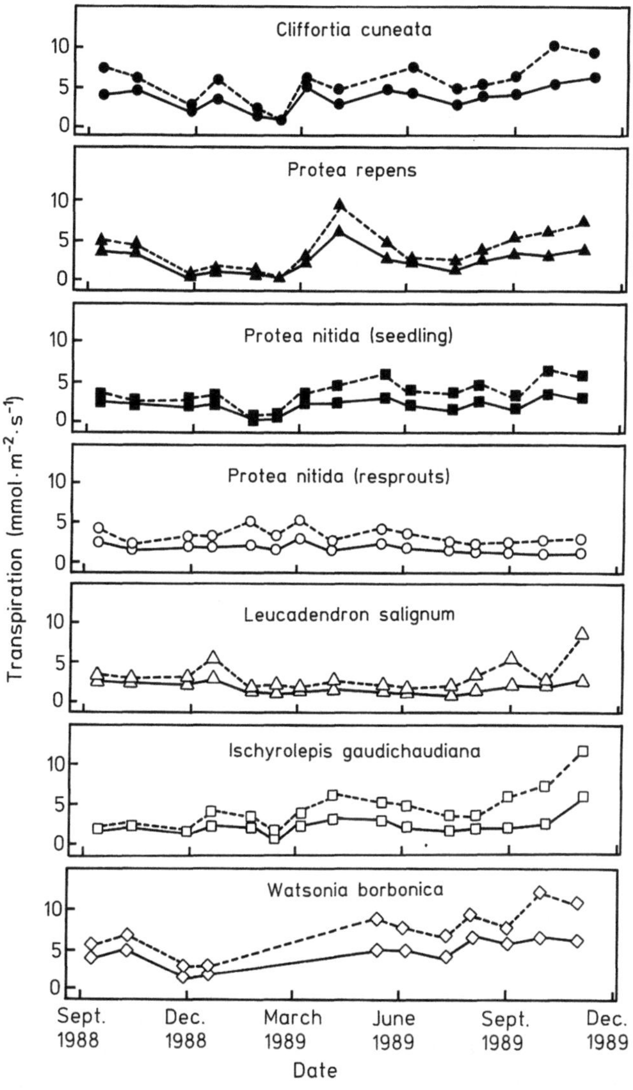


**Fig. 6.2.** The median (*solid line*) and peak (*dotted line*) transpiration rates for each measurement day (September 1988–December 1989), for selected seeding (*closed symbols*) and sprouting (*open symbols*) species

and in the open on fynbos soils (dry regime similar to Swartboskloof soils) (Fig. 6.4). Only non-sprouter seedlings survived in the shade. A few seedlings of both regeneration modes survived in the shade, probably because of the more favourable microclimate (more moisture, lower PAR, lower temperatures, more diffuse light, lower soil temperatures). Mortality was higher among the obligate sprouters (Fig. 6.4). The addition of water greatly enhanced the survival of (especially) non-sprouter and facultative sprouter seedlings in the open stand, but had little effect on seedlings of obligate sprouters. The sprouter seedlings seem, therefore, to be less sensitive to water stress than non-sprouter seedlings. There do not appear to be clear differences in the response of sprouter and non-sprouter seedlings to shade.

## 6.3 Comparison of Swartboskloof with Other Regions

### 6.3.1 Other Fynbos Areas

Ecophysiological data from other fynbos areas are limited. Most data come from the Cederberg near Clanwilliam (ca. 250 km north of Swartboskloof), a more xeric region which is probably more similar to the chaparral than to Swartboskloof. Even in the xeric Cederberg there were no marked differences in xylem pressure potentials and stomatal conductances between the regeneration types. As is the case at Swartboskloof, some of the seeders in the Cederberg (e.g. *Elytropappus gnaphaloides*) had deep roots, with high water potentials (ca. $-2.5$ MPa) and conductances (400–2000 $mmol\,m^{-2}\,s^{-1}$) (Miller et al. 1983). Some of the sprouters (e.g. *Agathosma giftbergensis*, *Diosma hirsuta*) had shallower roots with relatively low conductances ($\pm 80$ and 160 $mmol\,m^{-2}\,s^{-1}$ respectively) at low potentials ($\leqslant -7$ and $-4$ MPa) (Miller et al. 1983; Miller 1985). Like Swartboskloof, maximum stomatal conductances were variable between the regeneration types. Sprouters like *Agathosma giftbergensis* and *Nylandtia spinosa* stopped carbon uptake (stomatal conductance) at $\leqslant -6$ MPa, whereas *Protea glabra* (also a sprouter) stopped at $-2.5$ MPa. The seeders showed a similar lack of consistency.

Fynbos species at Swartboskloof (mesic) and the Cederberg (xeric) show no clear separation into sprouting and non-sprouting categories on the basis of ecophysiological characteristics. There do appear to be comparatively clear differences in the water relations of shallow- and deep-rooted fynbos plants. Deep-rooted species experience less water stress at both high and low stomatal conductances than shallow-rooted plants, and respond more moderately to the onset of drought or the first winter rains (Miller et al. 1983; Moll and Sommerville 1985; Smith 1990; Smith and Richardson 1990). Fynbos species should perhaps be classified in terms of functional attributes like rooting depth rather than regeneration or life history types.

### 6.3.2 Californian Chaparral

Most of the chaparral paradigms concerning sprouters and non-sprouters refer to the two extreme groups – the obligate sprouters and the non-sprouters. Facultative sprouters assume an intermediate position, but tend towards the non-sprouter syndrome. In the fynbos, obligate sprouters are most common in the forest-fynbos fringe or in the moist riverine forests (see Chap. 5). Reference to seeders and sprouters in fynbos shrublands usually implies seeders and facultative sprouters. Under short fire frequencies, similar to those found in the fynbos, chaparral obligate sprouters may also be restricted to riverine sites in California (J.E. Keeley pers. comm.).

In the chaparral, seeders (e.g. *Arctostaphylos glauca*, *A. stanfordiana*, *Ceanothus greggii*) are usually shallow-rooted (Hellmers et al. 1955; Kummerow et al. 1977). The low water potentials during the summer drought reflect the shallow rooting depths (Miller and Poole 1979; Davis and Mooney 1985; Hart and Radosevich 1987; Davis 1989; Thomas and Davis 1989; Table 6.1). As is the case in fynbos, there is no clear difference between water potentials of seedlings of sprouters and those of seeders in the chaparral (Davis 1989; Hastings et al. 1989; Thomas and Davis 1989). Stomatal conductances and transpiration rates of seeders in the chaparral are mostly higher than those of facultative and obligate sprouters (e.g. *Adenostoma fasciculatum*, *Heteromeles arbutifolia*, *Quercus dumosa*, *Rhus* spp.) during spring and early summer (Parker 1984; Hart and Radosevich 1987). Seeders appear to be less sensitive to water stress, and their stomata stay open at lower water potentials (Poole and Miller 1975; Poole et al. 1981; Keeley 1986; Oechel 1988; Davis 1989; Thomas and Davis 1989). They may, however, exhibit complete stomatal closure during the dry summer months. The higher survival of non-sprouter seedlings when compared to sprouter seedlings in the chaparral is ascribed to their lower sensitivity to water stress. Non-sprouter seedlings were also found to be comparatively less tolerant of shade (Thomas and Davis 1989). These observed physiological patterns appear to correspond approximately with the life history categories of chaparral species. Seeders usually have the lowest seasonal water potentials and presumably the shallowest roots, and obligate and facultative sprouters have the highest potentials and the deepest roots (Davis 1989). Shallow-rooted seeders are usually less sensitive to water stress (low potentials at which stomata shut), whereas sprouters are relatively sensitive to water stress (Miller and Poole 1979).

Although chaparral species with different regeneration modes often do respond differently to changes in soil moisture and other environmental factors, there are notable exceptions. For example, the water relations of *Rhamnus californica*, a resprouter, are more similar to those of other seeders than "typical" sprouters. This is probably because *R. californica* has relatively shallow roots and low water potentials (Davis and Mooney 1986; Davis 1989). A species like *Adenostoma fasciculatum* is effective as a seeder and as a sprouter. Water potentials are often low (−5 MPa; Table 6.1), similar to those of non-sprouters (Parker 1984). Transpiration rates are often well within the range of non-sprouters, and conductances are also often closer to those of seeders than sprouters (Table 6.1) on occasions (Poole et al. 1981). In other instances, photosynthesis rates have been found to be at the upper end of the range for sprouters (Parker 1984). The flexibility of characteristics displayed by this species may account for its widespread distribution and its dominance in low elevation chaparral (Parker 1984).

Although differences in regeneration mode commonly indicate differential water status and root stratification among chaparral species, the picture is not always so clear. Species like *A. fasciculatum* (facultative sprouter) which is relatively adaptable, confuse the issue somewhat. If, as it seems, facultative sprouters generally respond more like seeders than sprouters in the chaparral, many of the fuzzy patterns observed for the fynbos may make more sense. The fact that many sprouters in xeric fynbos are facultative sprouters rather than obligate sprouters may explain the apparent lack of consistent differences between the sprouting and non-sprouting faculties in the fynbos. It does appear, however, that fynbos seeders and sprouters do not fit the aforementioned ecophysiological paradigms of the chaparral very well. One of the reasons for this may be that chaparral and fynbos (especially mesic fynbos like Swartboskloof) are not climatically equivalent mediterranean-type ecosystems (Miller et al. 1983; see also Chap. 2). The lowest xylem pressure potentials at Swartboskloof (−4 MPa) were recorded in the shallow-rooted seedlings. Kruger (1987) also comments that fynbos xylem pressure potentials are rarely, if ever, as low as those of the chaparral. Xylem pressure potentials two to three times lower were recorded in plants from Californian chaparral, indicating much higher levels of drought stress. Miller et al. (1983), working in Swartboskloof, tested the hypothesis that the fynbos of South Africa was functionally equivalent to the chaparral of California (i.e. that a period of plant water stress, similar to the period measured in chaparral species, should occur in fynbos species in the dry summer). Their results indicated little or no plant water stress in most fynbos species in the more mesic fynbos regions (like Swartboskloof). In more arid fynbos areas (e.g. the Cederberg; Miller 1985), some species behaved more like chaparral plants (low potentials and low conductances), due to the significantly lower rainfall and higher vapour pressure deficits in these arid areas (Miller et al. 1983), although stomatal conductances of the these more arid species are generally still far higher than those of the chaparral non-sprouters. Miller et al. (1983) concluded that while arid fynbos (precipitation less than 650 mm) could be viewed as equivalent to chaparral, the vegetation at Swartboskloof would be functionally equivalent to coniferous and broad-leaved evergreen forest in western North America.

### 6.3.3 Australian Kwongan

Shallow-rooted species in the kwongan (e.g. *Hibbertia aurea*, a sprouter) usually experience greater water stress (⩽−6 MPa) during summer (and at an earlier stage in the season) than deeper-rooted species (Dodd et al.

1984). Some of the deep-rooted species (phreatophytes) were able to tap groundwater and maintained high transpiration rates without experiencing low water potentials (e.g. *Stirlingia latifolia*, a sprouter). Others with shallower roots may not reach the water table, and maintain relatively high potentials by controlling water loss (e.g. *Adenanthos cygnorum*, a non-sprouter). Some seeders are deep-rooted (e.g. *A. cygnorum*) and certain sprouters are shallow-rooted (e.g. *Hibbertia aurea*).

A study of the water relations of *Banksia hookeriana* (a seeder) and *B. menziesii* (a sprouter) also showed that the hypothesis that species with different life histories will have different rooting patterns, water relations and phenologies, does not always hold. Both these banksias have deep roots that give them access to groundwater during the summer drought. Both showed high water potentials throughout summer and autumn ($>-1.6$ MPa) in comparison with shallow-rooted chaparral and kwongan species (e.g. Dodd et al. 1984; Hart and Radosevich 1987) with pre-dawn potentials also remaining above $-0.8$ MPa. Both species exhibited stomatal closure over the mid-day period during the mid-autumn and conductances declined from 200 to 60 mmol $m^{-2} s^{-1}$ over summer and early autumn. The two species were therefore basically similar, and differences between the two regeneration modes will probably only be evident immediately after fire when *B. hookeriana* seedlings will be forced to adapt a more conservative strategy than the pre-fire *B. menziesii* adults (Bergl and Lamont 1988).

## 6.4 Community Composition Along a Moisture Gradient

Data from the vegetation monitoring plots (see Chap. 4) were used to examine trends in the ratio of seeders:sprouters along a moisture gradient. We tested the hypothesis that xeric sites have a higher proportion of seeders than mesic sites. Because soil moisture measurements were not available, plots were ranked according to a moisture index. This was based on mean annual rainfall on the plot, estimated using the Thiessen polygon method (Viessman et al. 1977), soil depth, soil type, aspect and the slope of the site. Ranked values were given to each of these characteristics on the strength of their contribution towards a more mesic environment, with low values indicating xeric conditions. Rainfall between 1800 and 2100 mm $yr^{-1}$ (lowest) was given a rank of 1, while the class 2700 to 3000 mm $yr^{-1}$ (highest) was given a rank of 6. Sandstone soils were ranked as 1 and granite soils as 3. Northerly aspects were ranked as 1 and southerly aspects as 2. Steep gradients (>45°) are usually well drained and were ranked as 1 (highest), while flat slopes (0–15°) were ranked as 4. Deep soils were ranked as 3 and shallow soils as 1. The proportion of sprouting species contributing to the total biomass at 30 months post-fire on each plot was then plotted against the moisture index (Fig. 6.3).

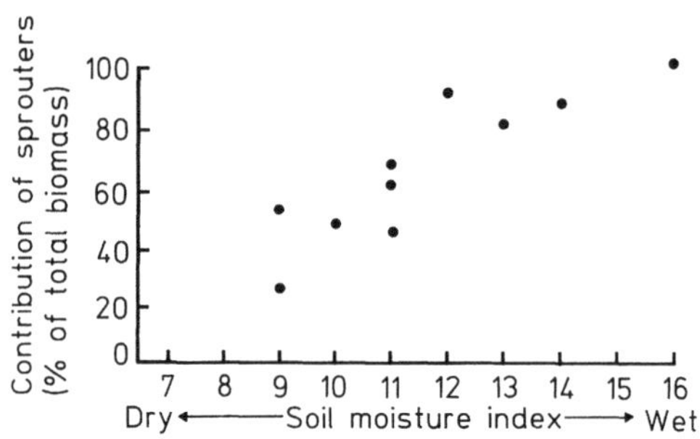


**Fig. 6.3.** The contribution of sprouting species to total plot biomass on ten fynbos plots in Swartboskloof 30 months after fire, plotted against the soil moisture index (see text)

The relative recovery rates of these species after the fire (in terms of biomass) were also measured over a period of 30 months. The relative contributions of seeders and sprouters to the vegetation canopy prior to burning, and during the first 3 years after fire were examined on ten plots (see Chap. 4) in Swartboskloof. They were surveyed once 3 months prior to the fire, and at 9, 18 and 30 months post-fire. The plots were subdivided into 50 sub-plots of 1 $m^2$ each. At each survey, the maximum height of the vegetation, and a visual estimate of the percentage canopy cover was made on each of the 50 sub-plots within each plot. In addition, the three dominant species (in terms of visual estimates of live biomass) were listed and ranked on each sub-plot. Species were allocated importance values in proportion to their rank, as follows: rank 1 = 1, rank 2 = 0.66 and rank 3 = 0.33. These importance values were totalled per species for each plot. The mode of post-fire regeneration was recorded for each species (see Chap. 4). Seeder: sprouter ratios were calculated from the importance values on each plot.

A surrogate measure of live biomass was calculated for each plot. This was done by multiplying the total importance value of each of the species per plot by the mean cover and height of the vegetation for that plot. These values were totalled for all species on the plot to arrive at a surrogate biomass estimate. The biomass development of a relatively xeric and a mesic site (according to the moisture indices) show that sprouters recover comparatively quickly on more mesic sites, and thereby exclude seeders to a certain extent (Fig. 6.4). The most xeric site, on the other hand, shows a greater recovery of seeder biomass than the most mesic site (Fig. 6.4).

Soil moisture does play a role in determining the proportion of seeders and sprouters in the fynbos of Swartboskloof. Wettest sites have higher percentages of sprouters (Fig. 6.3). Consistent with this is the predominance of obligate sprouters in ravines and riparian sites (see Chap. 5). There is also a comparatively greater seeder biomass recovery on xeric sites than on mesic sites, probably due to the formation of gaps between sprouter canopies. Xeric sites have lower pre-fire sprouter biomass ratios than mesic sites. These results support Specht's (1981b) contention for the heathland and Californian chaparral (see also Keeley 1986) that moisture gradients do

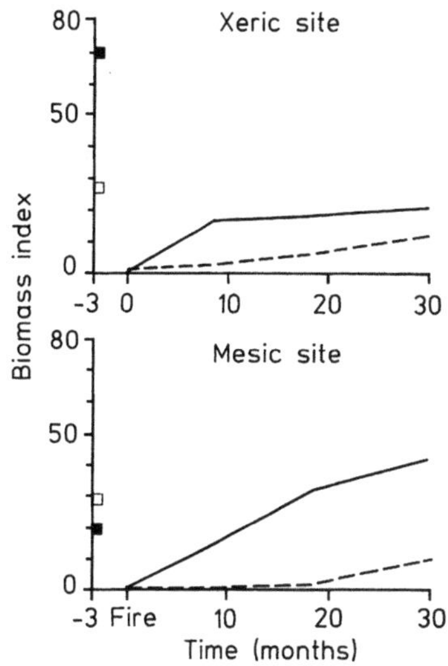


**Fig. 6.4.** The increase in biomass of seeders (*dotted lines*) and sprouters (*solid lines*) after fire on a xeric and a mesic site in Swartboskloof. Pre-fire estimates of biomass for seeders (■) and sprouters (□) are indicated on the y axis. The data comprises the mean biomass for all the plots occurring on these sites (see text)

exist in some mediterranean-climate regions, and that xeric sites generally have more non-sprouting species than mesic sites.

## 6.5 Conclusions

We set out to test two hypotheses. Firstly, we determined whether seeders and sprouters are separated by a suite of ecophysiological factors that may explain their spatial distribution and relative success in fynbos communities. Our second hypothesis was that soil moisture plays a role in determining the relative abundance of seeders and sprouters in fynbos communities.

Our data did not support the first hypothesis. There does not appear to be a suite of ecophysiological characteristics that separate seeders and sprouters in the fynbos at Swartboskloof, as was proposed for the chaparral by Miller (1981b) and Keeley (1986). This question has been addressed in the chaparral by comparing seeders and obligate sprouters. We compared seeders and facultative sprouters, and found few real differences. This may be because the behaviour of facultative sprouters is more similar to that of non-sprouters than to that of obligate sprouters. Another reason for the lack of clear differences between regeneration modes may be that mountain fynbos in Swartboskloof is not strictly comparable with xeric mediterranean-climate regions like the chaparral. Swartboskloof is more mesic than the chaparral (see Chap. 2; Fig. 2.2) and the selective pressure for water-conservative features is probably less intense, making differences less pronounced.

Our results provide strong support for the second hypothesis. The relative importance of seeders and sprouters in the mesic mountain fynbos of Swartboskloof changes along a soil moisture gradient. In agreement with Specht's (1981b) hypothesis, seeders were more prevalent on xeric sites. This relationship, although clear, may have been even stronger had we included riverine sites in the study. Post-fire biomass recovery of sprouters also appears to be lower on xeric sites than mesic sites, presumably as a result of the lower pre-fire sprouter biomass on these sites.

*Acknowledgements.* We thank D. Druce, G. Moses, G. Mostert and E. Theron for assistance with fieldwork, and D.C. Le Maitre for useful suggestions.

# 7 The Relative Advantages of Seeding and Sprouting in Fire-Prone Environments: a Comparison of Life Histories of *Protea neriifolia* and *Protea nitida*

D.C. Le Maitre

## 7.1 Introduction

Woody shrubs that are killed by fire and that regenerate exclusively from seed are often dominant in shrublands in mediterranean-climate regions of the world (Kruger 1983; Keeley 1986). Seed-regenerating plants (hereafter termed seeders) of the genera *Arctostaphylos* and *Ceanothus* dominate extensive communities in chaparral (Keeley and Keeley 1988), whereas pines (*Pinus* spp.) often dominate communities in the Mediterranean Basin (e.g. Mazurek and Romane 1986). Reseeding taxa of *Banksia* and *Hakea* spp. are important components of kwongan (Beard 1990), and *Protea* and *Leucadendron* are important in fynbos communities (Kruger 1979). Tall woody sprouters dominate landscapes in chaparral, e.g. *Adenostoma fasciculatum* (Stohlgren and Rundel 1986), and kwongan, e.g. *Banksia attenuata* (Beard 1984), but are seldom dominant in fynbos (see Chap. 4).

Three regeneration syndromes are evident among woody plants in mediterranean shrublands (Table 7.1): seeding, facultative sprouting, and obligate sprouting. Seeders are killed by fire, accumulate seed banks between fires, and regenerate only after fires. Facultative sprouters regenerate after fires from both dormant buds and seeds accumulated between fires, and may experience high mortality rates in intense fires. Obligate sprouters produce few seedlings (almost exclusively between fires), survive fires by sprouting, and have bird-dispersed seeds.

Recruitment of seed-regenerating shrubs, notably serotinous Proteaceae, is strongly influenced by fire season (Jordaan 1965; Le Maitre 1987a, 1988a; Midgley 1989), post-fire seed predation (Bond 1984; Bond et al. 1984), and post-fire climate (Le Maitre 1988b). This means that probabilities of local extinction are high (Bond 1989a). Effective conservation management of fynbos requires a sound understanding of the factors that favour regeneration. One way to understand the factors that favour regeneration from seed is to identify the factors that favoured the evolution of obligate seed regeneration in mediterranean shrublands.

Sprouting is a widespread trait in woody angiosperms (Blake 1983). It is common in, but is not confined to, fire-prone environments. Sprouting

**Table 7.1.** Generalized sprouter and seeder syndromes in woody chaparral plants (After Keeley 1986; with additional data from Carpenter and Recher 1979; Keeley 1981, 1987; and Zedler et al. 1983; and from inference = boldface type)

| Trait | Seeder (fire-killed) | Facultative sprouter | Obligate sprouter |
|---|---|---|---|
| Annual seed production | Moderate-high | Low-moderate | Temporally variable |
| Accumulation of seed banks | High | Low-moderate | None |
| Reproductive allocation | High | Lower | **Low?** |
| Germination | | | |
| between fires | None | None | Low |
| after fires | High | High | None |
| cues | Heat/charate | Heat/charate | None |
| | Fire-induced gaps | Fire-induced gaps | |
| Population expansion | | | |
| post-fire | High | Moderate-high | None |
| between fires | None | None | Low-moderate |
| Resilience of adults to length | | | |
| of fire interval – short | No tolerance | High tolerance | High tolerance |
| – long | High tolerance | Low tolerance | High tolerance |
| Juvenile period:[a] | | | |
| primary | 5–25 yrs | **Short-moderate?** | **Long?** |
| secondary | None | Short | 2 yrs |
| Mortality between fires | | | |
| seedlings | Low | Moderate-high | **High?** |
| juveniles | Low-moderate | Moderate-high | **High?** |
| adults | Low-high near end of life span | Low | Very low |
| Adult longevity | Short-moderate (<**100 yrs**) | Moderate-long (>**100 yrs**) | Very long (>**200 yrs**) |
| Shoot renewal by basal sprouts | None | Yes | **Yes** |
| Examples: chaparral | *Ceanothus greggii* | *C. leucodermis* | *Quercus dumosa* |
| | *C. crassifolius* | *Arctostaphylos glandulosa* | *Heteromeles arbutifolia* |
| | *Arctostaphylos glauca* | *Adenostoma fasciculatum* | *Rhus ovata* |
| | | | *Rhus* (Malosma) *laurina* |
| fynbos | *Protea repens* | *P. nitida* | *Kiggelaria africana* |
| | *P. neriifolia* | *Leucadendron salignum* | *Maytenus oleoides* |
| | | | *Olea europaea* |

[a] Keeley did not distinguish between primary (seed to maturity) and secondary (resprout to maturity) juvenile periods.

provides resilience to a variety of disturbances other than fire, such as insect defoliation or windthrow (Putz and Brokaw 1989; Boucher et al. 1990). Sprouting trees and shrubs are important or dominant in eucalyptus forest and mallee communities in Australia (Gill 1975; Blake 1983), oak woodlands in California (Wells 1969), tropical forests (Boucher et al. 1990) and savannas (Rutherford 1981). The prominence and frequent dominance of seeders in mediterranean shrublands distinguishes them from most other shrubland or woodland communities.

### 7.1.1 Evolution of Seed Regeneration in Mediterranean Shrublands

Few studies have examined whether sprouting or seeding is the more evolutionarily advanced trait. Midgley (1987a) found that sprouting was a derived (advanced) trait in the fynbos genus *Leucadendron*. Epicormic sprouting is a primitive trait, and seed regeneration is a derived trait in the genus *Banksia* (George 1981). Seed regeneration is considered a derived trait in many genera (Wells 1969; James 1984), including *Ceanothus* and *Arctostaphylos*, which are important genera in chaparral (Wells 1969; Keeley 1986), and *Eucalyptus* (Blake 1983).

The distinction between seeders and sprouters is comparable to the dichotomy between semelparity (one reproductive episode per lifetime) and iteroparity (many reproductive episodes per lifetime) (Keeley 1986; Bond 1987). Selective pressures for the evolution of semelparity and iteroparity have been explored in life-history theory (e.g. Charnov and Shaffer 1973). These analyses provide insights into the evolution of seeding. For seeders to have an evolutionary advantage over sprouters, they must recruit enough seedlings to compensate for the seedling recruitment of the sprouters, *and* the survival of established plants (i.e. the resprouts). In simple mathematical terms:

$$SB\_Sdr \times NR\_Sdr > SB\_Spr \times NR\_Spr + Spr\_survivors,$$

where Spr = sprouter; Sdr = seeder; NR = net recruitment rate; SB = seed bank.

Thus, seeders can gain an advantage by producing more seeds, or by having longer-lived seeds (greater seed banks), seeds with a lower mortality during fires, or seedlings and juveniles with a higher survivorship (Bond 1987; see also Keeley 1986). Another evolutionary advantage may lie in the greater population turnover in seeders because genetic overlap between generations is limited to seeds (Wells 1969). Regeneration from seed may have evolved in chaparral shrubs because of the opportunity for more rapid evolution (more generations and greater selection per generation through self-thinning) and adaptation to changing environments (Wells 1969).

### 7.1.2 Trade-Offs Linked to Seeding and Sprouting

Studies of congeneric pairs of sprouting and seeding chaparral shrubs have shown that the ability of sprouters to survive fire is negatively correlated with seed production and seedling establishment (Keeley and Zedler 1978). Woody sprouters in kwongan and fynbos usually produce fewer seeds and seedlings than co-occurring seeders (Specht et al. 1958; Carpenter and Recher 1979; Midgley 1987b) although some sprouting species may produce many seeds (Lamont 1985).

Studies of a sprouting and a seeding *Banksia* species showed that the sprouter released its seeds more rapidly and its seed germination was more rapid (Zammit and Westoby 1987). These traits may give sprouter seedlings the competitive advantage of early recruitment. The high risk of recruitment failure associated with complete seed germination in a single episode is offset by the survival of the established plants of sprouters. Seeders reduce this risk by spreading seed germination over several episodes (Zammit and Westoby 1987).

Seedlings of vigorous sprouters may also have higher mortality rates during drought than weak sprouters and seeding species (Keeley and Zedler 1978; Zammit and Westoby 1987). Juvenile plants of sprouting species may also require long fire-free periods to achieve fire tolerance and to mature, so frequent fires can have a negative influence on population sizes in the long-term (Carpenter and Recher 1979; Bradstock and Myerscough 1988).

There is some evidence that seeding shrub species in mediterranean shrublands show senescence (Kruger 1983); an increase in the mortality rate with increasing age, especially near the end of the lifespan (Leopold 1961). Senescence is important because recruitment between fires is rare and insignificant (Kruger 1983). Recruitment in senescent stands is poor, and long fire-free periods can result in extinction (Bond 1980, 1984). The principle of limited resource availability in life-history theory predicts that there is a trade-off between allocation to reproduction and to growth and longevity (Stearns 1976). Thus, seeders should have shorter life spans than sprouters because they allocate more resources to reproduction than sprouters (Kruger 1983). Because serotinous species lack seed banks that will outlive the parent plant, there is strong selective pressure for both maximum longevity, to ensure that they survive till the next fire, and for sustained high fecundity (Midgley 1987b). The mortality patterns of serotinous seeders should, therefore, show evidence of senescence; those of sprouters should not.

### 7.1.3 Hypotheses

The aim of this study was to test several hypotheses concerning the ecological and evolutionary trade-offs of seeding and sprouting. I compare the

**Fig. 7.1.** *Protea nitida* shrubs in a firebreak below Swartboskloof. The species is a facultative sprouter that sprouts from epicormic buds or a lignotuber. Seeds are retained for 1–2 years in persistent woody inflorescences (Photo: A.D.A. Simmers)

demography and reproductive biology of two congeneric shrubs, a seeder (*Protea neriifolia*) and a sprouter (*P. nitida*), between and after fires. Population responses to fire are given special attention. Other environmental and ecophysiological factors that govern the coexistence of seeders and sprouters in Swartboskloof are discussed in Chapter 6.

*Protea nitida* is a facultative sprouter that sprouts from epicormic buds or a lignotuber (Fig. 7.1) and *P. neriifolia* is a seeder (Fig. 7.2). Both species store their seeds in persistent woody inflorescences but *P. nitida* only retains its seeds for 1–2 years (Bond 1985), compared with 3–7 years in *P. neriifolia* (Le Maitre 1990). Populations of both species were studied in 0.25 ha (50 × 50 m) permanent plots at different sites in Swartboskloof and elsewhere in the Jonkershoek valley. I used data from these plots and from the literature, and simple models to test the following hypotheses:

1. Net population growth after fires is greater for *P. neriifolia* than for *P. nitida*.
2. Age-specific mortality is higher for juveniles of *P. nitida* than *P. neriifolia*, but vice versa for adults.
3. *Protea nitida* takes longer to mature than *P. neriifolia*.
4. *Protea neriifolia* produces more seeds than *P. nitida*.
5. Seeds of *P. nitida* germinate more rapidly than those of *P. neriifolia*.
6. Recruitment between fires is not a significant factor in the population dynamics of either species.

**Fig. 7.2.** *Protea neriifolia* in 12-year-old fynbos adjacent to Swartboskloof. This species is killed by fire and regenerates only from seeds which are retained for 3–7 years in the canopy or until the adult plant dies (Photo: A.D.A. Simmers)

**Table 7.2.** Recruitment of *Protea neriifolia* and *P. nitida* after a fire in the Jonkershoek valley

| Species | Post-fire age (yrs) | Stand[a] density (plants $ha^{-1}$) | Number of | | Plant mortality[c] (%) | Seedlings per parent |
|---|---|---|---|---|---|---|
| | | | Parents[b] | Seedlings | | |
| *P. neriifolia* | 12 | 76 | 14/5 | 1045 | 100 | 74.64 |
| | 29 | 1376 | 278/53 | 5899 | 94 | 21.22 |
| *P. nitida* | 3 | 1036 | 140/119 | 166 | 7 | 1.19 |
| | 29 | 904 | 95/131 | 1038 | 12 | 10.93 |

[a] Includes *P. repens*, another tall reseeding shrub species.
[b] Number with/number without inflorescences, seedlings/parent calculated for plants with inflorescences.
[c] Mortality of the pre-fire population.

## 7.2 Population Growth After Fire

The net seedling recruitment of *Protea neriifolia* after the first winter following the fire in March 1987 was determined at two 50 × 50 m plots in Swartboskloof. In a sparse population last burnt in 1973 there were 75 seedlings per parent plant (Table 7.2). This was about four times the estimated seed bank of the parent plants (D.C. Le Maitre unpubl. data) so seeds must have been dispersed from *P. neriifolia* plants in the surroundings. In a dense 28-year-old population there were about 21 seedlings per parent after allowing for about 22% mortality during the first summer (Jonkershoek F.R.C. unpubl. data). These recruitment ratios for *P. neriifolia* are markedly higher than those recorded elsewhere in fynbos: mean seedlings per parent = 1.39 (SD = 2.39, n = 39, maximum 14), with about half the samples having a net recruitment of about one seedling per parent (Bond 1989a).

*Protea nitida* populations had fewer seedlings per parent in the same fire (Table 7.2). When the established plants are included, this gives a net increase after fire of 148 plants (57%) in the 3-year post-fire population and 1011 plants (447%) in the 28-year post-fire population. The corresponding values for *P. neriifolia* were 5400% (12-year-old) and 1688% (28-year-old), one to two orders of magnitude greater. These findings support the hypothesis that seeders recruit more seedlings after fires than similar co-occurring sprouters.

## 7.3 Mortality Rates

### 7.3.1 *Protea neriifolia*

Data on long-term trends in populations of *Protea neriifolia* are available from two sites in Swartboskloof and one from the nearby Langrivier catchment (Chap. 1, Fig. 1.1; Le Maitre 1989). These data were analyzed using the life table approach of Leverich and Levin (1979; Fig. 7.3). Mortality rates during the first summer ranged from 20.1% (Table 7.3, Swartboskloof plot 1 after a fire in 1973) to 7.9–8.9% at two other sites in the valley. Additional records of mortality rates during the first summer after the fire in March 1987 range from about 22% (F.R.C. Jonkershoek unpubl. data) to 45% (R.E. Smith unpubl. data). The mean annual mortality rate from 1–13 years after a fire in 1973 (Table 7.3, Swartboskloof plot 1) was about 30%. This was significantly higher than that between 15 and 29 years of 8.5% (Table 7.3, Swartboskloof plot 2) ($P < 0.05$, Tukey test, GLM procedure; SAS Institute Inc. 1985b) but not significantly different from the rate between 30 and 44 years of age (19.3%, Langrivier, Table 7.3). A similar analysis for a *P. neriifolia* population at Jakkalsrivier (30 km SE of Jonkershoek near Grabouw; Kruger 1987) from the age of 0.9 to 8.8 years (Fig. 7.3) gave a mean mortality rate of 2.3%, which was significantly lower than the rates recorded at all the sites in Jonkershoek. These results taken together suggest that the annual mortality rate in the Swartboskloof (plot 1) population after the 1973 fire was unusually high. It is reasonable to conclude that the annual mortality rate increases with age, as found by Kruger (1987), but the

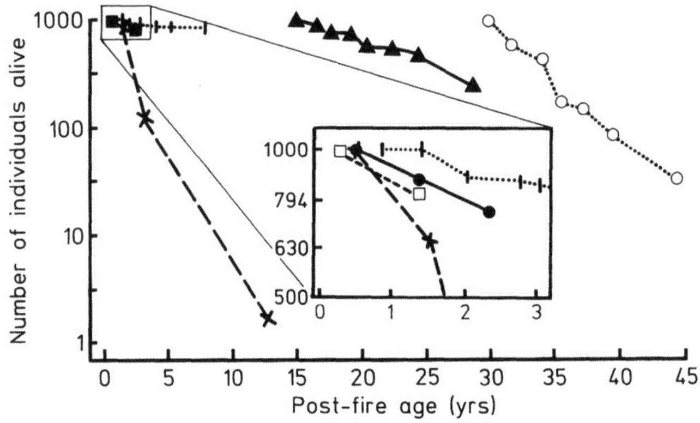


**Fig. 7.3.** Trends in the numbers of *Protea neriifolia* plants with increasing age at different sites. All populations were adjusted to an initial cohort of 1000 plants. The sites are the same as those in Table 7.3 plus one site at Jakkalsrivier (30 km SE of Swartboskloof). Note the log scale for numbers. ○, Langrivier before fire in October 1987; □, same site after fire; X, Swartboskloof (plot 1) before fire in March 1987; ●, same population after fire; ▲, Swartboskloof (plot 2); ▮, Jakkalsrivier

increase is gradual and without a marked rise towards the end of the life span, which is about 50 years at the study sites (Fig. 7.3).

### 7.3.2 *Protea nitida*

There are few records of *P. nitida* seedling mortality during the first summer. Haynes (1976) recorded 32.8% mortality, and 8.9% mortality was noted at one study site after the fire in March 1987 (R.E. Smith unpubl. data). Mortality data are also available for *P. nitida* populations of greater post-fire ages (structure and maturity classes are defined in Table 7.4; Table 7.5). In the 29-year post-fire population in Swartboskloof a total of 68 plants (25.4%) died between 1976 and 1987, 18 to 29 years post-fire. The highest mortality rates were for plants in class i and class ii (juveniles; Table 7.5). In the Eerste River population (east of Swartboskloof, see Chap. 1) 11 plants (5.2%) died between 34 and 45 years post-fire. The mortality rate was markedly lower than that in the Swartboskloof population, but small plants experienced the greatest mortality (Table 7.5). In both populations the new recruits added between the first and last surveys experienced lower mortality rates than plants in class i in 1976, but it is not clear why this happened. Many of these class i plants may already have been moribund in 1976 and eventually succumbed to competition or pathogens. The more vigorous new recruits were probably more successful. Juvenile plants also experienced higher mortality rates than mature plants, with the mortality rates of class iii (mature) being only 2–3% over 10 or more years (Table 7.5). No class iv

**Table 7.3.** A life table analysis of the mortality trends in three populations of *Protea neriifolia* in the Jonkershoek valley. The actual numbers have been converted to a cohort of 1000 plants of facilitate comparisons. Plant ages have been calculated using 15 June of the first winter after the last fire as the mean germination date

| Site and date of last fire | Date | Number of plants in cohort | Cohort age (years) | Length of interval (years)[a] | Annual mortality rate ($Q_x$) |
|---|---|---|---|---|---|
| Langrivier; October 1987 | 05/10/88 | 1000 | 0.31 | | |
| | 10/11/89 | 902 | 1.40 | 1.10 | 0.089 |
| Swartboskloof (plot 1); March 1987 | 18/12/87 | 1000 | 0.51 | | |
| | 14/11/88 | 928 | 1.42 | 0.91 | 0.079 |
| | 13/10/89 | 862 | 2.33 | 0.91 | 0.078 |
| Swartboskloof (plot 1); September 1973 | 09/01/75 | 1000 | 0.57 | | |
| | 10/12/75 | 816 | 1.49 | 0.92 | 0.201 |
| | 01/07/77 | 125 | 3.04 | 1.56 | 0.543 |
| | 16/11/78 | 66 | 4.42 | 1.38 | 0.346 |
| | 04/02/87 | 2 | 12.64 | 8.22 | 0.118 |
| Swartboskloof (plot 2); February 1958 | 22/05/73 | 1000 | 14.94 | | |
| | 14/01/75 | 879 | 16.58 | 1.65 | 0.121 |
| | 07/01/76 | 768 | 17.56 | 0.98 | 0.126 |
| | 07/07/77 | 724 | 19.06 | 1.50 | 0.038 |
| | 16/11/78 | 567 | 20.42 | 1.36 | 0.160 |
| | 16/09/80 | 542 | 22.26 | 1.83 | 0.024 |
| | 20/10/82 | 482 | 24.35 | 2.09 | 0.053 |
| | 03/02/87 | 245 | 28.64 | 4.29 | 0.114 |
| Langrivier; December 1942 | 01/03/73 | 1000 | 29.71 | | |
| | 20/01/75 | 600 | 31.60 | 1.89 | 0.212 |
| | 30/06/77 | 428 | 34.04 | 2.44 | 0.117 |
| | 16/11/78 | 172 | 35.42 | 1.38 | 0.433 |
| | 28/08/80 | 149 | 37.20 | 1.78 | 0.076 |
| | 12/11/82 | 84 | 39.41 | 2.20 | 0.198 |
| | 23/10/87 | 33 | 44.36 | 4.94 | 0.124 |

[a] From previous to current date.

**Table 7.4.** Classification of *Protea nitida* plants (after Haynes 1976; K.B. Higgins, P.T. Manders unpubl. data). The plants were further subdivided into adults which had flowered, and saplings (only class i) or juveniles which had not yet flowered

| Class | Description |
|---|---|
| i | Plant with a single dominant shoot, <0.5 m tall |
| ii | Multi-stemmed plant with no dominant shoot and without a rounded adult form |
| iii | Plants with a mature, rounded canopy and a single stem, <2.5 m tall |
| iv | As for class iii but >2.5 m tall |

**Table 7.5.** Changes in the proportion of *Protea nitida* populations in different structure and maturity classes (for definitions see Table 7.4, sap = sapling, juv = juvenile, ad = adult) during long intervals

Eerste River population. Changes from June 1976 to October 1988 (12.3 years). Last fire December 1942.

| Class 1976 | Maturity 1976 | Class and maturity in 1988 | | | | | | | Total number |
|---|---|---|---|---|---|---|---|---|---|
| | | i | ii | | iii | | iv | Dead | |
| | | sap | juv | ad | juv | ad | ad | | |
| i | Sapling | 0.13 | 0.62 | | | | | 0.25 | 8 |
| ii | Juvenile | | 0.65 | 0.17 | 0.02 | 0.04 | | 0.11 | 46 |
| ii | Adult | | | 0.50 | | 0.50 | | | 2 |
| iii | Juvenile | | 0.31 | 0.13 | 0.06 | 0.44 | | 0.06 | 16 |
| iii | Adult | | | 0.05 | | 0.72 | 0.21 | 0.02 | 123 |
| iv | Adult | | | | | 0.07 | 0.93 | | 15 |
| New recruits | | 0.84 | 0.13 | 0.03 | | | | | 38 |

Swartboskloof population. Changes from May 1976 to March 1987 (10.8 years). Last fire February 1958.

| Class 1987 | Maturity 1987 | Class and maturity in 1988 | | | | | | | Total number |
|---|---|---|---|---|---|---|---|---|---|
| | | i | ii | | iii | | iv | Dead | |
| | | sap | juv | ad | juv | ad | ad | | |
| i | Sapling | 0.02 | 0.18 | | | | | 0.80 | 56 |
| ii | Juvenile | | 0.61 | 0.02 | 0.13 | 0.04 | 0.02 | 0.18 | 104 |
| ii | Adult | | | 0.05 | | 0.68 | 0.23 | 0.05 | 22 |
| iii | Juvenile | | 0.16 | 0.03 | 0.31 | 0.41 | 0.03 | 0.06 | 32 |
| iii | Adult | | | 0.03 | | 0.50 | 0.44 | 0.03 | 32 |
| iv | Adult | | | | | 0.05 | 0.95 | | 22 |
| New recruits | | 0.36 | 0.60 | | | | | 0.05 | 42 |

Swartboskloof population. Changes from June 1976 to February 1987 (10.7 years) in a population in a firebreak burnt in September 1969, August 1978 and September 1984.

| Class 1976 | Maturity 1976 | Class 1987 | | | | | | | Total number |
|---|---|---|---|---|---|---|---|---|---|
| | | i | ii | | iii | | iv | Dead | |
| | | sap | juv | ad | juv | ad | ad | | |
| i | Sapling | 0.01 | 0.16 | | | | | 0.83 | 128 |
| ii | Juvenile | | 0.46 | 0.07 | 0.04 | 0.18 | | 0.25 | 89 |
| ii | Adult | | | 0.36 | | 0.48 | 0.09 | 0.06 | 33 |
| iii | Juvenile | | | 0.25 | 0.75 | | | | 4 |
| iii | Adult | | | 0.11 | | 0.54 | 0.31 | 0.03 | 61 |
| iv | Adult | | | 0.10 | | | 0.83 | 0.07 | 30 |
| New recruits | | 0.06 | 0.73 | 0.01 | 0.01 | | | 0.18 | 71 |

**Table 7.6.** Changes in the proportion of *Protea nitida* populations in different structure and maturity classes (for definitions see Table 7.4, sap = sapling, juv = juvenile, ad = adult) as a result of fires

*High intensity fire*: Eerste River population. Changes from October 1988 to October 1989 (1 year) following a fire in December 1988. Previous fire December 1942. Unidentifiable dead = 8.

| Class 1988 | Maturity 1988 | Class and maturity in 1989 | | | | | | | Total number |
|---|---|---|---|---|---|---|---|---|---|
| | | i | ii | | iii | | iv | Dead | |
| | | sap | juv | ad | juv | ad | ad | | |
| i | Sapling | | 0.35 | | | | | 0.65 | 43[a] |
| ii | Juvenile | | 0.61 | | 0.07 | | | 0.32 | 41 |
| ii | Adult | | | 0.63 | | | | 0.37 | 19 |
| iii | Juvenile | | 1.00 | | | | | | 2 |
| iii | Adult | | | 0.66 | | 0.05 | | 0.29 | 99 |
| iv | Adult | | | 0.65 | | 0.25 | | 0.10 | 40 |
| New recruits | | 1.00 | | | | | | | 354 |

*Moderately intense fire*: Swartboskloof population. Changes from March 1987 to September 1988 (1.6 years) following a fire in March 1987. Previous fire February 1958.

| Class 1987 | Maturity 1987 | Class and maturity in 1988 | | | | | | | Total number |
|---|---|---|---|---|---|---|---|---|---|
| | | i | ii | | iii | | iv | Dead | |
| | | sap | juv | ad | juv | ad | ad | | |
| i | Sapling | | 0.50 | | | | | 0.50 | 16 |
| ii | Juvenile | | 0.85 | 0.01 | | | | 0.14 | 103 |
| ii | Adult | | | 1.00 | | | | | 5 |
| iii | Juvenile | | 0.79 | | 0.08 | 0.04 | | 0.08 | 24 |
| iii | Adult | | | 0.51 | | 0.43 | | 0.06 | 49 |
| iv | Adult | | | 0.37 | | 0.12 | 0.37 | 0.14 | 43 |
| New recruits | | 1.00 | | | | | | | 1038 |

*Low intensity fire*: Swartboskloof population in a regularly burnt firebreak. Changes from February 1987 to September 1988 (1.57 years) following a fire in March 1987. Previous fire September 1984.

| Class 1987 | Maturity 1987 | Class 1988 | | | | | | | Total number |
|---|---|---|---|---|---|---|---|---|---|
| | | i | ii | | iii | | iv | Dead | |
| | | sap | juv | ad | juv | ad | ad | | |
| i | Sapling | | 0.67 | | | | | 0.33 | 6[b] |
| ii | Juvenile | | 0.86 | | 0.01 | | | 0.13 | 116 |
| ii | Adult | | | 0.88 | | 0.12 | | | 24 |
| iii | Juvenile | | 0.83 | | 0.17 | | | | 6 |
| iii | Adult | | | 0.34 | | 0.66 | | | 68 |
| iv | Adult | | | 0.15 | | 0.02 | 0.83 | | 48 |
| New recruits | | 1.00 | | | | | | | 67 |

[a] Includes seven plants first recorded after the fire.
[b] Includes one plant first recorded after the fire.

plants died during this period. The changes in a population in a firebreak in Swartboskloof differed in that 7% of the class iv plants died during the fires in 1978 and 1984 (Table 7.5). Most of the class i plants present in 1976 had died by 1987, but the same is true of the unburnt plot in Swartboskloof where there were no fires during the same period. Again, the new recruits experienced lower mortality rates than the class i plants (Table 7.5).

Mortality of *P. nitida* during fires is similar to that between fires. Class i plants experience the highest mortality, and mortality increases with increasing fire intensity (Table 7.6). The proportion of plants that drop one or more classes lower also increased with increasing fire intensity, because most plants resprout only from their lignotuber after intense fires (Haynes 1976). An intense fire killed 33% of the Eerste River population, with only 5% of the class iii, and no class iv plants maintaining their status (Table 7.6). No plants in this population had flowered by September 1989, 1 year after the fire, compared with 8% of the mature plants that survived the intense fire and 19% of those that survived the low intensity fire in Swartboskloof. Fire intensity can, therefore, have a significant impact on the population structure of *P. nitida*. Projections, using the transitions and time steps from Table 7.5 as a transition matrix with an initial cohort of 1000 class i plants, support this conclusion. There would be 200 and 750 plants left in the unburnt Eerste River and Swartboskloof populations, respectively, after the first transition interval (10–12 years; Table 7.5) compared with 26 in the twice burnt Swartboskloof firebreak population. After the second transition interval the corresponding numbers were 152, 643 and 12 respectively. These projections involve several assumptions which would need to be tested for a rigorous analysis (see Manders 1987). They do, however, show that fire-induced mortality in class i plants can restrict the population growth of *P. nitida*.

These analyses revealed no consistent differences between the seedling and juvenile mortality rates of these species. Mortality rates are low for mature *P. nitida* plants; large plants (class iii and iv) usually survive even

**Table 7.7.** Reproductive development of a population of *Protea neriifolia* at Jakkalsrivier (30 km SE of Swartboskloof) from 1984 to 1989. The site was burnt in March 1980. The data are the percentages of the population with the given number of flowering seasons

| Year | Number of flowering seasons | | | | | |
|---|---|---|---|---|---|---|
| | 0 | 1 | 2 | 3 | 4 | 5 |
| 1984 | 97 | 3 | 0 | 0 | 0 | 0 |
| 1985 | 86 | 14 | 0 | 0 | 0 | 0 |
| 1986 | 71 | 24 | 5 | 0 | 0 | 0 |
| 1987 | 56 | 32 | 11 | 1 | 0 | 0 |
| 1988 | 31 | 39 | 29 | 11 | 1 | 0 |
| 1989 | 16 | 31 | 22 | 20 | 9 | 1 |

intense fires (Table 7.6). Mortality rates in this species appear to decrease with age. This is in marked contrast to *P. neriifolia* where the mortality rates increase with increasing age, except the first summer when mortality rates of *P. neriifolia* seedlings may be 20% or more.

The mortality rates of seedlings and juvenile plants did not differ between the two species, contrary to the predictions of Keeley and Zedler (1978; Table 7.1). The mortality rates of adult *P. nitida* between fires, and even during intense fires (Table 7.6), were lower than those of *P. neriifolia*; this fits the predictions (Table 7.1). *Protea nitida* also showed no evidence of senescence, unlike *P. neriifolia*. Mortality rates decreased rapidly with increasing age in *P. nitida*, which supported the predictions.

## 7.4 Age at Maturity

*Protea neriifolia* plants begin flowering at 4 years of age (Table 7.7). Many plants take longer to mature; the population was 8 years old when more than half had flowered at least once. At least 12 years would be required for more than half to flower a minimum of three times, the criterion for the minimum age for successful recruitment proposed by Kruger and Lamb (1979, see also Kruger 1982).

There are no direct observations for *P. nitida* but one (3%) of the plants recruited between 1976 and 1988 in the Eerste river population (Table 7.5) reached maturity and one (1%) in the Swartboskloof firebreak population over a period of 11 years. None of the class i plants present in 1976 reached maturity at any of the sites. A preliminary projection, using the transition matrices of Table 7.5, suggests that maturation of half the class i plants would take 20 to 30 years. Most of these mature plants would still be in class ii and would not have reached maximum fecundity. *Protea nitida* clearly requires far longer than *P. neriifolia* to reach maturity. It may take 50 or more years without any fires to reach full maturity (class iii or iv) by which time most *P. neriifolia* plants in the same area will have died.

## 7.5 Seed Production and Seed Banks

### 7.5.1 *Protea neriifolia*

Seeds of *Protea neriifolia* take about 7 months to mature (van Staden 1978). This means that the first mature seeds are produced at an age of about 5 years. Only one or two inflorescences are produced in the first flowering season. Because of the sympodial branching pattern, the annual production of inflorescences increases rapidly (Table 7.8), and probably reaches a

**Table 7.8.** An analysis of the annual inflorescence production of *Protea neriifolia* plants with different numbers of flowering seasons. The data are the number of shrubs with a given number of inflorescences and the mean inflorecsence production of the sample

| Number of flowering seasons | Number of inflorescences per shrub | | | | | | | | Number of inflorescences (mean ± SD) |
|---|---|---|---|---|---|---|---|---|---|
| | 1 | 2 | 3 | 4 | 5 | 6 | 7 | 8 | |
| 1 | 53 | 6 | 3 | 0 | 0 | 0 | 0 | 0 | 1.19 ± 0.51 |
| 2 | 12 | 12 | 6 | 2 | 3 | 1 | 0 | 0 | 2.31 ± 1.37 |
| 3 | 5 | 9 | 5 | 1 | 1 | 1 | 0 | 1 | 2.61 ± 1.75 |
| 4 | 2 | 0 | 2 | 0 | 1 | 1 | 0 | 0 | 3.17 ± 2.04 |

maximum at between 15 and 20 years of age. As many as half of a *P. neriifolia* population may skip one or more flowering seasons (Le Maitre 1987b; D.C. Le Maitre unpubl. data) so that the proportion of a population with a given number of flowering seasons is not a simple linear function of age (Table 7.7).

Annual seed production and total seed bank sizes per *P. neriifolia* plant were skewed (coefficients of variation >100%, Table 7.9), with maxima very much larger than the means. Many plants had no inflorescences or had inflorescences without seeds (Table 7.9). The mean number of full seeds (with a firm white embryo) per inflorescence in the current crop declines from 9.6 ± 1.8 (SD) at 7 years of age to 2.7 ± 0.7 at 28 years of age (D.C. Le Maitre unpubl. data). The mean of 4.9 seeds per inflorescence recorded by R.S. Adamson (unpubl., cited in Collins and Rebelo 1987) falls within this range, but it is markedly lower than the 17.4 seeds per inflorescence recorded by Collins and Rebelo (1987).

*Protea neriifolia* can retain seeds in its persistent inflorescences for up to 7 years. However, up to 20% of 3-year-old inflorescences had released their seeds, and few viable seeds were retained for longer than 4 years. The accumulated seeds form a buffer against fluctuations in annual seed output and against seasonal variations due to seed predation and loss of viability of seeds in the capitulum (Bond 1985; Coetzee and Giliomee 1987). The contribution of the current crop in *P. neriifolia* declined from about 67% at 7 years of age to between 27 and 38% at 28 years of age (D.C. Le Maitre unpubl. data; Table 7.9). The proportion of seed-bearing plants and the numbers of seeds per 12-year-old plant (Table 7.9) were lower than those of the 7-year-old plants. These low values were at least partially due to the smaller size of the 12-year-old plants (D.C. Le Maitre unpubl. data).

### 7.5.2 *Protea nitida*

Little is known of the inflorescence and seed production of *P. nitida*. A random sample of 8-year-old *P. nitida* plants in the Swartberg in the

**Table 7.9.** Annual inflorescence and full seed production of *Protea neriifolia* populations of different ages in Swartboskloof. Percentage barren = plants which either had no inflorescences or no seeds in their inflorescences

| | Current crop: number of | | Total number of | |
|---|---|---|---|---|
| | Inflorescences | Seeds | Inflorescences | Seeds |
| Population A, effective age = 7 years, n = 74 | | | | |
| Mean | 2.89 | 26.80 | 8.11 | 53.4 |
| Coefficient of variation (%) | 117 | 155 | 125 | 142 |
| Maximum | 19 | 193 | 48 | 328 |
| Median | 8.5 | 7 | 4 | 25 |
| Percentage barren | 23 | 34 | 9 | 24 |
| Population B, effective age = 12 years, n = 30 | | | | |
| Mean | 6.16 | 22.57 | 19.40 | 34.33 |
| Coefficient of variation (%) | 116 | 208 | 121 | 174 |
| Maximum | 29 | 209 | 95 | 275 |
| Median | 2.5 | 4.5 | 10.5 | 4.5 |
| Percentage barren | 33 | 47 | 23 | 40 |
| Population C, effective age = 28 years, n = 140 | | | | |
| Mean | 3.09 | 8.91 | 22.26 | 43.99 |
| Coefficient of variation (%) | 259 | 292 | 171 | 234 |
| Maximum | 60 | 212 | 316 | 787 |
| Median | 0 | 0 | 9 | 9 |
| Percentage barren | 51 | 61 | 18 | 28 |
| Population D, effective age = 28 year, n = 20 | | | | |
| Mean | 15.65 | 37.90 | 87.95 | 99.40 |
| Coefficient of variation (%) | 107 | 177 | 102 | 151 |
| Maximum | 50 | 303 | 260 | 649 |
| Median | 8.5 | 22.5 | 45 | 47 |
| Percentage barren | 20 | 20 | 10 | 10 |

Southern Cape had 7.6 (SEM = 1.2, n = 20) current and 0.3 (SEM = 0.2, n = 20) 15-month old inflorescences which still retained seeds (Bond 1985). A mean of 15.2 (SD = 13.2, n = 119) full seeds per inflorescence was recorded in samples from two Swartboskloof populations (R.A. Haynes unpubl. data). This is higher than the mean of 2.6 recorded by R.S. Adamson (unpubl., cited in Collins and Rebelo 1987), 5.5 by Wiens et al. (1983) and 9.5 by Collins and Rebelo (1987). Unlike *P. neriifolia*, *P. nitida* does not retain its seeds in old inflorescences. Some inflorescences may retain seeds for up to 3 years (Haynes 1976) but most seeds are shed shortly after maturation (Bond 1985), with seed release peaking in autumn (Haynes 1976). *Protea nitida* has a long flowering season with most inflorescences opening from May to August and a few flowering inflorescences can be found at other times (Haynes 1976; Rourke 1982) so mature plants generally have some seeds in the canopy throughout the year.

*Protea nitida* produces more seeds per inflorescence than *P. neriifolia* but its annual inflorescence production is apparently lower. These findings provide only tentative support for the hypothesis that sprouters produce fewer seeds than similar seed-regenerating plants (Carpenter and Recher 1979). More detailed studies of the seed production and seed-bank dynamics of *P. nitida* are needed for a thorough test of this hypothesis.

## 7.6 Seed Germination

Seed germination of *Protea neriifolia*, *P. nitida* and *P. repens* was studied in a greenhouse (P.T. Manders unpubl. data). The seeds were planted in March 1987 in seed trays, 13 per species, which had 24 individual 50 × 50 × 75 mm deep holders filled with a mixture of coarse sand and a loamy soil. One seed was planted in each container at a depth of 10–20 mm. The trays were regularly watered.

Seed germination was calculated as a percentage of the total number of seeds germinating in each replicate because the accuracy of seed sorting differs between species. Seeds of *P. nitida* and *P. repens* have a thick woody testa. Many seeds, which appear to be "full", based on their external appearance, contain no embryo. In *P. nitida* a mean of 61.9% (SD = 24.1%, n = 114) of the apparently full seeds contained no embryo (R.A. Haynes unpubl. data). *Protea neriifolia* seeds lack the woody testa, and "full" seeds are easily distinguished by feel and sight. More than 95% of sorted seeds contained a firm white embryo (Le Maitre 1990).

*Protea nitida* seeds germinated first, beginning on day 23, while germination of *P. neriifolia* and *P. repens* began 7 days later (Fig. 7.4). By the 35th day the percentage germination of *P. nitida* was significantly higher than *P. neriifolia* and *P. repens* and by the 65th day of germination *P. neriifolia* seeds was complete (Table 7.10; Fig. 7.4). Fifty percent germination was achieved after 38 days by *P. nitida*, 41 days by *P. neriifolia* and 55 days by *P. repens*. The mean number of days elapsed per germinating seed was significantly greater for *P. repens* than for the other two species, with *P. nitida* showing a greater variability than *P. neriifolia* (Table 7.10).

Abrupt and rapid germination of *P. neriifolia* was also recorded in other studies (e.g. Le Maitre 1990). In growth chamber experiments seed germination in *P. neriifolia* began after about 21 days and was almost complete by 30 days (Le Maitre 1990). Germination of planted seeds in seed trays was slower, with a greater initial delay due to the extra time required for the seedling to emerge from the soil compared with radicle emergence from the testa, the criterion used in growth chamber experiments (Lombaard 1971; Le Maitre 1990). Similar seed germination rates for *P. nitida*, with germination beginning on day 25 and essentially complete by day 60, were recorded by Haynes (1976). Lombaard (1971) planted *P. repens* seeds at a

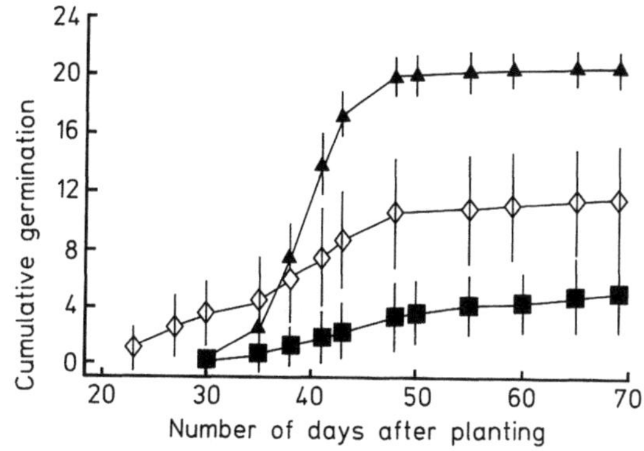


**Fig. 7.4.** Germination patterns of *Protea neriifolia* (▲), *Protea nitida* (◊), and *Protea repens* (■) (mean ± SD)

**Table 7.10.** A comparison of seed germination rates (mean ± SD) in three *Protea* species found in Swartboskloof. Values in columns with different letters are significantly different ($P < 0.05$; n = 13; Tukey's studentized range test)

| Species | Mean percentage germination after: 35 days | 65 days | Mean number of days elapsed per germinating seed |
|---|---|---|---|
| *Protea neriifolia* | 13.83 ± 8.90 b | 100.00 ± 0.00 a | 40.96 ± 1.01 a |
| *P. nitida* | 35.37 ± 17.84 a | 91.89 ± 5.71 a | 42.16 ± 5.89 a |
| *P. repens* | 8.38 ± 11.78 b | 58.36 ± 24.50 b | 79.53 ± 23.29 b |

similar depth in seed boxes and found that germination began after about 30 days and continued till about the 74th day.

Seeds of *P. neriifolia* germinated more rapidly than those of *P. nitida* but *P. repens*, another seeder, had the slowest germination rate. These findings do not support the hypothesis that seed germination in sprouters is faster than that of seeders (Zammit and Westoby 1987). Seeders can also spread the risk of recruitment failure by slow and intermittent seed release (Zammit and Westoby 1987). Seed release rates of *P. neriifolia* were slow, with some inflorescences retaining their seeds for 3 to 4 months after the fire (Le Maitre 1990; Le Maitre pers. observ.). Seed release rates of *P. nitida* were not examined but would probably be rapid as the open capitulum is shallow and the bracts open wide (Haynes 1976) unlike the cup-shaped capitulae of *P. neriifolia* and *P. repens*.

## 7.7 Recruitment Between Fires

Long-term studies of *Protea neriifolia* populations in Langrivier and Swartboskloof have shown that seedling recruitment of *P. neriifolia* increased from about 24–28 years after the previous fire. The increased

**Table 7.11.** Fire survival of *Protea nitida* plants in different age classes after fires of different intensities in the Jonkershoek valley. See Tables 7.5 and 7.6 for the fire history of the sites (sap = sapling, juv = juvenile, ad = adult, – = no plants in this class)

| Site | Recruitment period | Pre-fire[a] age (yr) | Number of survivors | | Percentage mortality by pre-fire class | | | |
|---|---|---|---|---|---|---|---|---|
| | | | pre-fire | post-fire | i sap | ii juv | ad | iii juv |
| *High intensity fire December 1988*: | | | | | | | | |
| Eerste River | 1980–88 | 1–7 | 38 | 1 | 97 | 100 | 100 | – |
| *Intense fire March 1987*: | | | | | | | | |
| Swartboskloof | 1976–80 | 6–10 | 5 | 4 | – | 20 | – | – |
| (old veld) | 1980–87 | 1–6 | 35 | 28 | 47 | 0 | – | – |
| *Low intensity fire March 1987*: | | | | | | | | |
| Swartboskloof | 1976–80 | 6–10 | 12 | 11 | – | 10 | 0 | 0 |
| (firebreak) | 1980–84 | 3–5 | 21 | 20 | 0 | 5 | – | – |
| | 1984–87 | 1–2 | 25 | 19 | 77 | 28 | – | – |

[a] Calculations based on seedling recruitment during the first and last winter between the sampling dates.

recruitment was apparently linked to the opening up of the stands following the senescence and mortality of the cohort that established after the previous fire (van Wilgen 1981, 1982; K.B. Higgins, P.T. Manders unpubl. data). When the Langrivier catchment was burnt in October 1987, 45 years after the previous fire, most shrubs from the first generation had died (Le Maitre 1989). Most shrubs from the second generation had, however, matured and there were sufficient seeds to get a seedling to parent ratio of 8.7 (K.B. Higgins, P.T. Manders unpubl. data), which is high for a spring fire (Bond 1989a). Successful development of a second generation of proteas without fire is uncommon (Bond 1980; Kruger 1983); the stand is vulnerable to extinction at the intermediate stage when few first generation adults remain and most or all of the second generation plants are still immature (K.B. Higgins, P.T. Manders unpubl. data). Seedling recruitment is generally much lower in senescent fynbos than in mature fynbos and there is a high probability of local extinction if the vegetation remains unburnt (Bond 1980).

Seedling recruitment of *P. nitida* also occurs between fires (Table 7.5). Mortality rates of these seedlings can be high during fires, especially in high intensity fires, and mortality rates are highest in the youngest plants (Table 7.11). Few of these plants reached maturity even during long fire-free intervals but the survivors probably would reach maturity before the next fire. Nevertheless, seedling recruitment between fires, even during long intervals, is an order or magnitude lower than recruitment after fires (Tables 7.6, 7.11). Post-fire recruitment is far more important than inter-fire recruitment in the demography of *P. nitida*. Adult plants are long-lived and mortality is low during long fire-free intervals so that there is no vulnerable senescent stage, but mortality during the intense fires that may follow can be high (Table 7.6). The results of these studies contradict the prevailing paradigm that recruitment between fires is absent or negligible in the demography of woody shrubs in fynbos. Indeed, inter-fire recruitment was a key factor in the persistence of *P. neriifolia* populations in Langrivier.

There is an important difference between the chaparral species, used by Keeley (1986) to define the recruitment syndromes, and the two *Protea* species. The chaparral species have long-lived soil-stored seed banks, and seed germination is dependent on fire, so that germination between fires is absent (Keeley 1986). *Protea* species store seeds in woody cone-like inflorescences. Seed dormancy is imposed by retention in the persistent inflorescence and the seeds will germinate in the first winter after release. Seed germination is, therefore, not as strongly coupled to the occurrence of fires. Recruitment of the chaparral species is much more tightly coupled with disturbance than is the case for *P. neriifolia* and *P. nitida*. The germination syndromes of the proteas are similar to Keeley's (1986) obligate sprouters that recruit seedlings between fires (Table 7.1) and other factors, such as herbivory and seed predation by small mammals, control seedling recruitment (Bond 1984; Bond and Breytenbach 1985). Senescence of the adult *P. neriifolia* plants opened up the stand and increased the seed rain when small mammal densities were low (Willan and Bigalke 1982; see Chap. 10) resulting in seedling recruitment. Further studies will be needed to show whether fynbos species with soil-stored seeds fit the predicted recruitment syndromes.

The results of these studies support the hypothesis that the reseeding shrub, *P. neriifolia*, should show a greater population growth after fire than the sprouting *P. nitida* (Keeley and Zedler 1978; Keeley 1986; Bond 1987).

## 7.8 Effects of Fire Frequency

*Protea nitida* is more resilient to both frequent and infrequent fires than *P. neriifolia*. Fires at 3-6-year intervals exclude the latter (van Wilgen 1981), and it generally dies during long fire-free intervals. A second generation of *P. neriifolia* is rarely recruited between fires (Kruger 1983). The mortality pattern of *P. nitida* is similar to that of *Adenostoma fasciculatum*, a common facultative sprouter in chaparral. Both show a constant mortality rate without evidence of senescence (see Stohlgren and Rundel 1986 for *A. fasciculatum*). In accordance with predictions, *P. nitida* is more resilient to changes in fire frequency than *P. neriifolia*. Frequent fires do, however, kill many juvenile plants, and increase the mortality rates of adults. These factors cause a slow decline in the population; the same trend is evident for woody shrubs in analogous shrublands in Australia (Bradstock and

Myerscough 1988; Bradstock 1990). The declines or extinctions of seeding Proteaceae with high fire frequencies (van Wilgen 1981; Bradstock and O'Connel 1988) are simply more obvious than the slow attrition of populations of co-occurring sprouting species (Bradstock and Myerscough 1988).

## 7.9 Coexistence of Seeders and Sprouters

Recent theories suggest that variable recruitment is the key to the coexistence of ecologically similar species (Chesson and Warner 1981; Shmida and Ellner 1984; Comins and Noble 1985; Fagerstrom 1988). Thus, coexistence of seeders and sprouters in fynbos may depend on the long-term consequences of variation in recruitment and mortality under variable fire regimes. This perspective raises some important questions. For example, how do seeders and sprouters differ in the way they recover from population crashes? A high probability of population crashes in seeding species would also be linked to the formation of small populations that favour speciation. This may explain the high species:genus ratio of seed-regenerating shrubs in fynbos (Cowling 1987).

Bond (1989a) developed a stochastic model of recruitment in serotinous Proteaceae, based on the observed mean recruitment ratio (seedlings: parent). He suggested that unless the mean ratio exceeded four seedlings per parent, the probability of population extinction was high (>40%). His model incorporates, but does not explicitly allow for some important factors such as the effect of fire season. High seedling recruitment occurs after fires in summer or autumn (Bond 1984; Midgley 1989); large fires of natural origin are most likely in these seasons (Horne 1981; Kruger and Bigalke 1984). I suggest that Bond's (1989a) model may underestimate recruitment. Nevertheless, his approach highlights the significant impact of stochastic variations, whether in the behaviour of fires, the post-fire climate, or the impact of granivores, on minimum viable population sizes for serotinous proteas.

When recruitment of a seeder is successful, it can recruit very high numbers of seedlings, but when recruitment is poor it may become locally extinct. A sprouter, even under optimum conditions, only recruits a few seedlings, but local extinction is rare. If, however, the best measure of fitness is the geometric mean growth rate of the population (Felsenstein 1976 cited in Zammit and Westoby 1987), then recruitment failure implies that the seeder strategy fails because extinction = zero fitness. I suggest that there is a two-fold explanation for this apparent anomaly: firstly, the increased fitness conferred by high net recruitment rates in seeders is greater than that of persistence by sprouting at the expense of recruitment. Secondly, extinction is generally only local so that recolonization via dispersal is possible (see Chap. 9).

## 7.10 Implications for Management

The mechanism of recolonization cannot operate once natural communities become fragmented, as has happened in much of the lowland fynbos (Cowling 1987). This means that stochastic factors are vital in determining whether populations persist or become extinct. This has important implications for management policies and practices concerned with the maintenance of species diversity. Managers need to ensure that their actions result in a high probability of successful recruitment to compensate for the uncontrollable factors such as rainfall patterns. This is especially so if global climate changes decrease the likelihood of successful regeneration (van Wilgen et al. 1991). Adult proteas are little influenced by summer drought, but seedlings may experience severe moisture stress (Chap. 6). High seedling mortality rates, particularly during the first summer after a fire, may provide the first indications of the adverse impact of climatic change. Managers can only anticipate changes of this kind if they have sufficient data to develop predictive models of seedling mortality and other key components of the life cycle. The development of these models should have a high priority in future fynbos research.

## 7.11 Conclusions

Findings of this study lend support to some of the hypotheses and refute others. The net population growth of *Protea neriifolia* after fires is greater than that of *P. nitida*. This supports the hypothesis that increased seedling recruitment, and thus fitness, may have favoured the evolution of seeders (Bond 1987). Contrary to the predictions of Keeley and Zedler (1978), mortality rates of seedlings and juvenile plants of the two species showed no marked difference. *P. neriifolia* mortality rates increased with age whereas mortality rates of adult *P. nitida* are very low. This supports the idea that there is a link between reproductive investment and longevity (Kruger 1983). *Protea neriifolia* produces fewer seeds per inflorescence but more infloresences than *P. nitida*. *Protea neriifolia* also retains its seeds in the canopy longer and accumulates larger seed banks. This supports the hypothesis that seeders accumulate larger seed banks than ecologically similar sprouters (Keeley and Zedler 1978). Contrary to the prediction that sprouters should have more rapid seed germination than seeders (Zammit and Westoby 1987), the seeds of *P. neriifolia* germinated more rapidly than those of *P. nitida*. Recruitment between fires is important in *P. nitida* populations, and was a key factor in the persistence of *P. neriifolia* at one site. This contradicts the prevailing paradigm that recruitment between fires is rare and insignificant (Bond 1980, 1989a; Kruger 1983).

Seed germination and seedling recruitment are the key phases in the life history of proteas. Of the major factors that control regeneration of seeders (fire season, post-fire seed predation, post-fire climate), only fire season can be manipulated given our current knowledge. The current policy of burning fynbos during late summer or autumn at 12–15-year intervals seems to be the most appropriate strategy for managing both reseeding and resprouting shrubs. Unplanned fires, which are inevitable, introduce an element of stochasticity which may favour a particular guild of plants in the short term. However, both reseeding and resprouting fynbos shrubs appear to be remarkably resilient, except when intervals between fires are very short.

*Acknowledgements.* I thank A.J. Lamb, K.B. Higgins, G. Bailey, G. Hockey and A.D.A. Simmers for directing and assisting with fieldwork. I also thank K.B. Higgins and P.T. Manders for allowing me to use their unpublished manuscript. I thank J.J. Midgley and W.J. Bond for stimulating discussions which clarified my ideas on the evolution of seeders in mediterranean shrublands. Finally, I thank D.M. Richardson and B.W. van Wilgen for their useful comments and suggestions during the editing of this chapter.

# 8 Life Cycles and Fire-Stimulated Flowering in Geophytes

D.C. Le Maitre and P.J. Brown

## 8.1 Introduction

Fire-stimulated flowering (hereafter FSF) is a common trait in herbaceous plants in fire-prone vegetation (Gill 1981a; Rundel 1981a). FSF of geophytes occurs in all the mediterranean shrublands of the world (Naveh 1974; Gill and Groves 1981; Trabaud 1981; Keeley 1986). The Cape flora is renowned for its rich and diverse geophytic flora (Table 8.1), and there are many anecdotal accounts of the profuse flowering of geophytes for one or more years after fire (Levyns 1966; Martin 1966; Goldblatt 1978; Frost 1984), but few quantitative studies.

Fire stimulated flowering in more species and more individuals of geophytes than hoeing, suggesting that flowering was dependent on fire rather than the removal of cover (Levyns 1929). Flowering of orchid species increased after fires but some species only responded in the second flowering season after the fire (Hall 1959). These fires occurred in mid-winter in the middle of the growing season, which may explain why several species responded only in the second year. Mass flowering of *Watsonia borbonica* (Pourret) Goldblatt (Iridaceae), formerly *W. pyramidata*, after autumn fires (Fig. 8.1) resulted in successful seedling recruitment (Kruger 1978; Kruger and Bigalke 1984; Le Maitre 1984). Fire-stimulated mass flowering led to mass seedling recruitment by *Bobartia* (Iridaceae; Richardson et al. 1984).

### 8.1.1 Factors that Stimulate Flowering After Fires

There have been few experimental studies of the proximate (secondary) cues for post-fire flowering. Increased light levels, similar to those after shrub cover is removed by fire, stimulated flowering of *Brodiaea* (Liliaceae) in chaparral (Stone 1951). Species of the Orchidaceae in maquis shrubland in Israel show similar responses (Naveh 1974). Bean (1962) studied the responses of several geophytic species, including three *Watsonia* hybrids and *W. borbonica*, to a number of factors including increased nutrient levels and

**Table 8.1.** Comparison of geophytic floras of mediterranean shrublands. Monocotyledonous geophytes belong primarily to families in the Liliiflorae, notably Liliaceae (sensu lato), Iridaceae, Amaryllidaceae and Orchidaceae

| | |
|---|---|
| Mediterranean region | Flora Europea contains ca. 600 species (6%) of geophytic monocots (collated from Tutin et al. 1980). Mediterranean Israel has 225 geophytic species (10%) (Danin and Orshan 1990). Geophytic species comprised 7.3% of a flora of 1554 species (Shmida 1981) and 6–17% of the flora in 1000 $m^2$ plots in Israeli shrublands (Naveh and Whittaker 1979). |
| California chaparral | The Californian Floristic Province has about 240 species (5%) of geophytic monocots (data collated from Raven and Axelrod 1978). Geophytic species comprised 2.5% of a flora of 402 species (Shmida 1981) and 2–8% of the species in 1000 $m^2$ plots (Naveh and Whittaker 1979). |
| Chilean matorral | Geophytic species comprised 6–9% of the flora in 1000 $m^2$ plots (Naveh and Whittaker 1979). |
| Australian heath | Geophytes comprised 5–12% of species in Western Australian heath samples (George et al. 1979) and 8.8% of 429 kwongan species examined (Pate et al. 1984). |
| Cape fynbos | The Cape Flora contains about 1300 species (15%) of monocotyledonous geophytes and there are also numerous geophytic dicots, e.g. 120 *Oxalis* and 41 *Pelargonium* species (Bond and Goldblatt 1984). Geophytes comprised about 18% of the 638 species in the Swartboskloof flora (McDonald and Morley 1988 and additions). Geophytes comprised 10–25% of the species in 1000 $m^2$ plots in fynbos shrublands (Kruger 1979; Cowling 1983a). |

heating of corms. Heating of corms increased flowering but this was confounded with the effect of lifting and disturbing corms, which also stimulates flowering (Bean 1962). Studies of two *Watsonia* species showed that three treatments (burning, clearing the vegetation and mulching) did not increase flowering relative to unburnt vegetation. Mulching significantly decreased flowering (Waher 1974). However, the study sites had been mowed each year, a treatment likely to inhibit flowering (Kruger 1978). Increased water availability after fires may also stimulate flowering (Waher 1974). Ethylene released from leaf bases damaged by fire, stimulated flowering in *Xanthorrhoea australis* (Liliaceae; Gill and Ingwersen 1976), but subsequent studies showed a strong interaction with the season of fire or leaf removal which suggested that leaf removal per se is a strong stimulus (Gill

**Fig. 8.1a–c.** Fire-stimulated flowering in a population of *Watsonia borbonica* in a frequently burnt firebreak adjacent to Swartboskloof. **a** Resprouting 3 weeks after fire; **b** profuse flowering 6 months after fire; **c** seed production 10 months after fire (Photos: G.G. Forsyth)

a
b
c

1981a). Studies of grasses suggest that the removal of cover stimulates flowering in many temperate species through raised soil temperatures and thus a longer growing season (Daubenmire 1968; Old 1969). Increased temperatures in burnt tussocks after removal of old foliage stimulated flowering of *Chionochloa rigida* (Poaceae; Mark 1965, 1969; Rowley 1970). The evidence from these studies support the hypothesis that flowering is not stimulated directly by a fire (heat from flames, tissue damage), but by the indirect effects of fire on the microclimate.

### 8.1.2 Factors Favouring the Evolution of Fire-Stimulated Flowering

Few studies have addressed the ultimate (fundamental) reasons for the evolution of FSF. Fires create open seed beds and lead to increased light levels at the surface, increased soil moisture and nutrient levels (Rundel 1981a; Kruger 1983) and decreased seed predation and herbivory (Bond 1984; see also Chap. 10). The simplest hypothesis is that FSF is simply a consequence of the enhanced vigour and shoot growth of sprouting species in the post-fire environment (Zedler and Zammit 1989). An alternative hypothesis is that FSF leads to mass seed production and the satiation of seed predators and maximum seedling recruitment in the open post-fire environment (Gill 1981a; Kruger and Bigalke 1984; Le Maitre 1984; Bond 1987). According to this hypothesis, FSF is analogous to mast fruiting in temperate and tropical forest trees which may also have evolved as a means of satiating seed predators (Janzen 1971; Waller 1979; Chan 1980; Silvertown 1980; Ims 1990). Mass flowering also acts as a signal to attract pollinators. A pre-condition for the evolution of this strategy is that the pollinator(s) should not be satiated while the seed predators are (Augspurger 1981; Ims 1990).

The flowering responses of geophytes fall into two distinct groups. A few species flower within weeks of a fire at any time of the year, e.g. some *Cyrtanthus* (Amaryllidaceae) species (Gordon-Gray and Wright 1969; Olivier and Werner 1980). Most species delay flowering until the next flowering or growing season. Species with delayed flowering also fall into two groups. Flowering of some species may be shifted by weeks or months depending on the timing of the fire, e.g. *Protasparagus* (Asparagaceae; Michell 1922) or *Bobartia* (Strid 1974). Most species flower in the normal season, e.g. *Haemanthus canaliculatus* (Amaryllidaceae; Michell 1922; Levyns 1966; Snijman 1984) and *W. borbonica* (Le Maitre 1984; Goldblatt 1989).

This chapter reports on a series of studies carried out at Swartboskloof on two geophyte species, *W. borbonica* and *Cyrtanthus ventricosus*. The studies were aimed at describing the regeneration patterns of the species after fire, in accordance with one of the central aims of the Swartboskloof experiment (see Preface). They were also designed to test the findings of earlier studies on *W. borbonica* (Kruger and Bigalke 1984; Le Maitre 1984). The following questions were addressed:

- What factors control the variation in the percentage of *W. borbonica* ramets (individual plants) in flower after fires in different seasons?
- What are the proximate cues that stimulate flowering after fires?
- What factors determine the type of fire response – immediate flowering in *C. ventricosus* versus flowering in its normal spring season in *W. borbonica*?
- What are the ultimate reasons for the evolution of FSF?

## 8.2 The Life Cycle of *Watsonia borbonica* (Iridaceae)

*Watsonia borbonica* is a large geophyte with leaves up to 1 m long and an annually replaced corm up to 70 mm in diameter (De Vos 1977; Goldblatt 1989). It is common on dry, rocky slopes in sandy or loamy soils in the mountains of the southwestern Cape. The inflorescence is up to 1.4 m tall with many large, pink flowers in spring (Goldblatt 1989). The flowering time in the Jonkershoek valley varied from September–October at 250 m to November at 350 m and December–January at 600 m above sea level. The flowers are protandrous and the primary pollinators are bees of the genus *Apis* (Goldblatt 1989). Inflorescences enclosed in 1-mm gauze bags did not set any seeds, suggesting that self-fertilization is not possible (D.C. Le Maitre, P.J. Brown unpubl. data). The seeds are large (14.5 mg) and have rudimentary wings. The plant multiplies vegetatively, forming clones of up to 28 individual plants (ramets) which can live for up to 60 years (Kruger and Bigalke 1984).

In a non-flowering year the apical bud of the old corm forms a new corm. The reserves in the old corm are used to support leaf and root growth (Waher 1974) and the old corm is reduced to a hard woody disc that resists decay for many years (De Vos 1977; Fig. 8.2a). In a flowering year the apical bud develops into an inflorescence and, almost invariably, two new corms grow in the axils of the lower leaves (De Vos 1977; Fig. 8.2b). The history of a clone can be determined because the persistent discs remain attached (dichotomous branchings marking flowering years). The occurrence of simultaneous branchings in the same clone, or different clones from the same site, can be used to determine the dates of previous fires. In years between fires and after spring or winter fires less than 10% of the ramets will flower but after summer or autumn fires 50–80% of the ramets will flower (Table 8.2), producing a spectacular display. Studies of populations in fire-breaks have found that seedling recruitment is rare or absent during intervals between fires but abundant after mass flowering events (Kruger and Bigalke 1984; Le Maitre 1984).

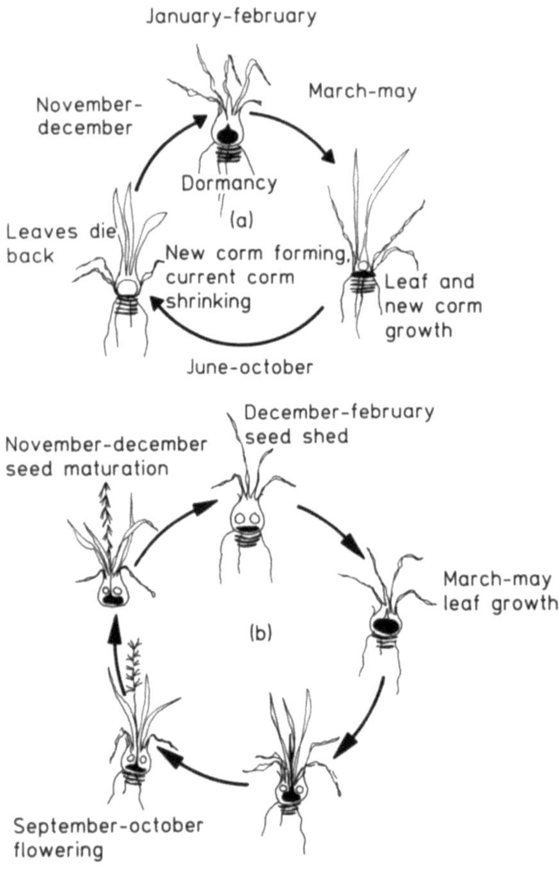


**Fig. 8.2.** The annual corm replacement cycle of a *Watsonia borbonica* plant in: **a** a non-flowering year, showing the reduction of the old corm to a disc (*shaded*), and the development of the new corm (*unshaded*); **b** a flowering year, showing the development of the terminal bud into an inflorescence and the development of two daughter corms

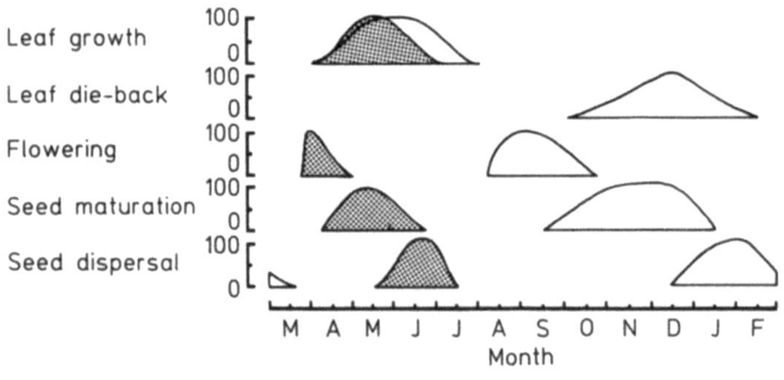


**Fig. 8.3.** A phenology diagram for populations of *Watsonia borbonica* (*open*) and *Cyrtanthus ventricosus* (*shaded*) after a fire in Swartboskloof in March 1987

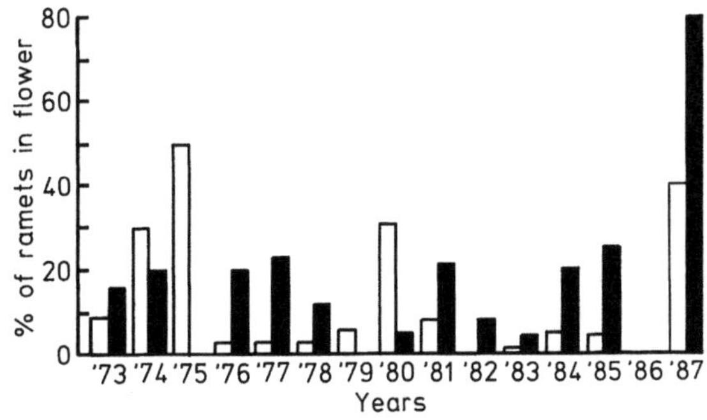


**Fig. 8.4.** The flowering history of excavated clones of *Watsonia borbonica* in a firebreak (*unshaded bars*) burnt in 1975, 1980 and 1987, and in 29-year-old vegetation burnt in 1987 (*shaded bars*). No plants flowered in the 29-year-old vegetation in 1986, and the percentage flowering in the firebreak in 1986 was not determined

**Table 8.2.** The reproductive responses of two populations of *Watsonia borbonica* to fires in autumn and spring in the Jonkershoek valley (After Kruger 1978)

| Date of fire | Number of clones tagged | Number and percentage of ramets in flower in spring in the given year | | | | | | | |
|---|---|---|---|---|---|---|---|---|---|
| | | 1974[a] | | 1975 | | 1976 | | 1977 | |
| | | No. | % | No. | % | No. | % | No. | % |
| 6 April 1975 | 81 | – | – | 485 | 49.9 | 615 | 0.0 | 609 | <0.1 |
| 13 October 1975 | 49 | 231 | 7.0 | 248 | 8.9[b] | 248 | <0.1 | 255 | <0.1 |

[a] Estimated from a count of the persistent 1974 spikes.
[b] Pre-burn assessment.

Root and leaf growth begin after the first autumn rains (Fig. 8.3). The new corm is initiated in autumn and it continues growing until the onset of dormancy in the following summer. Inflorescences emerge from the leaves in late winter and the flowers open in spring, with an inflorescence taking 2 to 3 weeks to complete flowering (Fig. 8.2). By early summer the fruits have matured and seed dispersal occurs from December to February (Fig. 8.3). The dormant seeds remain on or in the soil until the start of the wet season in late autumn and winter, when germination takes place. There were no studies of seed germination of *W. borbonica*. The primary cue is probably the low temperatures (<15°C), associated with winter rains, as found in *W. fourcadei* (Esterhuizen et al. 1986).

An analysis of past flowering events from *W. borbonica* clones excavated at two sites (a firebreak and a 29-year-old shrubland) showed that years with high percentage flowering in the firebreak were coincident with past fires, except 1974. No marked synchronization of flowering was apparent in a sample from a shrubland last burnt in 1958 (Fig. 8.4). This study found that seedling establishment can occur between fires in certain circumstances. Many young plants were found in the 29-year-old shrubland before the fire in March 1987, but young plants were rare in the firebreak. Several small plants were excavated during the study. Ramets frequently take at least 15, and occasionally more than 35 years (Kruger and Bigalke 1984) to reach maturity from seed. Ramets can flower at intervals of 1–40 or more years, but the mean interval between flowering events was 6–7 years in both firebreak and shrubland samples (D.C. Le Maitre, P.J. Brown unpubl. data).

Annual mortality rates of young plants were about 14.5% from 1–5 and 6.2% from 5–11 years of age (F.R.C. Jonkershoek unpubl. data). Mortality rates of adult plants are apparently low, about 1.6 and 19.7% in two samples of daughter corms and about 4% of the ramets in a spring fire (Kruger, 1978). The corms of several excavated ramets had been tunnelled out by larvae and it is likely that plants could be killed in this way. Rodent moles feed on *W. borbonica* corms; mole hills covered about 0.54 and 25.42% of the soil surface in two transect samples in September in *W. borbonica* populations on south- and north-facing slopes respectively (F.R.C. Jonkershoek unpubl. data).

## 8.3 The Life Cycle of *Cyrtanthus ventricosus* (Amaryllidaceae)

A few *Cyrtanthus* species flower shortly, and almost exclusively, after fires (Martin 1966; Gordon-Gray and Wright 1969; Olivier and Werner 1980). The distribution of *C. ventricosus* is poorly known and it has only been recorded from rocky slopes above 400 m in the Jonkershoek Valley. It has a perennial bulb about 50 mm long that produces one or two slender leaves about 160 mm long. It bears up to 15 orange-red flowers in an umbel on the end of a 150-mm-long peduncle. Flowering is apparently restricted to the period immediately following a fire. Vegetative reproduction via offsets is rare, unlike other species in this genus which consistently produce large numbers of bulbels (Olivier and Werner 1980; Reid and Dyer 1984). Reproduction is, therefore, virtually restricted to seedling recruitment after fires. The first inflorescence buds emerged within 2 weeks of the fire in March 1987 in Swartboskloof and over the following 5 weeks the individual flowers opened and were pollinated (Fig. 8.3). Leaf growth began as the flowers wilted and fruits developed but an examination of herbarium specimens shows that leaves and flowers may be present simultaneously in plants flowering after winter fires. The capsules ripened about 12 weeks after the fire and seeds were dispersed over a period of 3 weeks. Some plants only produced leaves in the second winter after the fire and some retained their leaves and remained evergreen for at least two seasons.

*Cyrtanthus* is the only genus in the southern African Amaryllidaceae that has flattened, black, hard-coated seeds. The remainder have globose seeds or fleshy berries (data from Dyer 1976; D. Snijman pers. comm. 1989). These seeds are dormant and can survive dry periods between dispersal and germination. The short-lived seeds of most other amaryllids may even germinate before dispersal (Markotter 1936; Snijman 1984). The standard horticultural method of inducing germination is to immerse the seeds in water, and change the water regularly to maintain high oxygen levels (Olivier and Werner 1980). Although seeds are long-lived compared to fleshy seeds, the life span is limited. Seeds kept in an envelope were germinated after storage for 2 and 3 years. Ninety-three percent of the 2-year-old seeds germinated after 2 months of soaking, but only 30% of the 3-year-old seeds germinated after the same treatment.

Plants which grew under what had been a dense, closed stand of *Protea neriifolia* tended to be smaller and produced fewer flowers and seeds after the fire than those in a more open environment (D.C. Le Maitre, P.J. Brown unpubl. data). Although only 40% of the flowers on a plant produced mature capsules, seed production was high. A seed count yielded a mean of 27 seeds per capsule, which, at a mean yield of two capsules per plant, gave a potential reproduction of 54 seedlings per plant (D.C. Le Maitre, P.J. Brown unpubl. data). Inflorescences enclosed in 1-mm plastic gauze bags did not set seeds. Self-fertilization did not occur, and the species appears to depend on pollinators for successful pollination. The individual flowers were protandrous but receptive periods of older flowers overlap with anthesis in younger flowers. *Aeropetes tulbaghia* (Linn.), the Mountain Pride butterfly, was the only potential pollinator seen visiting the species in Swartboskloof during several periods of observation which included early morning and late evening (Brown and Le Maitre 1990).

## 8.4 Fire Season and the Flowering Response of *Watsonia borbonica*

The effect of fires in different months was simulated experimentally by removing the leaves of plants monthly from April 1986 to February 1987 at two sites, one in a frequently burnt firebreak (3–6-year interval) and one in the adjacent 29-year-old shrubland (D.C. Le Maitre, P.J. Brown unpubl. data). There was a marked decline in the percentage of ramets in flower from a maximum after leaf harvesting from December to March to a minimum after harvesting in August and September at both sites (Fig. 8.5). A smaller proportion of the plants flowered and leaf harvesting had a more marked impact on the percentage of plants in flower in the firebreak (Fig. 8.5). Ramets in the firebreak had fewer flowers and produced fewer seeds per inflorescence than those in the shrubland. This, combined with the smaller proportion of ramets producing seeds, resulted in markedly lower seed production (seeds $m^{-2}$) in the firebreak (Fig. 8.5).

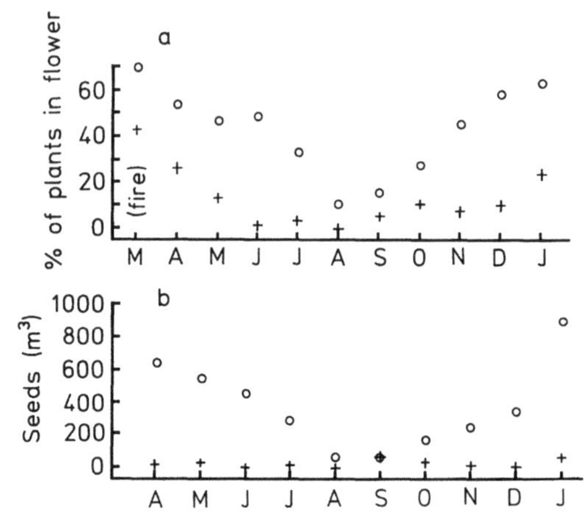


**Fig. 8.5.** The effects of leaf harvesting from April to January and a fire in March on the percentage of *Watsonia borbonica* plants in **a** flower and **b** seed production in a firebreak (burnt at 3–6-year intervals) and 29-year-old vegetation (+ and O respectively)

There was a direct link between the proportion of ramets in flower in October 1987 following leaf removal in different months (Fig. 8.5) and the corresponding stage in the corm replacement cycle (Fig. 8.2). It seems, therefore, that the percentage of plants in flower after the fire in March 1987 was controlled by timing of leaf removal in relation to the corm replacement cycle during the previous year. Leaf harvesting from April to June resulted in additional leaf growth that would have depleted the remaining reserves in the old corm or resulted in re-allocation from the new corm. Leaf harvesting after June did not result in additional leaf growth (D.C. Le Maitre, P.J. Brown unpubl. data) and the plants appeared to go dormant, completing their annual cycle with their remaining reserves. The depletion of the reserves following leaf removal seems to have left many plants with insufficient reserves to flower after the fire in March 1987.

## 8.5 Proximate Cues for Flowering

Potential cues for flowering in *W. borbonica* were studied in an experiment designed to test whether flowering was cued directly by heat from a fire, gases or ash from burning plant material or simply by removal of the vegetation. The experiment was conducted on three 5 × 5 m quadrats. One was burnt in March 1987 and two were brush-cut and cleared to exclude fire. In the first brush-cut treatment all vegetation except the *W. borbonica* plants was cut and raked out of the quadrat. In the second all vegetation was cut and cleared. Each experiment was replicated three times at each of three

**Table 8.3.** Effects of burning and vegetation clearing in March 1987 on the percentage of mature *Watsonia borbonica* ramets in 5 × 5 m quadrats that flowered by November 1987 in Swartboskloof. The vegetation clearing treatments were: brush-cutting and clearing of all vegetation, and cutting and clearing but leaving the *W. borbonica* plants

| Treatment | Sample size | *Percentage of ramets in flower* Mean ± SE | Median |
|---|---|---|---|
| Cutting, including *Watsonia* | 6 | 50.50 ± 19.16 | 51.67 |
| Cutting, excluding *Watsonia* | 6 | 43.67 ± 21.34 | 39.80 |
| Fire | 6 | 54.14 ± 12.08 | 52.40 |

sites, one in each of two firebreaks and one in an adjacent 29-year-old shrubland. The clearing simulated the removal of cover during the fire.

The clearing of vegetation did not exclude fire entirely. The fire in this area was intense, and the remaining *W. borbonica* plants, litter and stubble was scorched by the flames. However, the fire intensity on the treated quadrats would have been low because there was almost no fuel. Thus the fire would not have markedly heated the soil or the *Watsonia* corms. Little ash was deposited on the cleared quadrats (D.C. Le Maitre pers. observ.). The proportion of ramets in flower did not differ significantly between the burnt controls and the treated quadrats (Table 8.3, Chi-square = 1.631; DF = 2; $P$ = 0.44). This experiment supports Bean's (1962) hypothesis that flowering of *W. borbonica* is not stimulated directly by fire but by factors, such as soil temperature regimes, which are indirectly influenced by fire.

The results of the leaf-harvesting experiment also support this conclusion. Fires during April and May would normally stimulate flowering during the following spring, but leaf harvesting during April and May 1986 did not stimulate flowering in spring 1986. This suggests that the physical damage caused by leaf harvesting, without removing the surrounding vegetation, does not provide an appropriate flowering cue.

As *C. ventricosus* only appeared and flowered after the fire, the environmental changes that cue flowering could not be determined. No plants were seen in flower in similar but unburnt habitats adjacent to the burnt area. Some plants did, however, flower in a hoed and cleared, but unburnt, firebreak. Tagged plants in the field were visited during the second and third autumn after the fire but there was no evidence that any of the plants flowered again. Excavated bulbs were placed in pots in a shading experiment with 0, 30 and 50% shade. None of these plants have flowered in the three flowering seasons since the fire (D.C. Le Maitre, P.J. Brown unpubl. data). Many geophytic species need to reach a "critical mass" of reserves before they will flower (Rees 1972; Dafni et al. 1981a; Pate and Dixon 1982). The lack of flowering in this experiment may have been a result of insufficient time to replenish bulb reserves after flowering in March 1987. Flowering may, therefore, only occur at intervals of 3 or more years even under ideal conditions. It is difficult to stimulate flowering of *Cyrtanthus ventricosus* in horticulture, as it is for other *Cyrtanthus* species that flower only after fires (Olivier and Werner 1980; Reid and Dyer 1984). The same applies to ecologically similar species from other genera in the Amaryllidaceae, e.g. *Hessea stellaris* (D. Snijman pers. comm. 1989). The findings of this study, particularly the observation of flowering in an unburnt, hoed firebreak, support the hypothesis that flowering of *Cyrtanthus ventricosus* is stimulated by fire-caused changes in the environment rather than the direct action of fire and the by-products of combustion.

## 8.6 Factors Favouring the Evolution of Fire-Stimulated Flowering

Mass flowering of *W. borbonica* results in mass fruit set and production that satiates pre- and post-dispersal seed predators (Le Maitre 1984) and results in seedling recruitment (Kruger and Bigalke 1984; Le Maitre 1984). Plants in unburnt vegetation had approximately the same number of flowers per inflorescence as those in the burnt 30-year-old vegetation. Fruit set was higher in the mass-flowering burnt populations of *W. borbonica* (mean = 45.7%) than in a sparsely flowering population in an adjacent unburnt firebreak (mean = 36.4%), so there was no evidence of pollinator satiation in the mass flowering population (D.C. Le Maitre, P.J. Brown unpubl. data). Pre-dispersal predation of fruits by snout-beetle (Circulionidae) was significantly higher in the unburnt population (mean = 56.6%; n = 40; only one plant escaped predation) than the adjacent burnt populations (mean = 9.8%; n = 147; 60 plants escaped predation) (Wilcoxon two-sample test Z = 8.086; $P < 0.01$). Fruits damaged by beetles almost invariably contained no full seeds so pre-dispersal predation had a significant impact on seed production. The low percentage of damaged fruits in the burnt sites suggests that granivore populations were satiated by the high proportion of seed-producing plants in the burnt area, supporting the findings of Le Maitre

**Table 8.4.** A comparison of biological traits of *Watsonia borbonica* and *Cyrtanthus ventricosus*

| Trait | *W. borbonica* | *C. ventricosus* |
|---|---|---|
| Storage organ | Annually replaced corm | Perennial bulb |
| Vegetative multiplication | After flowering | Rare |
| Flowering | Not fire-obligate | Fire obligate |
| | Spring | Shortly after fire |
| | After leaf growth | Before leaf growth |
| Pollination | Bees | Butterflies |
| Seed type | Dormant | Dormant |
| Seed germination | Low temperatures | Prolonged soaking |

(1984). These findings are consistent with the hypothesis that mass flowering can only evolve where it does not result in pollinator satiation but results in mass fruiting and seeding that satiates fruit or seed predators (Augspurger 1981; Ims 1990). The results of these experiments also support Keeley's (1986) view that post-fire flowering and seed production is an alternative strategy to the accumulation of large reserves of seeds with fire-cued germination.

## 8.7 Geophyte Life Cycles and the Type of Flowering Response

The life cycles of *W. borbonica* and *C. ventricosus* differ in several respects (Table 8.4), the most important being in the timing of flowering after fire. *Watsonia borbonica* has a synanthous habit, i.e. leaves are present and functioning at the time of flowering. *Cyrtanthus ventricosus* has a hysteranthous habit, i.e. functional leaves are absent at the time of flowering. These different habits are the consequence of different phenological and developmental patterns (Dafni et al. 1981a). In synanthy, the general course of events in the annual cycle is: growth – reserve replenishment – flowering – seed production – dormancy. In hysteranthy, it is: growth – reserve replenishment – dormancy – flowering – seed production. The critical distinction is that in hysteranthy a dormant period separates the growth and flowering phases and that reserve replenishment precedes flowering. The flowering of synanthous species is constrained by the determinate developmental patterns to periods suitable for leaf growth either simultaneously with the flowers or preceding flowering. Hysteranthy is a derived condition in geophytes, releasing them from the fixed developmental sequence so that flowering can occur when leaf growth is constrained by low temperatures or moisture stress (Burtt 1970; Dafni et al. 1981a,b; Goldblatt 1983).

Two factors may have favoured the evolution of hysteranthy. One is that plants can reduce competition for pollinators or dispersers with other geophytes by flowering at a different time, and can time flowering to make maximum use of the ensuing wet season for seedling recruitment (Markotter 1936; Burtt 1970; Dafni et al. 1981a,b; Snijman 1984). The other is that the synanthous habit obliges the plant to first invest reserves in leaf growth and then in flowering and seed production. If the growing season is truncated, for example by an early drought, the plant may abort flowers or immature seeds, thus losing its investment. Flowering of hysteranthous plants is based on reserves already accumulated by the plant so that seed production is provided for and not dependent on conditions during the growing season (Dafni et al. 1981a,b). Hysteranthy is, therefore, a prerequisite for the classic fire-lily response of flowering within days of a fire, regardless of the season of the fire. Only a few *Cyrtanthus* species respond like true fire-lilies although genera with subterranean ovaries, e.g. *Gethyllis*, can potentially respond in the same way because they flower in response to rain (Burtt 1970; Markotter 1936).

Fire stimulates the flowering of several other amaryllids but flowering takes place in the normal period in summer and autumn (Levyns 1966; Frost 1984; Snijman 1984). Most of these species are hysteranthous, so the question arises: why don't they flower shortly after fires? The answer seems to lie in the biology of their seeds. These genera all have fleshy seeds that have no dormancy and may even begin to germinate before dispersal (Markotter 1936; Snijman 1984). If these species flowered shortly after a fire in spring or summer, the seeds would become dehydrated and die before the onset of winter rains. *Cyrtanthus* species have dry, dormant seeds that can survive the dry period before the winter rains. These dormant seeds are a prerequisite for seedling recruitment when flowering occurs shortly after fires whenever they occur.

## 8.8 Effect of Frequent Fires on *Wastonia borbonica*

While autumn fires promote flowering in *W. borbonica* there is evidence that fires at 3–6-year intervals may have a negative impact. The percentage of plants in flower was markedly higher in the 29-year-old shrubland than the firebreak (Fig. 8.5) and plants in the firebreak were smaller, had fewer

**Table 8.5.** Flowering responses of geophytes after fires and their relationship to other aspects of the life cycle (After Markotter 1936; Burtt 1970; Dafni et al. 1981a,b; Snijman 1984; and other references cited in the text)

**A.** Flowering shortly after a fire or preceding the next wet season. All species have a *hysteranthous* habit (no functional leaves present when flowers appear). Two kinds are found:

1. Species flowering within weeks of a fire – true fire-lilies:
   *Cyrtanthus* species – ovary aerial, seed dormant, wind-dispersed; *Gethyllis* species – ovary subterranean, seed dispersed following autumn rains[a].
2. Species flowering in their normal flowering season.
   *Haemanthus*, *Brunsvigia*, *Boophane*, *Amaryllis* (Amaryllidaceae), *Urginea* (Liliaceae) and possibly some *Pelargonium* species (section Hoarea).

**B.** Flowering during the following wet season and in their normal season. All these species have a *synanthous* habit (flowers are present together with the leaves). There are two groups:

1. Flowering early in the wet season, seed not dormant, plants all deciduous:
   *Oxalis* species
2. Flowering early on or later in the wet season, seed dormant, plants evergreen or deciduous:
   Iridaceae (e.g. *Watsonia borbonica*), Orchidaceae, Haemodoraceae, *Oxalis* species.

[a] The extent to which the flowering of *Gethyllis* species is influenced by fire is not known, flowering generally coincides with rainstorms in summer or autumn (Markotter 1936).

flowers and thus fewer seeds and seedlings (Fig. 8.5). The number of ramets per mature clone was lower at two sites in the firebreak (1.93 ± 1.33 and 1.61 ± 1.29) than in the 29-year-old shrubland (2.38 ± 2.01; Kruskal-Wallis test, Chi-square approximation = 21.93; D.F. = 2; $P < 0.01$) despite the greater potential for vegetative multiplication following the more frequent fires in the firebreak. This suggests that mortality rates of adult plants must be higher in the firebreak, and they are: a greater proportion of disc sequences in excavated clones ended abruptly in the firebreak sample (18.1%; n = 83) than in a sample from the 29-year-old shrubland (6.8%; n = 265). These findings support those of studies of woody sprouters that show that frequent fires do not benefit slow maturing species but actually result in a slow attrition of the population (Bradstock and Myerscough 1988; K.B. Higgins, P.T. Manders unpubl. data; see also Chap. 7). The plants are unable to develop rapidly enough between fires to replace the losses during fires or to reach maturity and recruit sufficient seedlings to compensate for fire-caused mortality (Bradstock and Myerscough 1988; K.B. Higgins, P.T. Manders unpubl. data).

## 8.9 Conclusions

This study was aimed at identifying the factors that control the variation in flowering of *Watsonia borbonica* after fires in different seasons, the proximate cues for fire-stimulated flowering, the factors that determine the timing of flowering after fires in *W. borbonica* and *Cyrtanthus ventricosus*, and factors that have favoured the evolution of fire-stimulated flowering.

Seasonal variation in the percentage of *W. borbonica* ramets in flower is directly related to its climatically and developmentally controlled growth pattern. Disruption of the annual corm replacement cycle by leaf removal in the growing season depletes the stored reserves which, in turn, reduces the percentage of plants in flower. Leaf removal in summer or autumn does not have this adverse effect, and results in mass flowering.

The stimulus for the flowering of *W. borbonica* and *C. ventricosus* is apparently not a direct effect of fire (i.e. heat-induced damage to leaves or apical buds), but is an indirect effect linked to changes in the environment such as altered soil temperature regimes.

The differences in the timing of flowering after fires in *W. borbonica* and *C. ventricosus* are imposed by their flowering habit (Table 8.5). The synanthous habit of *W. borbonica* imposes the developmental constraint that leaf growth must precede flowering. This delays flowering to the end of the first growing season after the fire. The evolution of a hysteranthous habit in *C. ventricosus* has freed it from this developmental constraint so that flowering can occur at times when leaf and root growth are not possible because of unfavourable environmental conditions (Dafni et al. 1981b). The evolution of this trait, plus that of dormant, non-fleshy seeds, enables *C. ventricosus* to flower shortly after fires at any time of the year. In these species developmental patterns have placed significant constraints on the evolution of their reproductive traits. Evolution of mass flowering of *W. borbonica* has only been possible because it is linked to two other factors: dormancy induced by water stress in summer and autumn, and the natural prevalence of fires at this time when the plants reserves are highest.

Fire-stimulated flowering results in decreased seed predation and successful seedling recruitment in *W. borbonica*. These findings support the hypothesis that mass flowering has evolved as a means of satiating seed predators and ensuring seedling recruitment (Janzen 1971; Gill 1981a; Kruger and Bigalke 1984; Le Maitre 1984).

Policies for managing fynbos are currently based on studies of the recruitment of serotinous Proteaceae (van Wilgen 1981; Bond 1984; Kruger and Bigalke 1984). Accordingly, the aim is to burn fynbos at intervals of 12 to 15 years in late summer or autumn. An unstated assumption underlying much of the research into plant ecology in fynbos is the notion that only seed-regenerating species are vulnerable to changes in the fire regime. Sprouting species, especially plants with plentiful reserves such as geophytes are considered highly resilient to changes in fire regime. This study has shown that this assumption is flawed. Although *W. borbonica* appears to flourish in frequently burnt firebreaks, frequent fires result in a reduction in ramet size, increase ramet mortality rates, and reduce seedling recruitment. Large, slow maturing synanthous geophytes like *W. borbonica* require fire regimes similar to the current prescriptions in order to ensure that their population structure is stable and recruitment balances mortality.

*Acknowledgements.* We thank Gregory Bailey, Adrian Simmers and Clare Jones for their assistance with fieldwork, the dismantling of excavated clones for reconstruction of their history and for putting data onto the computer for analyses. DLM wishes to thank William Bond for clarifying his ideas about the evolution of mass flowering in plants. We thank Dave Richardson and Brian van Wilgen for their comments and editing which helped to clarify the concepts presented in this chapter.

# 9 Why Is Mountain Fynbos Invasible and Which Species Invade?

D.M. Richardson and R.M. Cowling

## 9.1 Introduction

No account of the effects of fire in mountain fynbos would be complete without considering invasive alien plants. Introduced plants are a striking feature of fynbos landscapes, and they have a marked effect on the functioning of natural ecosystems (Richardson et al. 1991).

The alien flora of Swartboskloof (Table 9.1) includes many of the major weed taxa for mountain fynbos (sensu Taylor 1978; Kruger 1979; Campbell 1985) over the whole fynbos biome of South Africa (see later). Some important invaders in others areas of mountain fynbos, such as *Acacia cyclops*, *A. saligna* and *Leptospermum laevigatum*, have not been introduced to the catchment. Most of the alien species that are present are scarce, thanks to regular control operations over the past 20 years. Nearby areas with similar sets of alien species, and where control operations have not been carried out regularly, support dense thickets of alien trees and shrubs (Fig. 9.1). The herbaceous species *Briza maxima* and *Hypochoeris radicata*, although abundant in parts of Swartboskloof, are confined to paths and other highly disturbed areas; they are absent or of negligible importance in the principal vegetation communities of Swartboskloof (see Chap. 1). This is typical for mountain fynbos (Richardson et al. 1991). *Pittosporum undulatum* (Richardson and Brink 1985) and *Passiflora quadrangularis* (pers. observ.) are restricted to forest. Invasions of these species are little influenced by fire and will not be considered in this chapter. Rather, we concentrate on invasions in the fire-prone shrubland communities, and especially on invasive trees and shrubs with canopy-stored seeds dispersed by wind. This group, comprising mainly species of *Hakea* and *Pinus*, has been the most successful invasive guild in mountain fynbos (Macdonald and Richardson 1986; Richardson et al. 1990). Australian *Acacia* species that maintain banks of hard seeds in the soil (Holmes 1989) have generally been less successful in mountain fynbos. The invasion of fynbos by this guild is described elsewhere (Richardson et al. 1991).

Why have *Hakea sericea*, *Pinus pinaster*, *P. radiata* and several other species been so successful in mountain fynbos? The apparent ease with

**Table 9.1.** The alien flora of Swartboskloof, showing the distribution of species in the major vegetation communities (Chap. 1) and in disturbed areas; (*) indicates occasional occurrence

| | Disturbed areas (paths etc.) | Hygrophilous communities | Ericoid-restioid shrublands | Protea shrublands | Short riparian forest | Tall forest |
|---|---|---|---|---|---|---|
| **Trees and shrubs** | | | | | | |
| *Acacia longifolia* (Fabaceae) | | | | | * | |
| *Hakea sericea* (Proteaceae) | | | * | * | | |
| *Pinus halepensis* (Pinaceae) | | | | * | | |
| *P. pinaster* (Pinaceae) | | | * | * | | |
| *P. radiata* (Pinaceae) | | | * | * | | |
| *Pittosporum undulatum* (Pittosporaceae) | | | | | * | * |
| **Creepers** | | | | | | |
| *Passiflora* sp. (cf. *quadrangularis* Passifloraceae) | | | | | | * |
| **Herbs** | | | | | | |
| *Conyza bonariensis* (Asteraceae) | * | | | (*) | | |
| *Hypochoeris radicata* (Asteraceae) | * | | | | | |
| **Grasses** | | | | | | |
| *Briza maxima* (Poaceae) | * | | | | | |

**Fig. 9.1.** A moderately dense stand of the invasive species *Hakea sericea* (dark shrub in the *centre*), *Pinus halepensis* (tall trees in the *background* and *right foreground*) and *P. pinaster* (*left foreground*) in mountain fynbos in the Jonkershoek Valley, Stellenbosch (Photo: A.D.A. Simmers)

which these species have invaded tends to mask the numerous obstacles that an introduced species must overcome in order to invade (e.g. Roughgarden and Diamond 1986; Kruger et al. 1986). However, as in any other area (Elton 1958; Groves and Burdon 1986; Macdonald et al. 1986; Mooney and Drake 1986; Drake et al. 1989), the successful invaders in Swartboskloof are few relative to the total set of introduced organisms in the region. To develop strategies for managing current invaders and preventing future invasions, it is necessary to understand the process of invasion.

In assessing the processes governing invasions one must deal with fundamental issues concerning the determinants of community structure. We address two key questions: what makes fynbos communities open to invasion (invasible), and what type of species can invade? To answer the first question, we examine resource utilization and aspects of the structure and dynamics of fynbos communities that render them prone to invasion. Special attention is given to the role of fire. We then examine the life history attributes of species that have already become major weeds and generate a biological profile of a successful invader. We draw on information from Swartboskloof and other fynbos sites to explore these invasions. Comparisons between fynbos and other mediterranean-type ecosystems are used to emphasize salient invasion processes and determinants of invasibility.

## 9.2 What Makes Terrestrial Plant Communities Invasible?

Is it possible to predict, from community properties, the potential of a community to accept additional members? Several recent studies have addressed this question at a regional or global scale. Two important issues are examined: the role of disturbance in initiating and sustaining invasions, and the relationship between species richness and vulnerability to invasion. The role of extrinsic factors that influence the invader's ability to enter the community has also received some attention.

The disturbance regime is an important, but complex, determinant of invasibility. There is strong evidence in support of an "intermediate disturbance hypothesis" which argues that moderate levels of disturbance allow invasion of a community by the widest range of species, thus maintaining the highest species richness (e.g. Grime 1973; Abele 1976; Connell 1978; Hubbell 1979; Peet et al. 1983; Baker 1986; Crawley 1987; Richardson and Bond 1991). Exclosure experiments in grazed grasslands (e.g. Grime 1973) provide the clearest evidence for this phenomenon. In these cases, enhanced invasibility (due to reduced competition for resources resulting from selective grazing) *may* result in increased species diversity, provided there is a source of potential immigrants. Disturbance may create safe sites for germination and establishment of invading species by freeing otherwise unavailable resources. Disturbance will enhance invasibility only if it increases the availability of a limiting resource (Hobbs 1989); not all disturbances do this (Sutherland 1990). Nevertheless, most syntheses evoke a very general "disturbance/resource" explanation as a major "cause" of invasions (e.g. Elton 1958; Harper 1965; Baker 1986; Fox and Fox 1986; Orians 1986; Crawley 1987; Rejmanek 1989). Invasions tend to proceed more easily in sites subjected to ecosystem-level disturbances (Baker 1986; Crawley 1987; Lepart and Dubussche 1991; Richardson and Bond 1991). Furthermore, communities subject to recurrent disturbances are usually prone to invasion (Myers 1983; Fox and Fox 1986; Orians 1986, p. 133), especially where disturbances are catastrophic (Fox and Fox 1986; Rejmanek 1989). The role of disturbance in mediating invasions cannot be considered in isolation; its potency as a filter depends on many response modifiers, notably life history characteristics of the invaders and various biotic and abiotic constraints.

Species-rich communities may be less invasible than species-poor ones, presumably since resource niches are more fully occupied in the former (Elton 1958; Fox and Fox 1986; Moulton and Pimm 1986a,b). This generalization assumes that competition determines the structure of niche-differentiated communities, and conforms with the notion that more complex communities are more stable (reviews in Diamond and Case 1986, p. 71; Pimm 1986). These concepts were derived mainly from studies on oceanic islands, where communities are often "unsaturated" because of their isolation from regions with similar biota (Elton 1958; Mac Arthur and Wilson 1967; Moulton and Pimm 1986a,b). For many communities, however, especially those dominated by sessile organisms (e.g. plants), species richness is a poor reflection of the diversity of resource niches (Grubb 1977; Agren and Fagerstrom 1984; Yodzis 1986). Indeed, species richness may be unrelated to resource availability and may not govern invasibility (see also Pimm 1989).

Factors extrinsic to the community are often important determinants of invasibility, and sometimes override all other factors. For example, areas close to large sources of potential immigrants are most severely invaded (Crawley 1987). Historical factors such as the cultural ties between regions are often important determinants of the composition of invasive biotas (Kruger et al. 1989). This must be considered when evaluating the relative success of different organisms in invasive biotas of different parts of the world.

The above generalizations have to some extent improved our comprehension of the determinants of invasibility at a global scale. Besides providing some empirical support for Elton's (1958) pioneering hypotheses, these syntheses also emphasize the complexity of the determinants of invasibility due to the intermingling of intrinsic and extrinsic factors. For example, the susceptibility of a community to invasion by pines is usually the product of its "inherent susceptibility" (which for pines can be predicted from latitude and elevation) and the cumulative effects of various extrinsic factors that may enhance or reduce invasibility (Richardson and Bond 1991). The foregoing generalizations have mostly been gleaned from comparative studies of community patterns. As Pimm (1986) has pointed out, these are unsure ways of understanding what factors affect community persistence. In this chapter we have attempted to achieve an understanding of the mechanisms underlying plant invasions in mountain fynbos. We examine *why* the fynbos environment can support dense stands of invasive trees and shrubs, and also *how* the invaders gain entry to fynbos communities.

## 9.3 Plant Invasions in Mountain Fynbos: Why and How?

### 9.3.1 Invasions by Trees and Shrubs

Fynbos is sometimes considered to be remarkably susceptible to invasion by introduced plants when compared with other vegetation types (e.g. Taylor 1977). Macdonald (1984) has, however, argued that whereas the fynbos biome is *more severely invaded* than other southern African biomes, available data do not support the view that the fynbos is inherently *more invasible*. Certainly, the fynbos biome does not contain a disproportional *number of species* of major weeds. The few major weeds (sensu Baker 1986) (only about ten species of trees or shrubs: Table 9.2) have, however, invaded large areas of mountain fynbos. In many cases, these species have

**Table 9.2.** Major invasive plants in fynbos, listed in order of importance (ranked according to current magnitude of invasion, extent of potential habitat and potential rate of spread; Macdonald and Jarman 1984). Species marked with asterisks regularly invade mountain fynbos

| Species[a] | Origin | Life form | Juvenile period (years)[b] | Fresh seed mass (mg)[b] | Seed storage strategy[c] | Dispersal agent[d] | Fire tolerance (adults)[e] | Sprouting ability[f] |
|---|---|---|---|---|---|---|---|---|
| *Acacia saligna*** | Australia | Tree/Shrub | 3(1) | 16.4(1) | S | Wa/A | 0 | 2 |
| *Acacia longifolia**** | Australia | Tree/Shrub | 3(1) | 10.8(1) | S | Wa/B/A | 0 | 0 |
| *Acacia cyclops*** | Australia | Tree/Shrub | 3(1) | 35.0(1) | S | B/Wa | 0 | 1 |
| *Pinus pinaster**** | Mediterranean Basin | Tree | 6(2) | 45.0(2) | C1 | Wi | 1 | 0 |
| *Leptospermum laevigatum*** | Australia | Tree/Shrub | 5(4) | 0.3(5) | C1 | Wa/Wi | 0 | 0 |
| *Hakea sericea**** | Australia | Tree/Shrub | 2(3) | 29.7(3) | C2 | Wi | 0 | 0 |
| *Pinus radiata**** | California | Tree | 5(2) | 34.0(2) | C2 | Wi | 1 | 0 |
| *Pinus halepensis*** | Mediterranean Basin | Tree | 15(2) | 16.0(2) | C2 | Wi | 1 | 0 |
| *Paraserianthus lophantha** | Australia | Tree | 3(1) | 82.0(1) | S | Wa | 0 | 0 |
| *Acacia melanoxylon** | Australia | Tree | 3(1) | 16.7(1) | S | Wa | 0 | 2 |
| *Acacia mearnsii** | Australia | Tree | 3(1) | 12.8(1) | S | Wa | 0 | 2 |
| *Hakea gibbosa*** | Australia | Tree/Shrub | 2(3) | 37.1(3) | C2 | Wi | 0 | 0 |

[a] * = occasionally invades, but usually confined to other vegetation types in the biome; ** = forms dense stands over small areas; *** = a major weed in mountain fynbos.
[b] Numbers in parentheses indicate source of data; 1, P.J. Pieterse (unpubl. data); 2, Richardson et al. (1990): 3, Richardson et al. (1987); 4, Johnson (1978); 5, D.M. Richardson (unpubl. data).
[c] S = Soil-stored; C1 = canopy-stored (weakly serotinous); C2 = canopy-stored (strongly serotinous).
[d] A = Ants; B = birds; Wa = water; Wi = Wind.
[e] 0 = Poor (thin bark); 1 = moderate (adult plants occasionally survive fires).
[f] 0 = Does not sprout; 1 = occasionally sprouts, but sprouting does not ensure persistence after fire; 2 = good sprouting ability.

severely disrupted natural ecosystem processes (Macdonald and Richardson 1986).

The dominance of trees and shrubs, and the absence of forbs and grasses in the invasive flora of Swartboskloof (Table 9.1) and in mountain fynbos in general (Kruger et al. 1989) is unusual when compared with those of other southern Africa biomes (Macdonald and Richardson 1986) or with other mediterranean-climate regions (cf. Groves 1986; Mooney et al. 1986; Kruger et al. 1989). Historical and cultural differences that have influenced introductions to mediterranean-climate regions make the evaluation of the anomalous success of invasive trees in the Cape difficult (Kruger et al. 1989). Pines were introduced earlier and in greater numbers to the Cape (see Shaughnessy 1980, 1986) than to other mediterranean-climate regions (cf. Groves 1986; Mooney et al. 1986; Groves and Di Castri 1991) and this may explain, at least partially, their inordinate success as invaders (Kruger et al. 1989). Many other hypotheses have been put forward to explain the remarkable success of introduced trees and shrubs as invaders in fynbos (Table 9.3). We have consolidated these hypotheses with additional insights to propose a general explanation for the invasibility of mountain fynbos. Central to our argument are considerations of resource utilization patterns and the spatial and temporal dynamics of mountain fynbos communities.

### 9.3.2 Why Is Mountain Fynbos Invasible?

Fynbos plant communities differ in some important characteristics when compared with communities in other mediterranean-climate regions. A notable feature of fynbos is its low height (Cowling and Campbell 1980; Milewski 1981). This is perhaps best shown by comparing fynbos to vegetation in equivalent regions of Western Australia. Generally, Australian formations are an order of magnitude taller than the Cape for matched sites (Fig. 9.2). For example, proteoid fynbos in the Cape (Campbell 1985), represented in Swartboskloof by the *Protea* shrubland grouping (see Chap. 1), is the predominant formation over a wide rainfall range on sand. Corresponding formations in Australia range from structurally similar scrub heath at lower rainfalls (ca. $450\,mm\,yr^{-1}$) to woodlands and forests up to 18 times taller than fynbos at higher rainfalls. What is striking is the predominance of trees in Western Australia where rainfall exceeds about 600 mm on sands and granites. Trees are noticeably rare in the Cape (McKenzie et al. 1977; Campbell et al. 1979; Moll et al. 1980; see also Chap. 5). There is no conclusive evidence that the Cape, and mountain fynbos in particular, supported taller vegetation during the last glacial – interglacial cycle (Scholtz 1986). Fynbos communities are also characterized by much lower above-ground biomass than other mediterranean-climate regions for matched sites. The above-ground biomass in mature (but not

**Table 9.3.** A summary of frequently cited explanations for the severe invasion of fynbos by introduced trees and shrubs

**Properties of the Invader:**

*Life history attributes*

- Enhanced seed production in the absence of natural limiting factors (Neser 1968; Kluge 1983; Gill and Neser 1984) together with high germinability and rapid germination (Richardson and van Wilgen 1984; Jeffery et al. 1988) and the good heat tolerance for alien legumes (Jeffery et al. 1988) results in abnormally high recruitment.
- Enhanced seed production, together with good inherent dispersibility of individual seeds, results in exceptional dispersal which has permitted proliferation and spread (Middlemiss 1963; Kruger 1977c; Hall 1979; Richardson et al. 1987).
- Greater longevity (Kruger 1977c), and better fire survival (Kruger 1977c; Fugler 1983) facilitate superior resilience.

*Resource utilization*

- Invaders show earlier shoot growth (Sommerville 1981), faster height growth (Richardson 1985; Rutherford et al. 1986), and greater above- and below-ground absolute growth rates (Witkowski 1991) than the seed-regenerating shrubs that dominate most fynbos communities. The rapid growth rates result in the pre-emption of resources such as nutrients, moisture and light, reducing the availability of these to the slower-growing indigenous evergreen shrubs (e.g. Witkowski 1991).
- Invaders have nutrient-rich seeds which lead to enhanced germination and survival in the nutrient-poor fynbos soil (Mitchell and Allsopp 1984; Witkowski 1991).
- Invaders are better adapted to certain trace-element deficiencies (Schutte 1960).
- The nitrogen-fixing ability of some species (e.g. *Acacia* spp.; Roux and Warren 1963; Lamont 1982) has given them a competitive advantage (Milton 1979).

**Properties of the Receiving Environment:**

- The increased dispersion and size of transformed habitats means that most points within the fynbos biome are within "plant dispersal distance" of invasives (Hall 1979).
- Altered fire regimes (Kruger 1979), herbivore (Skead 1980) and insect communities (Skaife undated; Slingsby and Bond 1983), and sediment erosion and deposition patterns (Moll and Campbell 1976; Shaughnessy 1980; Brownlie 1982) have favoured the invasives.
- The occurrence of high velocity winds during summer when seeds are released from serotinous cones has added greatly to invasive potential (Richardson and Brown 1986).
- The elimination of indigenous large mammals from the fynbos biome (without replacement by introduced mammals) has reduced browsing levels, allowing certain introduced species to regenerate (Macdonald 1984).
- Invaders have replaced native species whose regenerative and maintenance capacities have been reduced through environment changes, i.e. fynbos is in a state of bioclimatic compression and the indigenous flora is not in equilibrium with the present environment (Cone 1973).
- The "vacant tree niche" hypothesis: trees that can cope with the fynbos environment (low nutrient status, summer drought, periodic fires) have failed to evolve in the biome, and this niche has been filled by introduced trees (Campbell et al. 1979; Moll et al. 1980).

senescent) mountain fynbos shrublands ranges from 15 000 to 51 000 kg ha$^{-1}$ (Kruger 1977a, van Wilgen 1982; van Wilgen et al. 1985). This is lower than in most structurally similar communities in the other regions which have richer soils but lower rainfall (cf. Ehleringer and Mooney 1983), and an order of magnitude less than areas within other mediterranean-climate regions with similar rainfall (cf. Le Houérou 1981; Ehleringer and Mooney

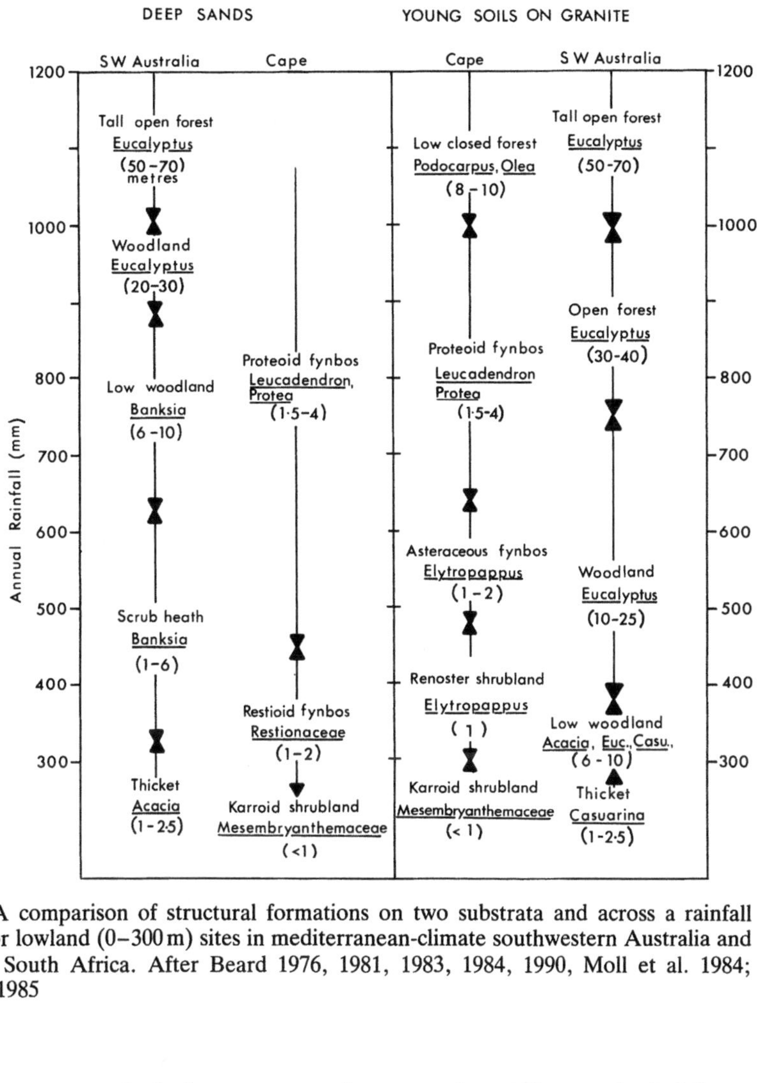


**Fig. 9.2.** A comparison of structural formations on two substrata and across a rainfall gradient for lowland (0–300 m) sites in mediterranean-climate southwestern Australia and the Cape, South Africa. After Beard 1976, 1981, 1983, 1984, 1990, Moll et al. 1984; Campbell 1985

1983). Most mountain fynbos areas receive more than 800 mm of rainfall per year. Equivalent habitats in the other regions support forest with much higher biomass. For example, above-ground biomass in dense pine forests in mediterranean-climate France (1100 mm rain $yr^{-1}$) is up to two orders of magnitude greater than that of fynbos communities in parts of the Cape with an equivalent annual rainfall (L. Trabaud unpubl. data; van Wilgen et al. 1985). Similarly, in lower rainfall (<800 mm $yr^{-1}$) regions (e.g. dry heaths in Australia, Specht 1969; *Ceanothus* chaparral in California, Gray 1982; and Chilean matorral, Kummerow et al. 1981) above-ground biomass of fynbos communities is markedly lower (cf. Kruger 1977a; Rutherford 1978; van Wilgen et al. 1985).

Annual biomass increments of between 1000 and 4000 kg $ha^{-1}$ in the first few years after fire in fynbos (Kruger 1977a; see also Chap. 3, Fig. 3.1) aproximate or slightly exceed those in California chaparral, European garrigue, scrub and heathland, and Australian heath and malee-broombush communities (Long et al. 1967; Specht 1969; Le Houérou 1981; Specht 1981a; Ehleringer and Mooney 1983; Merino et al. 1990). Shrubs at Swartboskloof have similar photosynthetic capacities to evergreen shrubs in other mediterranean-climate regions (Mooney et al. 1983), but are apparently restrained by the need to survive fire and to exploit the brief period after fire. The fynbos shrubs appear to lack the capacity (either as individuals or as a community) to optimize resource use later in stand development. Invasion of mountain fynbos by *Acacia saligna*, *A. longifolia* and *Hakea sericea* can double or treble the total above-ground plant biomass (Milton and Siegfried 1981; van Wilgen and Richardson 1985). The biomass in self-sown pine forests can be up to five times that of the pre-invasion fynbos (Versfeld and van Wilgen 1986). Introduced trees and shrubs show more rapid growth than those of fynbos shrubs. For example, 7 years after a fire in mountain fynbos near Swartboskloof the mean height of the dominant indigenous shrub *Protea repens* was 0.75 m, whereas the introduced species *Pinus pinaster* (1.75 m), *P. radiata* (2.05 m) and *Hakea sericea* (1.78 m) were much taller (Richardson 1985 and unpubl. data).

What are the implications of the low resource use in fynbos for invasibility? Clearly, fynbos communities at Swartboskloof lie below the steady-state biomass predicted from precipitation (see Miller 1981b, p. 382). The low biomass in fynbos communities cannot be attributed to the nutrient-poor soils; Australian soils are poorer than those of the Cape (Di Castri 1981, p. 22; Milewski and Cowling 1985; Lamont et al. 1985). It appears that resources, particularly moisture, are not fully exploited in fynbos ecosystems. The insular nature of the biome probably prevented migrations of pre-adapted trees from other regions. The evolution of fynbos plant communities has been a gradual in situ process, rather than a migration event (Deacon 1983c), and for some reason fynbos taxa have not adapted to exploit fully the available resources.

Invasion of fynbos by trees and shrubs has pushed the vegetation closer to the steady-state biomass predicted for the resource (precipitation) level (Fig. 9.3). The ready incorporation of invasive trees into fynbos, and the higher steady-state biomass of invaded compared with non-invaded communities, suggests a surplus of resources associated with a vacant niche in the fynbos for this life form (Moll et al. 1980). This contrasts with the mediterranean-climate regions of Australia (Groves 1986), California (Mooney et al. 1986) and Chile (Keeley and Johnson 1976), where plant invasions have not resulted in large increases in steady-state biomass. We suggest that the biomass of the natural communities in these regions approximates the steady-state biomass. For this reason, plant invasions in these regions usually involve the replacement of resident species by similar

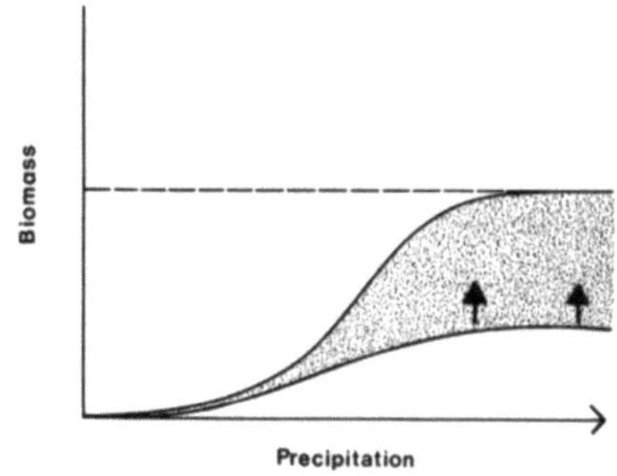


**Fig. 9.3.** Hypothetical relationship between precipitation and above-ground plant biomass in terrestrial ecosystems. The *upper curve* defines the steady-state biomass (sensu Miller 1981b) for any given rainfall. The *lower curve* defines the biomass/rainfall relationship for mountain fynbos. The *shaded area* represents the biomass that can still be incorporated into fynbos communities through replacement of indigenous by introduced plants at different precipitation levels

life forms or guilds [e.g. geophytes in Western Australian heath (kwongan) and Mediterranean grasses in chaparral].

Another factor that appears to have permitted invasion of fynbos by trees and shrubs from other mediterranean-climate regions is the relatively mild moisture deficits during summer which may alleviate seedling mortality. For example, mediterranean-climate California generally shows a distinct seasonal drought and plant xylem pressure potentials below −6 MPa are common (Poole et al. 1981; Davis and Mooney 1986; Hart and Radosevich 1987). A distinct seasonal drought is also common in mediterranean-climate Chile (Poole et al. 1981). In mountain fynbos xylem pressure potentials seldom fall below −2 MPa for deep-rooted proteoid shrubs and −4 MPa for shallow-rooted ericoid and restiod species (for Swartboskloof: Richardson and Kruger 1990; Smith and Richardson 1990; see also Chap. 6; for other mountain fynbos sites: see Miller et al. 1983; Miller 1985; Van der Heyden and Lewis 1989). In general, soil moisture in Swartboskloof is recharged rapidly following relatively light rainfall events (Richardson and Kruger 1990).

Another feature of fynbos community structure (at least at Swartboskloof and other western Cape sites), which is shared with other mediterranean-climate regions, is the poor representation in the field layer of vigorous and productive $C_4$ grasses (Vogel et al. 1978; Cowling 1983b). Of the 50 grass species in Swartboskloof (McDonald and Morley 1988), 36 (72%) are $C_3$. Those $C_4$ grasses that occur in the catchment are confined to soils derived from granite and shale (pers. observ.). Fast-growing $C_4$ grasses can potentially out-compete woody plants, especially the relatively slow growing gymnosperms such as pines, in the regeneration phase (references in Richardson and Bond 1991). Many $C_4$ grasslands could, like mountain fynbos, support woody vegetation with a much higher biomass. Unlike mountain fynbos, however, the incorporation of these woody species is often prevented by the competitive strength of the grasses (Bond 1989b), or by frequent fire (van Wilgen et al. 1990a).

We have identified reasons why mountain fynbos ecosystems such as Swartboskloof can support communities of plants with much greater resource-use efficiency than that exhibited by the natural communities. There remain, however, fundamental barriers for the incorporation of these alien organisms into fynbos communities.

### 9.3.3 How Is Mountain Fynbos Invaded?

The invasion of mountain fynbos could be explained by the dynamics of fynbos in relation to fire. There has been much comment on the floristic complexity of fynbos communities, particularly in the southwestern part of the biome (e.g. Taylor 1978; Kruger 1979; Campbell 1985). At the landscape level it is possible to recognize broadly defined structural and floristic formations that are predictably related to environmental factors (e.g. McDonald 1988; Chaps. 1 and 5 for Swartboskloof, and Taylor 1969; Boucher 1972; Cowling 1984; Campbell 1985; Cowling et al. 1988 for elsewhere in the fynbos). At the level of the community (or more specifically the phytosociologically defined association) the characterization of fynbos vegetation becomes intractable. Both Taylor (1969) and Boucher (1972), who have conducted detailed phytosociological studies in large areas (ca. 8000 ha and 24 000 ha respectively) of fynbos in the southwestern Cape, bemoan the high within-community floristic variation and associated difficulties in defining lower level phytosociological units; witness the large number of general and infrequent species (equivalent to noise) in the phytosociological tables for Swartboskloof (McDonald 1988) and other mountain fynbos sites (e.g. Boucher 1972; van Wilgen and Kruger 1985).

Much of the floristic complexity of fynbos could result from differential post-fire recruitment of coexisting species (Cowling 1987). Thus, each fire (or parts of the same fire; Hobbs and Atkins 1988) could be unique in its effects on recruitment levels as determined by pre-fire factors (e.g. sizes of seed banks), the fire itself (e.g. heat pulse for breaking seed dormancy or releasing seed), and post-fire conditions (e.g. drought-induced seedling mortality or predator dynamics). The result could be that over several fire cycles each species would encounter at least one set of conditions favourable for recruitment: competitive hierarchies would fluctuate with the changing relative abundances of species and coexistence would be maintained. This model incorporates elements of Chesson and Warner's (1981) lottery model with a storage effect (Warner and Chesson 1985).

All the evidence for post-fire variation in recruitment of fynbos species comes from non-sprouting, overstorey Proteaceae (*Protea* and *Leucadendron* spp.) (Bond et al. 1984; van Wilgen and Viviers 1985; Le Maitre 1988a,b;

Cowling and Gxaba 1990). Post-fire recruitment of this guild is highly variable (see Chap. 7). Cowling and Gxaba (1990) reported high spatial variation of *Leucadendron laureolum* recruitment after a single summer fire. They concluded that this species shows "drifting clouds of species abundance" (see Grubb 1986), the location, shape and density of the clouds varying after each fire. Moreover, dense stands of this species reduce significantly the understorey species richness (see also Campbell and van der Meulen 1980; Kruger 1987; Esler and Cowling 1990). Thus, a vagile and marauding population of an overstorey species imposes on the landscape a dynamic array of habitats which promotes community-wide floristic patchiness (Cowling and Gxaba 1990). Many of these overstorey proteoids can form dense thickets (or be driven to local extinction) in a wide range of communities, phytosociologically defined by the understorey flora. Examples in the southwestern Cape mountain fynbos are *Leucadendron laureolum* (Taylor 1969; Boucher 1972; Cowling and Gxaba 1990), *Protea repens* (Boucher 1972) and, in Swartboskloof, *P. neriifolia* (McDonald 1988; see also Chap. 7).

Mountain fynbos communities dominated by non-sprouting species (Kruger and Bigalke 1984) can show large fire-induced changes in population sizes of component species (see Chap. 7). In Yodzis's (1986) terms, the communities show a high degree of founder control where community structure is essentially a relict of the original colonization event (fire). Founder control would be most evident in the small-leaved understorey shrub component which includes many structurally (and possibly functionally) similar species which could be regarded as trophically equivalent (sensu Shmida and Ellner 1984, see Cowling 1987). Amongst these species, inter- and intra specific competition would be broadly similar, resulting in no dominance effects (cf. Yodzis 1986). Dominance control is introduced by the overstorey component which can suppress or even eliminate understorey species.

What are the implications of this non-equilibrium model of fynbos species population dynamics for invasion by introduced species? The fire-induced variation in the floristic structure of fynbos communities implies that local extinction of community components and invasion (or recolonization) is a normal and regular process. Fire introduces recurrent fluctuations in community connectance (sensu Pimm 1989), creating transient invasion windows when the non-selective botanical barrier is removed (Johnstone 1986). We argue that mountain fynbos communities, which are vulnerable to invasion and suppression by indigenous species, are readily invaded by certain introduced species, particularly if the latter exert strong dominance interactions. The effect of the aliens on reducing species richness (Cowling et al. 1976; Richardson and van Wilgen 1986a; Richardson et al. 1989) and on other aspects of community structure and functioning (Breytenbach 1986; Macdonald and Richardson 1986; Richardson and van Wilgen 1986a; Versfeld and van Wilgen persist because the aliens are *not* prone to fire-induced local extinction. The aliens initially behave much like the indigenous shrubs but, because they are well buffered against fire-induced population crashes (because of their short juvenile periods and large reserves of viable seeds), the non-equilibrium status of natural communities is disrupted, resulting in new depauperate steady states (Fig. 9.4). The only disturbance required to create invasion windows is fire, which is a naturally occurring phenomenon; invasions can occur without any exogenous disturbance (sensu Fox and Fox 1986). In the next section we consider in more detail the role of life history in determining invasive success.

## 9.4 Which Species Invade?

### 9.4.1 Describing a Good Invader

Several authors, most notably entomologists, have argued that it is almost impossible to predict which species will invade (Ehrlich 1986; Crawley 1986, 1987; Lawton and Brown 1986; Simberloff 1986). However, while accepting that biological invasions are highly idiosyncratic, we believe that attempts must be made to generalize and predict. This is especially crucial for mountain fynbos areas such as Swartboskloof where there will be a sustained need for the introduction of trees and shrubs for forestry and horticulture. Many species are likely to come from environmentally analogous regions.

A profile of a successful invader in a particular environment can be developed a priori from theoretical considerations and a posteriori from empirical data (Baker and Stebbins 1965; Baker 1986; Bazzaz 1986). A potential invader must be able to grow and reproduce in the target environment, and must possess life history attributes which permit population growth under the prevailing disturbance regime (Kruger et al. 1986). Empirical data for the biological characterization of successful invaders come from natural experiments involving many introductions over long periods. Using both approaches, we compiled a flow diagram to distinguish between species with low and high probabilities of becoming invaders in mountain fynbos (Fig. 9.5). The chart defines a research protocol for screening potential introductions.

What attributes would we expect a successful invader of mountain fynbos to have? The disturbance regime in Swartboskloof, as elsewhere in mountain fynbos, is overwhelmingly characterized by intense fires at intervals of 10 to 30 years (Kruger and Bigalke 1984). Invasive species must have life history attributes which can cope with this fire regime (Noble and Slatyer 1980). First, a successful invader would have a juvenile period of less than about 8 years to ensure adequate seed production between fires. Species which produce large seed crops should be more invasive than species with a low reproductive output. Relative to spontaneous seed release, serotiny (canopy storage of overlapping seed crops) maximizes seed avail-

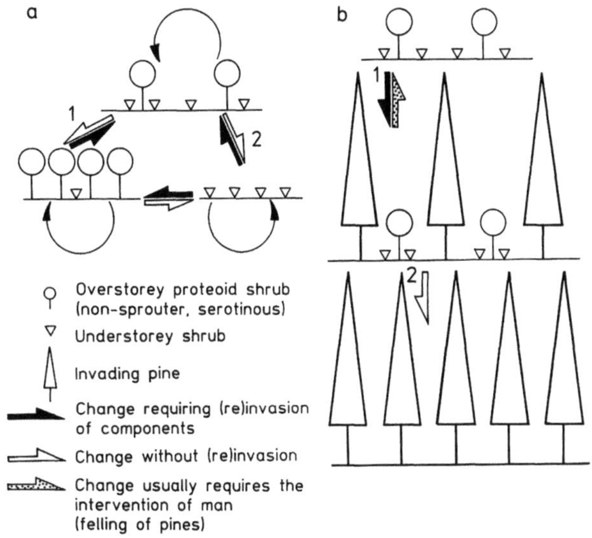


**Fig. 9.4.** Schematic representation of fire-induced changes in the structure of mountain fynbos shrublands: **a** in the absence of invasive pines (note cyclical replacement processes); **b** invaded by pines. In the absence of invasive pines, recruitment of overstorey shrubs is highly variable. They may proliferate and suppress understorey species (*a1*). Adverse conditions during or after a fire may result in poor recruitment or local extinction of overstorey shrubs (*a2*). Restoration of either condition requires recolonization by components. Once pines are established in fynbos communities, their superior fire resilience disrupts the prevailing non-equilibrium system. As the invaders proliferate (*b2*), competition with fynbos elements is intensified, eventually leading to their local extinction as residual seed stores are depleted. There is no cyclical replacement without abnormal conditions (e.g. felling of pines by man), and a depauperate steady state results

able for the next generation, provided fires occur at intervals less than the life span of the species (Lamont et al., in press). Since mountain fynbos fires occur at intervals far less than the potential life spans of all trees and shrubs, serotiny must enhance the invasiveness of these species. High seed dispersability would enhance colonization and hence invasive success (van Wilgen and Siegfried 1986; Richardson et al. 1987). Finally, for those species without complete serotiny, the ability to establish and reach reproductive maturity *between* fires would also promote invasiveness.

In the next section we describe the characteristics of invasive trees and shrubs in mountain fynbos. We evaluate the relative success as invaders of different species of *Hakea* and *Pinus* using the risk assessment model (Fig. 9.5) as a framework for our discussion.

### 9.4.2 Those That Have Made It: Pines and Hakeas

Taxa of the genera *Hakea* (Proteaceae) and *Pinus* (Pinaceae) have been particularly successful as weeds in mountain fynbos (Tables 9.1, 9.2). Several

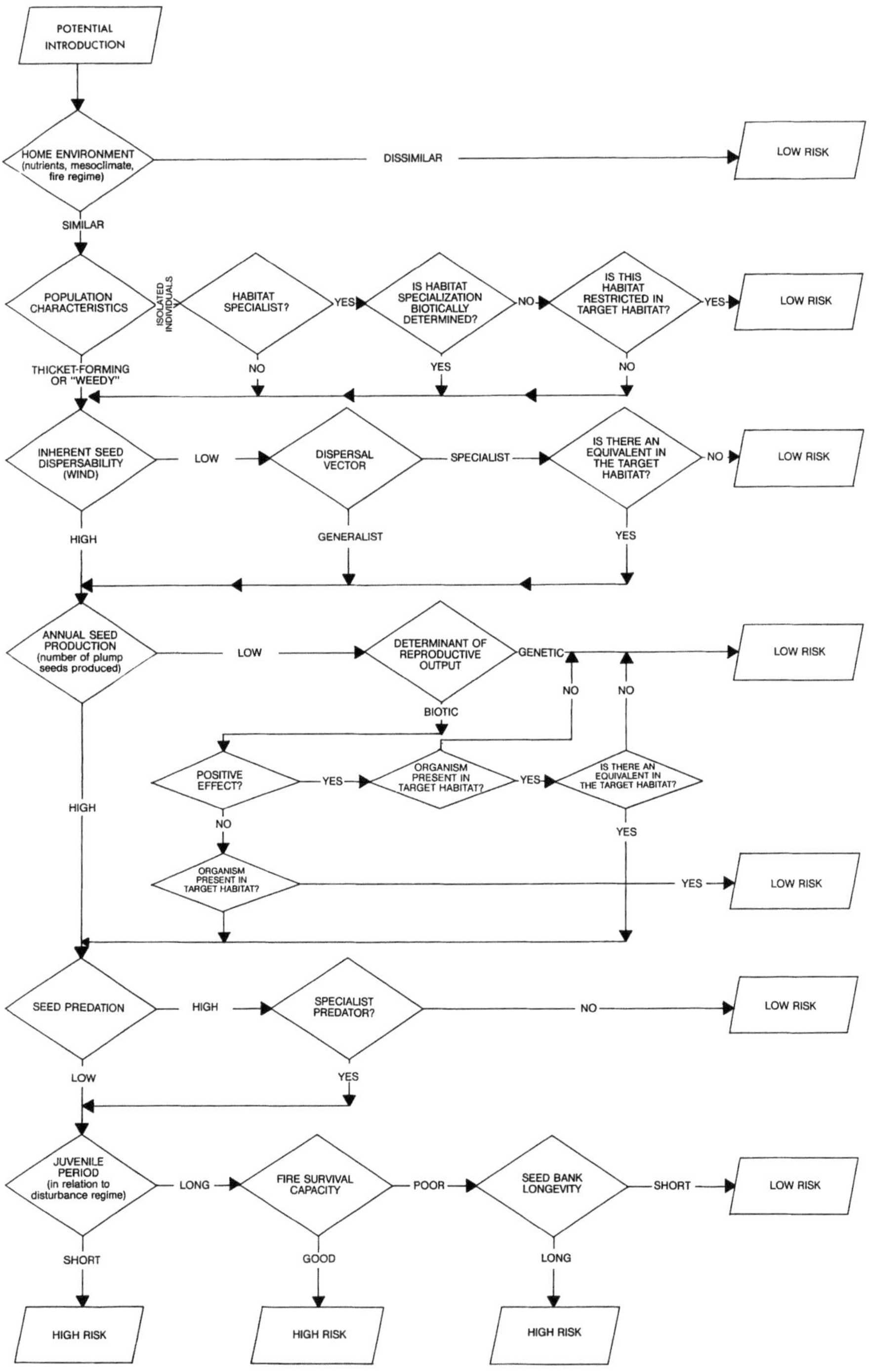
POTENTIAL INTRODUCTION
HOME ENVIRONMENT (nutrients, mesoclimate, fire regime)
DISSIMILAR
LOW RISK
SIMILAR
POPULATION CHARACTERISTICS
ISOLATED INDIVIDUALS
HABITAT SPECIALIST?
YES
IS HABITAT SPECIALIZATION BIOTICALLY DETERMINED?
NO
IS THIS HABITAT RESTRICTED IN TARGET HABITAT?
YES
LOW RISK
THICKET-FORMING OR "WEEDY"
NO
YES
NO
INHERENT SEED DISPERSABILITY (WIND)
LOW
DISPERSAL VECTOR
SPECIALIST
IS THERE AN EQUIVALENT IN THE TARGET HABITAT?
NO
LOW RISK
HIGH
GENERALIST
YES
ANNUAL SEED PRODUCTION (number of plump seeds produced)
LOW
DETERMINANT OF REPRODUCTIVE OUTPUT
GENETIC
LOW RISK
NO
NO
BIOTIC
POSITIVE EFFECT?
YES
ORGANISM PRESENT IN TARGET HABITAT?
YES
IS THERE AN EQUIVALENT IN THE TARGET HABITAT?
HIGH
NO
YES
ORGANISM PRESENT IN TARGET HABITAT?
YES
LOW RISK
SEED PREDATION
HIGH
SPECIALIST PREDATOR?
NO
LOW RISK
LOW
YES
JUVENILE PERIOD (in relation to disturbance regime)
LONG
FIRE SURVIVAL CAPACITY
POOR
SEED BANK LONGEVITY
SHORT
LOW RISK
SHORT
GOOD
LONG
HIGH RISK
HIGH RISK
HIGH RISK

**Table 9.4.** A comparison of available data on life history attributes of four introduced *Hakea* species. The order follows the flow chart in Fig. 9.5. Sources of data are referenced in Richardson et al. (1987)

* Of the four species, only *H. salicifolia* appears to lack the necessary pre-adaptation to the mountain fynbos environment.
* Only *H. sericea* is known to be weedy in *part* of its natural range (notably the population at Nerriga, New South Wales, 35° 14′ S; 150° 01′ E, D.M. Richardson unpubl data). We attribute the general absence of weedy tendencies to seed predation and other biotic factors. *Hakea sericea* has established adventive populations in New Zealand and Portugal.
* All four species have seeds that are well adapted for long-distance dispersal by wind.
* The relative annual seed production of the four species in Australia is unknown. In the Cape *H. salicifolia* and *H. sericea* are much more fecund than the other two species.
* All four species have host specific pre-dispersal seed predators. Pre-dispersal seed predation is generally low and generalist predators have a very small influence on the seed bank.
* *Hakea gibbosa* and *H. sericea* have short juvenile periods (2 years) compared to *H. salicifolia* and *H. suaveolens* (ca. 6 years).
* All species are fire-sensitive in that seedlings and adults are killed by fire. *Hakea salicifolia* differs markedly from the other three species in its capacity to persist in an environment subject to intense crown fires; the relatively small follicles provide insufficient insulation of seeds.
* All species are serotinous, but *H. salicifolia* much less so than the other three species. For the other three species, a large proportion of the seeds accumulated in the canopy are held in a viable condition for at least the average inter-fire period in mountain fynbos.

genera of both taxa have been introduced, some of which have become major weeds while others have shown moderate invasion or have not invaded at all. Are there general rules that govern the expression of invasive success? Four *Hakea* species (*H. sericea*, *H. gibbosa*, *H. suaveolens* and *H. salicifolia*) were introduced to the Cape between 1840 and 1860. *Hakea sericea* is highly invasive, *H. gibbosa* and *H. suaveolens* are moderately invasive and *H. salicifolia* does not invade mountain fynbos. The major weed, *H. sericea*, differs from the other species in some important characteristics, notably in its relatively high seed production, short juvenile period and good ability to persist in an environment subjected to intense crown fires (Table 9.4).

Many *Pinus* species have been introduced to the Cape. Most species have become naturalized but three species (*P. pinaster*, *P. radiata* and *P. halepensis*) readily invade mountain fynbos (all three species are present

**Fig. 9.5.** Proposed protocol of risk assessment for determining the likelihood of an introduced plant species invading mountain fynbos. The flow diagram distinguishes between *low risk* and *high risk* species (low and high probabilities of becoming major weeds; sensu Baker 1986; Kruger et al. 1986). High risk introductions may possess either a short juvenile period or a good capacity for surviving fires or long seed bank longevity (or a combination of these attributes)

in Swartboskloof; Table 9.1). The major invasive pines are all from environmentally similar regions and all show weedy behaviour in other parts of the world, inside and outside their natural ranges (Richardson and Bond 1991). These species, with several others, form a distinguishable group within the genus *Pinus*. They are fire-resilient pines with small seeds, low seed wing loadings, short juvenile periods, moderate to high degrees of serotiny and poor fire tolerance as adults (Richardson et al. 1990).

Serotinous pines and hakeas are prominent among successful invasive plants in mountain fynbos (Table 9.2). Fire stimulates the release of their seeds onto the nutrient-rich and competition-free ash bed that facilitates germination and establishment (Bond 1985; Lamont et al., in press). Seedling recruitment of these species is mainly in the immediate post-fire phase, and leads to the formation of even-aged thickets (Kruger 1977c; Richardson and Brown 1986; Richardson 1988, 1989). Sporadic inter-fire recruitment does occur for *Pinus* spp. (Kruger 1977c; Richardson and Brown 1986; Richardson 1989), and this contributes to spread and population growth through the establishment of new satellite seed banks. For most species, the time between germination and the production of viable seeds is less than 7 years. This facilitates the establishment of new seed sources between fires, which seldom occur at intervals of less than 8 years. As would be predicted, those species with longer juvenile periods (e.g. *Pinus halepensis*), are less widespread. In these species, persistence is achieved either through fire survival, which is rare, or through rapid immigration of seeds following local extinction (Richardson 1988). The serotinous invaders have winged seeds that are easily dispersed over several kilometres by wind (Kruger 1977c; Richardson and Brown 1986; van Wilgen and Siegfried 1986).

The perception of what makes a successful invader in mountain fynbos that emerges from the empirical analysis of past introductions corroborates the theoretical considerations presented above. Consequently, we can identify characteristics that are associated with invasive success for trees and shrubs. For some taxa, risk assessment based on life history attributes is relatively straightforward. For example, we consider any *Pinus* species with the attributes described above to be high risk introductions. *All* previous introductions of pines that possess this set of attributes (and that have enjoyed sufficient initial dissemination by people) have led to major invasions. Also, similar species are weedy elsewhere in the world (Richardson and Bond 1991). It follows that further introductions of similar species would constitute a high risk. Some high risk species would probably fail to become major weeds (sensu Baker 1986); we would attribute this to population responses and interactions in the target habitat which are difficult to quantify and impossible to predict (Richardson 1989; Richardson et al. 1990). We assign low risk status to non-serotinous pines with large seeds, high seed/wing loadings, long juvenile periods and moderate to good fire tolerance as adults; these traits are not conducive to rapid reproduction in fynbos. Some low risk taxa may invade moderately due to unpredictable interactions; such has been the case for *P. pinea* which has established an opportunistic mutualism with the introduced squirrel *Sciurus carolinensis* (Richardson et al. 1990). We would, however, be relatively confident of excluding potential major weeds among pines in the screening process. But pines are perhaps a special case; the biology and ecology of many species have been well studied, largely because of the economic importance of the genus. Few other genera of trees and shrubs have been studied this well. The thorough evaluation of decisions in the flow diagram (Fig. 9.5) requires a good data base of biological and ecological information, ideally for a representative sample of taxa in the relevant genus.

This approach provides the means to assess the risk potential of future introductions. The Australian genus *Banksia* (Proteaceae) contains trees and shrubs which grow in environments similar to mountain fynbos, and which share characteristics with pines and hakeas. At least ten species from Western Australia have recently been introduced to the Cape for horticulture. Richardson et al. (1990) defined a group of *Banksia* species that have a high potential for invading mountain fynbos. These species are tall, non-sprouting, serotinous shrubs with short juvenile periods, many small (highly dispersible) seeds per plant, and low fire tolerance.

## 9.5 Conclusions

We have attempted to achieve an understanding of plant invasions in mountain fynbos by examing *why* the environment can support dense stands of invasive trees and shrubs, and also *how* the invaders gain entry to natural plant communities. Recurrent intense fires in mountain fynbos cause large spatial and temporal fluctuations in population sizes of indigenous overstorey shrubs. We have argued that these stochastic fire-induced changes in community structure create transient invasion windows that permit not only invasion (or re-invasion) of indigenous elements, but also enable certain introduced tree and shrub species to invade. There is no evidence that invasion in mountain fynbos is dependent on a man-induced alteration of the fire regime. Populations of the introduced species essentially behave like the native marauders, but excel in two key facets: they have better dispersal abilities and are more fire-resilient. The first faculty improves vagility and facilitates invasion. Environmental conditions in mountain fynbos are sufficiently benign for the pre-adapted trees and shrubs. Once establishment barriers are overcome, there are sufficient resources for the transformation of fynbos into forests of introduced trees and shrubs. Superior fire resilience buffers the established populations against local extinction, thus disrupting the prevailing non-equilibrium system.

Plant invasions in mountain fynbos are influenced markedly by extrinsic factors such as large-scale afforestation and man-assisted dissemination of pines, the geographic location of the Cape and its cultural history. These factors have undoubtedly influenced the current status of invasive species in mountain fynbos. Nevertheless, we believe that mountain fynbos is inherently highly susceptible to invasion by some pines and other species that possess the suite of traits required to exploit the available invasion windows. Witness for example the relative ease with which species such as *P. halepensis* and *P. pinaster* establish in mountain fynbos (Kruger 1977c; Richardson 1988, 1989) when compared with recruitment patterns in the natural range of these species in the Mediterranean Basin. In France, both species apparently invade only communities currently dominated by pines (Trabaud 1991 for *P. halepensis*; Mazurek and Romane 1986 for *P. pinaster*). Outside these communities, competition from herbaceous species is a effective barrier to invasion (e.g. Acherar et al. 1984 for *P. halepensis*; Mazurek and Romane 1986 for *P. pinaster*). Biotic interactions, especially between plant competitors, are fundamental in determining the invasibility of communities for pines throughout the world (Richardson and Bond 1991). In fynbos, however, invasions are overwhelmingly mediated by fire and wind patterns; biotic interactions contribute little to observed patterns. The fynbos is certainly unique in that the naturally occurring disturbance regime permits invasion of pines. The severe ecosystem-level impacts of these invasions are also exceptional (cf. Vitousek 1990).

Pine and hakea invasions in mountain fynbos are particularly interesting for the questions they raise concerning the determinants of distribution of these taxa throughout their current range, natural and adventive. The pines of the Mediterranean Basin and of California are all believed to be native to specialized sites (Mirov 1967). However, these species are weedy in their natural ranges, readily invading abandoned fields, degraded scrublands and other disturbed sites (Richardson 1990; Richardson and Bond 1991). In the Mediterranean Basin, *P. halepensis* is restricted to calcareous soils with relatively high pH, whereas *P. pinaster* occurs on siliceous soils with low pH (Richardson 1988). In mountain fynbos these two species (and *P. radiata*) frequently co-occur (e.g. Richardson and Brown 1986; see also Fig. 9.1); the factors that usually restrict these species to specific sites in their home ranges clearly do not operate here. In contrast with these pines, most *Hakea* species are widely distributed (and often dominant) components of the vegetation in mediterranean-climate Australia (Beard 1984, 1990). And yet, *H. sericea* appears to have the same type of distribution as *P. pinaster* and *P. radiata* in mountain fynbos (alpha common, beta common, gamma common sensu Bond 1989a). These species conform with Baker's (1986, p. 51) concept of a major weed. The invasions in fynbos provide natural experiments with considerable potential for exploring the determinants of distribution of these taxa. We hope that our preliminary treatment will stimulate further research towards this end.

Through both theoretical and empirical analyses we have generated a biological profile of a successful invader, and a risk assessment model for screening new introductions. The model has the potential to identify with reasonable confidence those species with low invasive potential. Predicting which of the high risk species will become *major* weeds is more problematic, since idiosyncratic events sometimes confound predictions. Nevertheless, attempts at risk assessment, based on a biological profile of potential introductions, and an understanding of the structure of the target community are useful.

*Acknowledgements.* We are grateful to W.J. Bond, P.J. Grubb, R.J. Hobbs, B.B. Lamont, P.T. Manders, J.J. Midgley, W.D. Stock and B.W. van Wilgen for their helpful comments on the manuscript.

# 10 Do Small Mammals Govern Vegetation Recovery After Fire in Fynbos?

H.J. van Hensbergen, S.A. Botha, G.G. Forsyth,
and D.C. Le Maitre

## 10.1 Introduction

Fynbos vegetation is of low nutritive quality for mammals (Bigalke 1979b). Palatable grasses are scarce (Cowling 1983b), have low protein and phosphorus values and deficiencies in some trace elements (Louw 1969). The low productivity of fynbos vegetation, apart from the first few years after fire (Kruger 1977a; see also Chap. 9; Sect. 9.3.2), also limits the availability of food for mammals (Fox et al. 1985). In contrast with the foliage, the seeds of some fynbos plants are large and rich in both energy and nutrients and are therefore a valuable food resource for animals (Bond and Breytenbach 1985).

Because of limited food, small mammal densities in fynbos are generally lower than in savanna and grassland (Bigalke 1979b; Nel et al. 1980; Fox et al. 1985). The contribution to biomass of the larger mammals is also small. It is therefore unlikely that mammals play a significant role in nutrient cycling in these ecosystems (Bigalke 1979b). The leaves of sclerophyllous plants contain plant secondary compounds (Morrow 1983), probably as an adaptation to previous higher levels of herbivory (Fuentes and Etchegaray 1983). Low densities of small mammals in fynbos may indicate that their activities have little influence on the functioning of the system. However, plants employ two distinct strategies of seed dispersal in fynbos, myrmecochory and serotiny, which are thought to have evolved in response to mammalian granivory (Bond and Slingsby 1983; Bond 1984).

Many fynbos species produce seeds with a fleshy elaiosome that is attractive to ants. Ants quickly collect these seeds and carry them underground where the elaiosome is consumed and the seed abandoned (Bond and Slingsby 1983). This adaptation, which is thought to reduce seed consumption by rodent granivores in South Africa and granivorous ants in Australia (Hughes and Westoby 1990) is strongly correlated with nutrient-poor soils both in South Africa (Bond and Slingsby 1983) and Australia (Davidson and Morton 1981). In nutrient-poor areas the nutrient- and energy-rich seed is a valuable resource for granivores and must be protected by the plant. In the absence of ants, rodents consume most of the seed available but when ants are present they bury most of the seed before it is discovered by rodents, which have difficulty in finding seed once it has been buried (Bond and Breytenbach 1985).

Other fynbos plants are serotinous, storing seeds in closed cones in the canopy. Seeds are released when the plant dies, either of old age or in a fire. The cones offer adequate protection of seed reserves from rodents. These plants, mainly members of the Proteaceae, are often the dominant overstorey shrubs in fynbos communities (Bond et al. 1984). Bond (1984) and Breytenbach (1984) attribute the even-aged nature of these stands to high mortality of seeds released and seedlings germinating in established vegetation. However, recruitment of a second generation of *Protea neriifolia* plants between fires has been recorded at two sites in the Jonkershoek Valley (see Chap. 7). The question of whether rodents can maintain even-aged populations of proteas thus remains unanswered.

The impact of rodent granivory interacts with fire season to determine the degree of recruitment success after fire. Seedling recruitment of serotinous Proteaceae decreases with increasing time between the fire and the germination cue in winter (Bond et al. 1984; van Wilgen and Viviers 1985), due to the cumulative effect of seed consumption by rodents over time. Alternatively, prolonged exposure causes a loss of seed viability which can explain at least part of the difference (Midgley et al. 1989).

If rodents are still present to consume seeds immediately after fire then they may still have a significant effect on plant regeneration and therefore on post-fire vegetation structure (Bond et al. 1984). On the other hand, Midgley and Clayton (1990) consider that the numbers of small mammals remaining after fires in the southern Cape are too low to affect the regeneration of the vegetation. A drop in small mammal numbers after a fire may be sufficient to ensure adequate regeneration.

Although it is known what fynbos small mammal communities change in response to changes in the physical structure of the vegetation (Bond et al. 1980; Breytenbach 1982, 1987; Willan and Bigalke 1982; Botha 1988), it is not known what factors affect the distribution of small mammals after fire. Following a fire in the Duthie reserve in Stellenbosch, small mammals did not leave the burnt area immediately, but remained for several weeks (H.J. van Hensbergen unpubl. data). The pre-fire distribution of small mammals is therefore likely to affect the immediate post-fire distribution which in turn may determine the level of seed consumption.

In this chapter we examine the pattern of post-fire community succession in small mammals. We seek to determine if animals emigrate from the burnt area following the fire, and if repopulation is a result of slow immigration from the periphery of the catchment. We examine habitat selection by small mammals and relate this to the physical and floristic composition of the vegetation. We relate the density of small mammals to changes in the physical structure of the regenerating vegetation after fire. We also examined the effects of small mammals on the vegetation by

**Table 10.1.** Details of the layout of 12 small mammal trapping sites in Swartboskloof

| Site | Trapping dates | Layout (number of traps) | Grid size (m) | Vegetation |
|---|---|---|---|---|
| GR1 | March 1985–December 1989 | 10 × 10 | 90 × 90 | Proteoid shrubland and hygrophilous communities |
| GR2 | June 1986–June 1989 | 4 × 25 | 30 × 240 | Proteoid shrubland |
| GR3 Old Veld | June 1986–December 1989 | 2 × 50 | 10 × 490 | Proteoid shrubland |
| GR3 Firebreak | As above | As above | As above | Low herbland |
| SB1 | March 1986–October 1989 | 7 × 7 | 60 × 60 | Proteoid shrubland |
| SB2 | March 1986–October 1989 | 7 ×7 | 60 × 60 | Proteoid shrubland |
| SB3 | March 1987–October 1989 | 7 × 7 | 60 × 60 | Proteoid shrubland |
| R1–R4 | January 1987–October 1989 | 20 × 1 | 190 | Low forest |
| R5 | January 1987–October 1989 | 20 × 1 | 190 | Tall forest |
| R6 | April 1987–October 1989 | 20 × 1 | 190 | Low forest |

assessing whether granivory by small mammals prevents recruitment of *Protea neriifolia* between fires and after fires.

## 10.2 Methods

### 10.2.1 Factors Affecting the Distribution and Size of Small Mammal Populations

To determine patterns of small mammal succession after fire, and to detect immigration into, and emigration from the catchment, our experimental design comprised three trapping sites (Fig. 1.3). The sites had different purposes as reflected in their different shapes, sizes and positions (Table 10.1). The first site (GR1) was primarily designed to detect changes in community structure. A second site (GR2) crossed the boundary of the proposed fire so as to provide a control in unburnt vegetation, and to detect emigration into unburnt vegetation after the fire. A third site (GR3) consisted of two lines of traps running at right angles to the boundary of the proposed fire, to detect slow immigration from the edge of the burnt area.

GR1 falls entirely within the mature (29-year-old) vegetation, whereas half the grid points of GR2 fall in the mature vegetation, and half in vegetation 14 years old, which was not due to be burnt. GR3 was half in the old vegetation and half on a fire break burnt 6 years before the 1987 fire; the entire site was burnt in the 1987 fire.

To determine the role of unburnt patches of vegetation in providing shelter for small mammals in the post-fire environment we also trapped a number of "refuge" areas which we did not expect to burn during the 1987 fire. Five refuges (R1 to R5) within the burnt area were trapped before and after the fire. A sixth refuge (R6) was selected after the burn and was thereafter trapped on the same dates as refuges R1 to R5.

We also determined the density of small mammals at the sites (SB1 and SB2 in Table 10.1) of seed survival experiments (see Sect. 10.2.3), so that the rates of seed removal could be related to small mammal density. A third site (SB3) was placed in an area of burnt fynbos after the fire. Most trapping was done with standard Sherman folding traps. A single trap at each point was baited with a mixture of peanut butter, bran, lard and candlewax (Willan 1986). All captured small mammals were marked by toe clipping. In the immediate pre- and post-fire sessions a large number of plastic drainpipe traps (Willan 1979) were also used to increase the trap density to two traps per trap point and to enable us to trap all sites simultaneously. Trapping was initially carried out for 4 days each month and traps were cleared in the morning and in the late afternoon. Trapping was carried out for 7 days immediately before and after the fire. In the final year of trapping the sessions were extended to 6 days duration bimonthly in an attempt to derive more reliable population estimates.

The data from sites GR1 to GR3 were analyzed using the method of enumeration (calendar of captures) of Petrusewicz and Andrzejewski (1962). The pattern of change in community structure was analyzed by ordination of these minimum number alive counts using correspondence analysis (Digby and Kempton 1987). The data from other trapping sites were expressed as an index (captures/100 trap nights).

### 10.2.2 Vegetation Factors Affecting Habitat Selection by Small Mammals

To determine whether small mammal populations respond to changes in the vegetation composition and structure we sampled vegetation in two ways. Firstly, Neethling and Boucher (1986) carried out a pre-fire floristic survey of GR1 using 34 quadrats ($2\,m^2$) in the site to coincide with every third trapping point. The numbers of captures of each small mammal species in each of the three largest vegetation communities identified in this survey were compared to assess their preferences for particular vegetation communities.

Secondly, the vegetation was sampled after the fire. Ten permanently marked plots (5 × 10 m) were set out on the sites. Six of these were in GR1, two in each of the three major vegetation communities. Additional plots were placed in each half of GR2 and GR3. These plots were sampled 3 months after the fire and at six monthly intervals thereafter until 33 months after the fire, giving a total of six sampling sessions. The unburnt part of GR2 was sampled 33 months after the fire only. The plots were subdivided into 50 subplots of 1 $m^2$ each. A portable catwalk allowed observers to walk over the plot without damaging the vegetation. The percentage cover and maximum height of the vegetation was estimated in each subplot at each survey. Rock cover was estimated once only. In addition, a 5-m line was passed through the vegetation across the plot, and the distances from the edge of the plot to the start and end of all vegetation intercepts on the line were measured and the cover density within the intercept was estimated. This procedure was repeated on each plot at two positions 2.5 m from each end of the plot and for five heights (0.1, 0.3, 0.5, 0.75 and 1 m). These data were used to calculate percentage cover at each height and two indices (Shannons H' and Simpsons D) of horizontal cover diversity. These measures differ in their sensitivity to departures from evenness (Magurran 1988). The value of D decreases much more rapidly in response to the dominance of a single intercept than H', whereas H' is more dependent on the number of intercepts along the line.

Analysis of the relationship between the number of individuals in the small mammal populations and vegetation structure was carried out using multiple regression with stepwise variable selection of the measured parameters. Groups of variables were sought which would account for a reasonable proportion of the variance between sites and sampling occasions. Many variables in the models were significantly correlated with numbers of animals but accounted for little of the variance; these were rejected. Doyle (1990) found similar high correlations with habitat variables for small mammals in Oregon and used a high level of significance of $P < 0.005$. We retained only sets of variables where the additional contribution to the variance of the last variable entered was at least 10%. The use of the percentage rock cover in the analyses was a pseudo-replication as this could not change between surveys.

### 10.2.3 The Effect of Small Mammals on Plant Regeneration

Two experiments were carried out to determine the factors affecting seed and seedling survival both before and after the fire.

#### 10.2.3.1 Factors Affecting Plant Establishment in Mature Fynbos

We determined whether small mammals were an important cause of seed removal and seedling mortality before the fire by means of an exclosure

**Table 10.2.** Layout of each replicate of small mammal exclosure treatments in the pre-fire seedling recruitment experiment in Swartboskloof

| | | | | |
|---|---|---|---|---|
| Exclosure number | 1 | 2 | 3 | 4 |
| Exclosure type | Rodents not excluded | Rodents not excluded | Rodents excluded | Rodents excluded |
| Substrate | Litter removed | Litter left | Litter removed | Litter left |
| | sown / planted | sown / planted | sown / planted | sown / planted |

experiment. At two of the trapping sites (SB1 an SB2) a set of four 14-mm mesh exclosures (1 × 1 × 0.2 m high), each subdivided into two to provide eight treatments (Table 10.2), was established in March 1986, 1 year before the fire. Each experiment was replicated three times at each site. In each enclosure, ten *Protea neriifolia* seeds were hand-sown onto the surface in one half of the exclosure and ten seeds were planted 3–5 mm deep in the other half. The surface litter was either removed from the exclosure or left in place, to establish whether this affected germination success. Two of the exclosures in each experiment were left with the sides open to give access to small mammals and the other two were closed. Exclosures were included in the small mammal access treatment to ensure equality of microclimates between treatments. The number of seeds germinating and later the number of seedlings surviving in each enclosure were counted at regular intervals. Where seedlings had died an attempt was made to establish the cause of death.

#### 10.2.3.2 Factors Affecting Plant Establishment After Fire in Fynbos

We determined the effect of small mammals on *Protea* regeneration shortly after fire in fynbos. We also tested whether the distance from a refuge (unburnt) site that offered cover to rodents, affected regeneration. Seven parallel lines of ten small mammal exclosures were placed at varying distances from an unburnt refuge (R6). One line was situated in the refuge (a low forest), one in the partly burnt ecotone and the others in burnt fynbos. Four *Protea neriifolia* seeds were planted inside and outside enclosures about 3 mm below the surface 3 weeks after the fire. The germination and survival of seedlings was evaluated at regular intervals for 1 year. The data were analyzed by fitting a hierarchical log linear model (Dobson 1983) to the frequencies of seeds germinating and not germinating in each treatment. Seedling mortality was classified as herbivory, fungal attack and other.

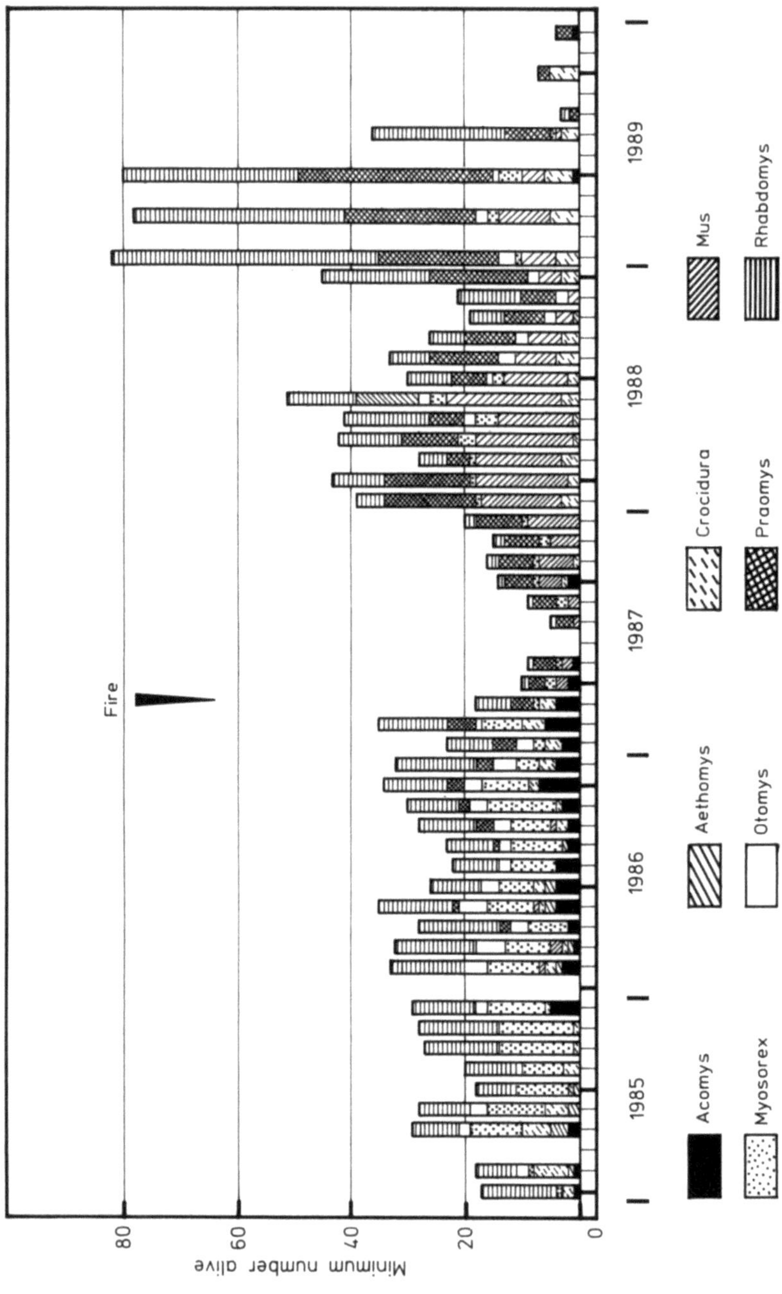


**Fig. 10.1.** Minimum number of small mammals alive on site GR1 in Swartboskloof in each month of trapping

## 10.3 Results

### 10.3.1 Post-Fire Development of Small Mammal Communities

Nine small mammal species (3155 individuals) were captured on the three main sites (1950 on GR1, 453 on GR2 and 752 on GR3). The species caught were *Acomys subspinosus* (Cape spiny mouse), *Aethomys namaquensis* (Namaqua rock mouse), *Crocidura flavescens* (greater musk shrew), *Dendromys mesomelas* (Brant's climbing mouse), *Mus minutoides* (pygmy mouse), *Myosorex varius* (forest shrew), *Otomys irroratus* (vlei rat), *Praomys* (*Myomyscus*) *verreauxi* (Verreaux's mouse, one of the multi-mammate mice) and *Rhabdomys pumilio* (striped mouse). Of these *D. mesomelas* was trapped five times only. Seven of the remaining species were caught at some time on all the sites but *P. verreauxi* was never caught on GR3. Although species richness was lower at some sites after the fire, the overall species richness of small mammals in Swartboskloof did not change after the fire. All species present before the fire were recaptured within 9 months of the fire.

Numbers of small mammals caught fluctuated seasonally on sites GR1 and GR3 (Figs. 10.1 to 10.3). No similar pattern was detected for GR2 (Fig. 10.2). These population fluctuations were greater following the fire on all sites, with higher peaks and lower troughs, than before the fire. The peaks 2 years after the fire were higher than first-year peaks in all areas except the burnt section of GR2. Fire did not affect the pattern of fluctuations, as patterns on the unburnt portion of GR2 were similar to those on burnt sites.

The first two axes of the correspondence analysis accounted for 54% of the inertia. The primary difference was between communities dominated by *A. namaquensis* (on the extreme right) and all other species (Fig. 10.4). The second axis separated communities dominated by *P. verreauxi* and *M. minutoides* (at the bottom of the plot) from those containing *A. subspinosus* and *M. varius* (at the top of the plot). Communities moved down the plot after the fire in most cases, reflecting a general increase in *M. minutoides* and *P. verreauxi* after the fire. Two sites in proteoid shrublands (GR3 and SB2) behaved quite differently from the other communities. Here, the numbers of *A. namaquensis* were high before the fire, and even higher after the fire. These two sites were on sandstone substrates, and were dominated by *Protea repens*, both factors which influenced vegetation structure.

### 10.3.2 Vegetation Factors Affecting Habitat Selection by Small Mammals

#### 10.3.2.1 The Relationship Between Floristic Communities and Small Mammal Occurrence

Five vegetation communities were identified on site GR1 (Fig. 10.5). Different small mammal species selected different vegetation communities both

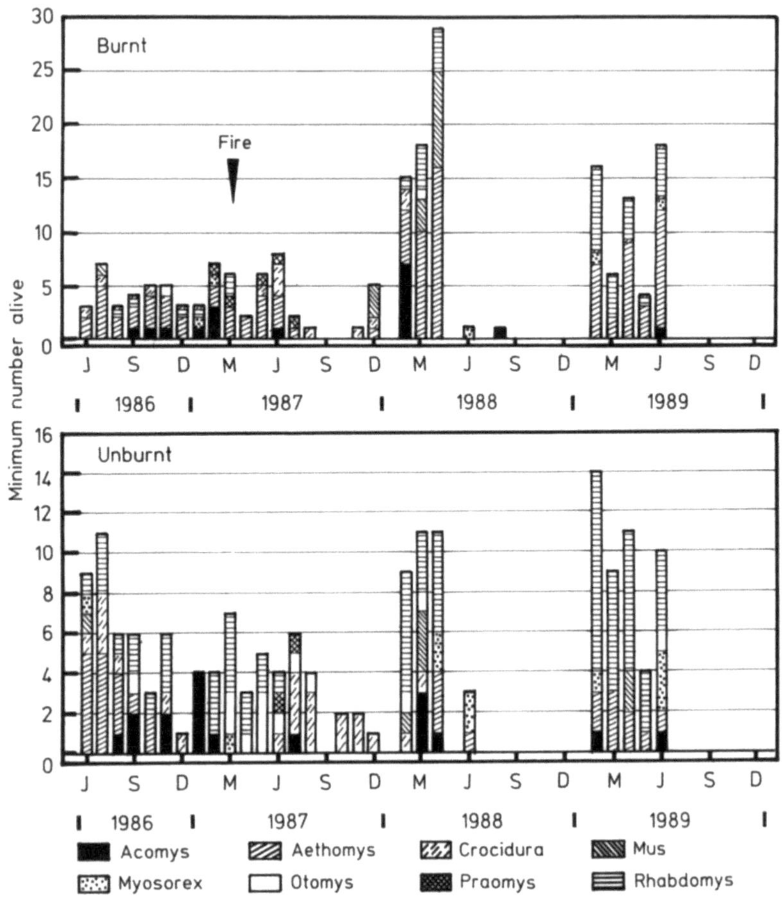


**Fig. 10.2.** Minimum number of small mammals alive on burnt and unburnt halves of site GR2 in Swartboskloof in each month of trapping

before ($\chi^2 = 53$; $P < 0.0001$) and after the fire ($\chi^2 = 214$, $P < 0.0001$) (Fig. 10.6). Furthermore, the preference for a given vegetation community changed after the fire. *Rhabdomys pumilio* made greater use of communities A and D, whereas *P. verreauxi* moved into communities D and E where it had been almost absent before the fire.

### 10.3.2.2 The Relationship Between Structure of the Vegetation and Small Mammal Densities

The numbers of small mammals were meaningfully correlated with vegetation structure. For most species, an increase in live cover, cover density, or diversity index 0.30 m above the ground was positively correlated with animal numbers. Cover higher in the canopy was negatively correlated.

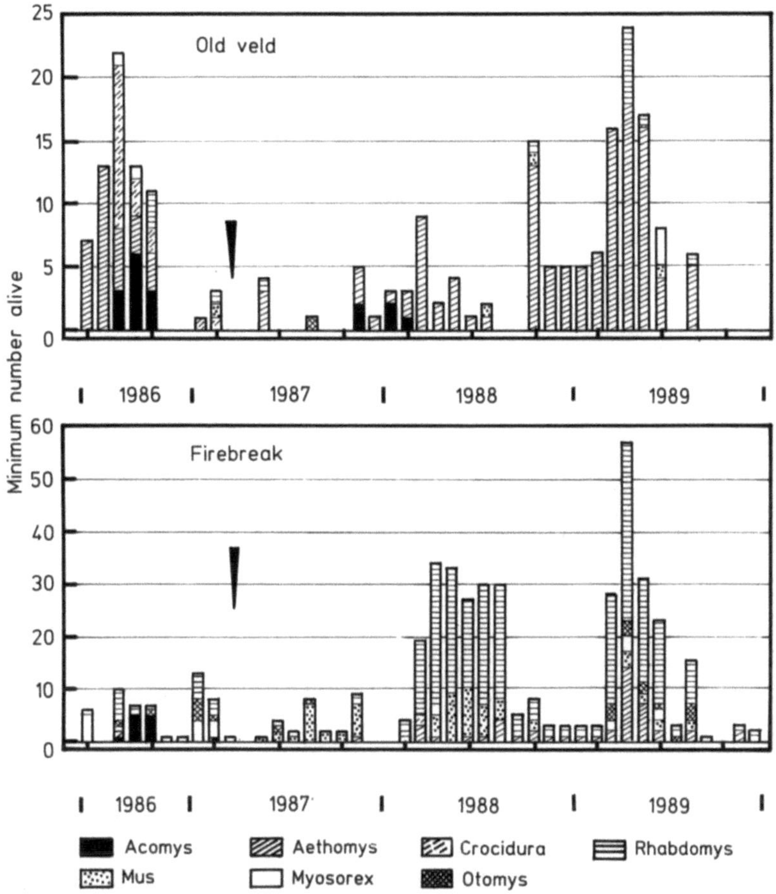


**Fig. 10.3.** Minimum number of small mammals alive on two halves of site GR3 in Swartboskloof in each month of trapping

*Aethomys namaquensis* differed from other species in that it preferred less cover below 0.3 m, with greater cover above. Models in which one variable accounted for at least 10% of the variance could be found for all species except *A. subspinosus* as there was almost no variation in the post-fire numbers of the latter (Table 10.3). It was also not possible to find a model for the populations of all species combined.

#### 10.3.2.3 Survival and Recolonization Strategies of Small Mammals

There was little evidence that the fire had any direct effect on small mammals. Many animals caught before the fire were trapped alive on the same sites immediately after the fire, although they had disappeared by the next month.

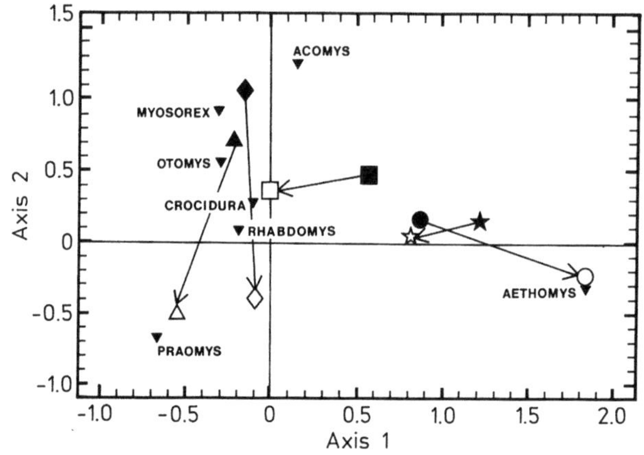


**Fig. 10.4.** Plot of means of sites (before and after fire) and species locations on the first two axes of a correspondence analysis ordination. The sites are GR1 (△), GR2 burnt (□), GR2 unburnt (★), GR3 proteoid shrubland (○), GR3 firebelt (◊). *Filled symbol* is before and *empty symbol*, after the fire

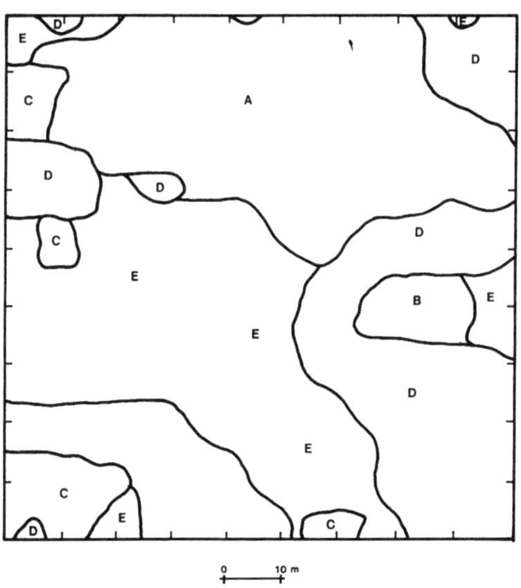


**Fig. 10.5.** Distribution of five vegetation communities on site GR1 in Swartboskloof. The communities were; *A* a hygrophilous herbland differentiated by the shrubs *Osmitopsis asteriscoides* and *Berzelia lanuginosa*; *B* a hygrophilous herbland differentiated by the grass *Merxmuellera cincta* and the shrub *Cliffortia polygonifolia*; *C* a hygrophilous herbland differentiated by *M. cincta* and the restioid herb *Platycaulos depauperatus*; *D* a community ecotonal between hygrophilous herblands and proteoid shrublands, differentiated by *M. cincta* and the shrub *Rhus angustifolia*; *E* a proteoid shrubland differentiated by the shrubs *Diospyros glabra* and *Protea repens*

**Table 10.3.** Coefficients of best-fitted regression models for small mammal populations in Swartboskloof. For *Mus minutoides* and *Praomys verreauxi* the square root of the numbers gave the best fit. Independent variables are: cover (% live plant cover), rock (% rock cover), maxht (mean of the 50 maximum plant heights in the plot), Px (% of line at height x in contact with vegetation), Hx (Shannon's H' for line at height x), D (Simpson's D for line at height x). For x, 1 = 100 mm, 3 = 300 mm, 500 mm and 7 = 750 mm. The inverses of, and ratios between percentage cover at different heights were also used. % Var = % of variance accounted for by the model

| Predictor | COVER | ROCK | MAXHT | 1/P1 | P3 | P5 | P5/P1 | P7/P1 | H3 | H5 | D1 | D5 | D7 | % Var |
|---|---|---|---|---|---|---|---|---|---|---|---|---|---|---|
| Species | | | | | | | | | | | | | | |
| *Acomys* | | | | | | | | | | | | | | None |
| *Aethomys* | | | −0.003 | | | | | | | | −0.57 | +1.98 | +1.84 | 55 |
| *Crocidura* | +0.02 | | | | | | | −1.26 | | | | | | 71 |
| *Mus*$^{0.5}$ | +0.036 | | | | | | | | | −0.298 | | | | 40 |
| *Myosorex* | | | | +13.0 | | | | | | | | | | 22 |
| *Otomys* | | +0.012 | | | +0.013 | +0.021 | | | | −0.38 | | | | 58 |
| *Praomys*$^{0.5}$ | +0.029 | +0.034 | | | | | −0.56 | | | | | | | 55 |
| *Rhabdomys* | | −0.024 | | | | | | | +0.88 | | | | | 28 |
| All species | | | | | | | | | | | | | | None |

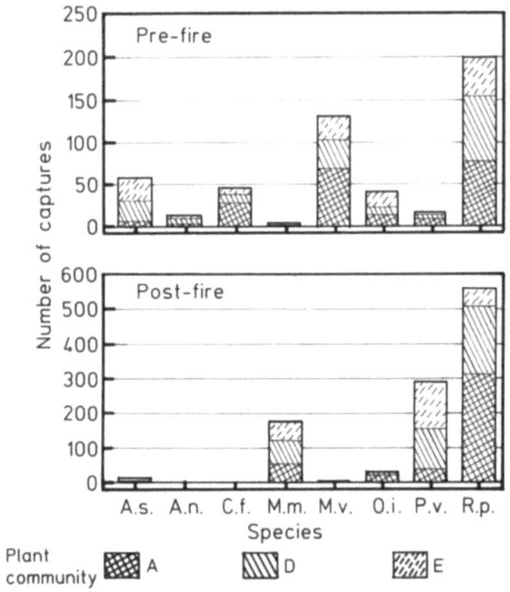


**Fig. 10.6.** Number of captures of small mammals in different vegetation communities (Fig. 10.5) on site GR1 before and after a fire in Swartboskloof. Small mammals are: **A.s** = *Acomys subspinosus*; **A.n.** = *Aethomys namaquensis*; **C.f.** = *Crocidura flavescens*; **M.m** = *Mus minutoides*; **M.v.** = *Myosorex varius*; **O.i.** = *Otomys irroratus*; **P.v.** = *Praomys verreauxi*; **R.p.** = *Rhabdomys pumilio*

Small mammals made use of unburnt refuge habitats after the fire, as shown by increases in trap success for 6 months after the fire (Table 10.4). The pre-fire trapping success in the two fynbos sites SB1 and SB2 was higher than that found in the five refuges, but fell in the burnt areas for approximately 6 months after the fire. Trapping success was similar in burnt and refuge areas 6 months after the fire. The refuges were not sampled again until 27 months after the fire, at which time the densities in the refuges were similar to those in the burnt areas.

There was little evidence for emigration from the burnt area as a result of the fire. A single juvenile *R. pumilio* caught initially in the burnt section of GR2 immediately after the fire was subsequently recaptured on the unburnt section of that site where it survived for a further 3 months. A number of unmarked *R. pumilio* adults appeared on the unburnt section of this site in the month after the fire and it seems likely that these immigrated from the burnt area. One *P. verreauxi* was caught on both halves of site GR2 after the fire and may have been resident in the unburnt area but foraged in the burnt half. Seventeen small mammals were recaptured on different sites after the fire having moved either between refuges or between refuges and burnt veld.

**Table 10.4.** The relative abundance (No. of captures/100 trap nights) of small mammals in refuges (tall and low forest, see text) and on fynbos sites shortly before and after an autumn burn in Swartboskloof. The number of trap nights per trapping session are given in parentheses in the heading. A dash (–) indicates no trapping conducted. The species are: An = *Aethomys namaquensis*; Cf = *Crocidura flavescens*; Mm = *Mus minutoides*; Mv = *Myomyscus verreauxi*; Mxv = *Myosorex varius*; Ospp = *Otomys* spp.; Pv = *Praomys verreauxi*; Rp = *Rhabdomys pumilio*

| Date of trapping session | Refuges Nos. 1–5 (400) | Refuge No. 6 (80) | Site 1 (196) | Site 2 (196) | Site 3 (196) |
|---|---|---|---|---|---|
| **Before the fire** | | | | | |
| Average[a] | – | – | 6.5 (Rp)[b] | 3.3 (An, As)[b] | – |
| 12–16 January 1987 | 1.0 (Pv) | – | – | – | |
| **After the fire** | | | | | |
| 19–23 March 1987 | 8.3 (Pv) | – | 7.6 (Rp) | 1.5 (Rp) | 0 |
| 7–11 April 1987 | 5.5 (Pv) | 3.7 (Cf, Pv) | 4.1 (Rp, Mv) | 0.5 (Rp) | 1.5 (Mv, Pv) |
| 8–12 June 1987 | 8.8 (Pv, An) | 17.5 (Pv) | 1.5 (Rp, Cf, Ospp) | 0 | 4.1 (Rp) |
| 24–28 August 1987 | 3.9 (Pv) | 11.2 (Pv) | 1.0 (Pv) | 7.7 (Pv) | 1.0 (Rp, Ospp) |
| 12–16 October 1987 | 3.2 (Pv) | 2.5 (Pv) | 2.6 (Pv, Mm) | 2.6 (An) | 2.0 (Rp, An) |
| 7–11 December 1987 | – | – | 1.0 (Rp, Cf) | 3.1 (An) | 7.1 (An) |
| 8–12 February 1988 | – | – | 7.1 (Rp, An, Mm) | 7.7 (An) | 7.1 (An) |
| 13–17 June 1988 | – | – | 21.9 (Rp, An) | 13.8 (An) | 25.5 (An) |
| 3–7 October 1988 | – | – | 15.3 (Pv, Rp) | 13.3 (An) | 21.9 (Rp, An) |
| 9–13 January 1989 | – | – | 14.8 (Rp) | 18.4 (An) | 27.0 (Rp) |
| 10–14 April 1989 | – | – | 28.6 (Rp) | 22.4 (An) | 40.8 (Rp) |
| 17–12 July 1989 | 12.2 (Pv) | 11.3 (Pv) | 5.1 (Rp) | 8.7 (An) | 29.1 (Rp) |
| 16–20 October 1989 | 5.0 (Pv) | 8.7 (Pv) | 12.8 (Rp, Pv) | 5.1 (An) | 6.1 (Rp) |

[a] Average of relative abundance in the last year before the burn.
[b] No. of trap nights = 980.

**Table 10.5.** The total number of seeds which germinated during the winter before the fire in each of eight treatments on two permanent trapping sites and the average germination (%) within each exclosure treatment, on each site (see Table 10.2 for details)

| | | Rodents excluded | | Rodents not excluded | | Site mean |
|---|---|---|---|---|---|---|
| Site | Substrate | Sown | Planted | Sown | Planted | |
| 1 | Clean | 15.0 ± 1.4 | 6.0 ± 3.0 | 0.0 | 0.0 | |
| | Litter | 8.0 ± 1.5 | 8.0 ± 0.8 | 0.0 | 0.0 | |
| Mean germination | | 70.0 | | 0.0 | | 35.0 |
| 2 | Clean | 6.7 ± 1.5 | 7.2 ± 1.1 | 1.0 ± 0.8 | 6.7 ± 2.1 | |
| | Litter | 6.6 ± 1.3 | 7.5 ± 0.6 | 2.7 ± 2.4 | 3.0 ± 1.4 | |
| Mean germination (%) | | 71.7 | | 33.3 | | 52.5 |

**Table 10.6.** Hierarchical log-linear analysis of factors affecting germination frequencies in a post-fire exclosure experiment. Only significant factors ($P < 0.05$) are shown here

| Factor affecting germination | Degrees of freedom | $\chi^2$ |
|---|---|---|
| Protection from rodents | 1 | 21 |
| Distance from refuge | 6 | 119 |
| Time after planting | 5 | 676 |
| Distance and time interaction | 30 | 99 |

There was no evidence that population recovery occurred as a result of slow immigration from outside the burnt area. On site GR3 (set out to detect immigration) there was no relationship between the distance from the edge of the burn at which animals were caught and the time after the fire.

## 10.3.3 The Effect of Small Mammals on Plant Regeneration

### 10.3.3.1 Factors Affecting Plant Establishment in Mature Fynbos

Small mammals limited the regeneration of proteas in mature fynbos. In the pre-fire experiment, germination was higher inside exclosures than outside (Table 10.5), but was not affected by any other factor. Seed removal was complete only on site SB1 which also had the highest density of small mammals. Of the 210 seeds that germinated on sites SB1 and SB2, 55% survived their first year; survival was independent of any of the treatment factors. Seventy to 90% of the seedlings died after wilting due to water stress.

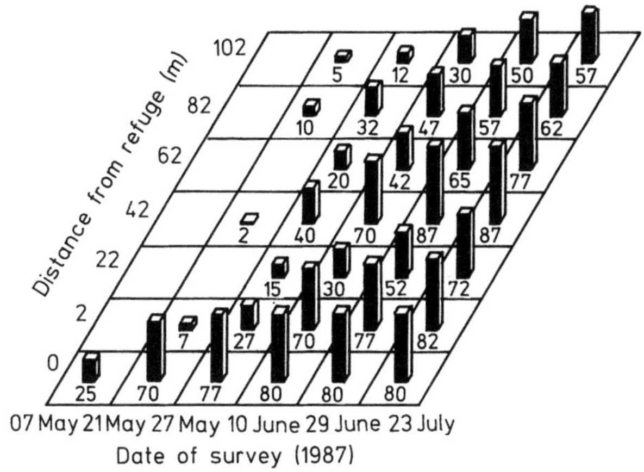


a

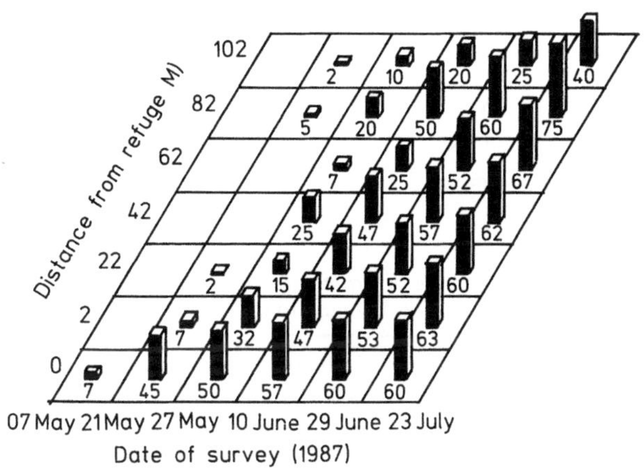


b

**Fig. 10.7.** The percentage germination of *Protea neriifolia* seeds at various stages after planting after fire, and at varying distances from a refuge site for small mammals in Swartboskloof. The treatments were **a** not accessible to small mammals and **b** accessible to small mammals

#### 10.3.3.2 Factors Affecting Plant Establishment in Burnt Fynbos

The effect of small mammals on *Protea* regeneration was smaller in burnt than in unburnt fynbos. Germination success in the post-fire experiment was 72%. Protection from rodents resulted in 20% more seeds germinating by the end of the trial (Fig. 10.7, Table 10.6). Seeds germinated sooner after planting in the refuge site (a low forest, see Fig. 10.7) and there was a decrease in germination as distance from the refuge increased (Table 10.6), although this had almost disappeared by the final survey. There was no difference in the effect of exclosures in different months or at varying distances from the refuge.

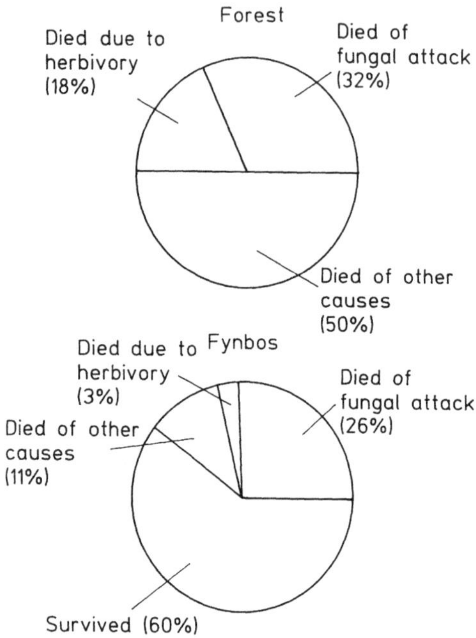


**Fig. 10.8.** The fate of *Protea neriifolia* seedlings germinating after a fire in forest (unburnt) and fynbos (burnt) vegetation in Swartboskloof

**Table 10.7.** Numbers of *Protea neriifolia* seedlings during the first year after a fire, inside and outside exclosures designed to exclude rodents in Swartboskloof

| Number of seedlings | Rodents excluded | Rodents not excluded | Total number of seedlings |
|---|---|---|---|
| Surviving | 120 | 86 | 206 |
| Dead | 94 | 103 | 197 |
| Total | 214 | 189 | 403 |
| Percentage germination | 76.4 | 67.5 | 72 |

*Protea neriifolia* seedlings which germinated in the forest refuge did not survive the first winter, whereas 60% of those germinating in fynbos did so (see also Chap. 6). The pathogenic fungus *Colletotrichum gloeosporioides* was the most important cause of death both inside and outside the forest (Fig. 10.8). Wilting without obvious fungal attack was the most important cause of death in the pre-fire trials. Herbivory was also important, but there was no difference between exclosure seed and control seed and it is likely that this was caused by insect herbivores. There was an improvement in first-year survival inside the exclosures (Table 10.7). The mean density of seeds on the soil 1 week after the fire was 340 $m^{-2}$, less than 3% of which were viable (D.C. Le Maitre, unpubl. data).

**Table 10.8.** Small mammal densities and species richness in relation to fire and precipitation from a variety of mediterranean scrublands. Peak densities were taken where seasonal data were available. Post-fire densities refer to the post-fire peak

| Site | Density (individuals per ha) | Mean annual rainfall (mm) | Species richness | Source |
|---|---|---|---|---|
| Swartboskloof | | | | |
| Pre-fire | 35 | 2328 | 9 | This study |
| Post-fire | 80 | 2328 | 9 | This study |
| Australia | | | | |
| Pre-fire | 5 | 1390 | 5 | Fox et al. (1985) |
| Post-fire | 12 | 1390 | 2 | Fox et al. (1985) |
| Cederberg (South Africa) | | | | |
| >9 years | 36 | 939 | 8 | Botha (1988) |
| <5 years | 18 | 939 | 8 | Botha (1988) |
| Pyrenees (Spain) | | | | |
| Pre-Fire | 34 | 930 | 3 | van Hensbergen (1984) |
| Outeniqua (South Africa) | | | | |
| Pre-fire | 14 | 700 | 9 | Breytenbach (1987) |
| Post-fire | 26 | 700 | 4 | Breytenbach (1987) |
| Swartberg (South Africa) | | | | |
| Post-fire | 28 | 662 | 4 | Breytenbach (1982) |
| Swartberg (South Africa) | | | | |
| Pre-fire | 16 | 450 | 6 | Breytenbach (1987) |
| Post-fire | 23 | 450 | 6 | Breytenbach (1987) |
| California (North America) | | | | |
| Pre-fire | 8 | 289 | 8 | Fox et al. (1985) |
| Post-fire | 14 | 289 | 4 | Fox et al. (1985) |

## 10.4 Discussion

### 10.4.1 Post-Fire Development of Small Mammal Communities

Populations of small mammals fell immediately after the fire and remained low for 6 to 9 months, before increasing to levels higher than those found before the fire. This pattern was similar to that found at other fynbos sites (Kruger and Bigalke 1984; Fox et al. 1985; Breytenbach 1987), but recovery was faster in Swartboskloof. The post-fire densities in Swartboskloof were highest in the second year after the fire and started to fall in the third. This is possibly a reflection of the rapid vegetation recovery (Fox et al. 1985) under the much wetter conditions in Swartboskloof than at the other sites. Under these conditions, vegetation canopy cover in the >500 mm layer would have recovered relatively rapidly, leading to a reduction in those small mammal species negatively correlated with this predictor (Table 10.3).

The absolute peak densities recorded in Swartboskloof before and after the fire are equal to or higher than those reported for most other sites in the fynbos and other mediterranean-type ecosystems (Table 10.8). The data suggest that small mammals densities are related to rainfall, as densities increase with increasing rainfall (Table 10.8). The data from Australia do not fit the pattern well ($R^2 = 0.38$ and 0.64 with and without Australian data respectively), as small mammal numbers remain low despite fairly high rainfall (Table 10.8). This supports the contention (Fox et al. 1985) that Australian ecosystems are different.

Breytenbach (1987) was able to show a consistent pattern of community succession after fire for a variety of sites. In Swartboskloof there were differences in the pattern of succession between sites. Sites with large numbers of *A. namaquensis* before the fire behaved differently to other sites. In sites where *A. namaquensis* was the dominant small mammal, all other species were rare after the fire. Other sites had more *M. minutoides* and/or more *P. verreauxi* after the fire, and *Rhabdomys pumilio* was common in these sites both before and after the fire.

### 10.4.2 Factors Affecting Habitat Selection by Small Mammals

Small mammals appear to select habitats on the basis of vegetation structure, and successional differences between sites can be explained in terms of the structure of the regenerating vegetation. *Aethomys namaquensis* numbers were highest in areas of sparse ground cover with a denser canopy at 0.5–0.7 m, on sandstones. All other species preferred dense cover at ground level or slightly above ground. This is in broad agreement with findings elsewhere in the fynbos biome for the small mammal species concerned (Bond et al. 1980; Botha 1989).

We found that small mammal communities dominated by *A. namaquensis* in Swartboskloof were little affected by the fire, presumably because the vegetation structure of areas selected by *A. namaquensis* was little affected by the fire. Site GR3 had an open understorey before the fire (McDonald 1985). After the fire this community was slowest in developing cover and therefore maintained an open understorey. After the fire the dominant plant species in this site were different from before the fire, suggesting that the habitat selection is based on physical structure and not plant species. The vegetation cover on the other site (GR2 burnt), which had a large population of *A. namaquensis* before the fire, recovered more quickly than GR3 following the fire and this site then had equal numbers of *A. namaquensis* and *R. pumilio*.

### 10.4.3 Do Small Mammals Limit *Protea* Regeneration from Seed?

There were large differences between pre- and post-fire seed removal rates. Granivores removed 50–100% of seeds in the pre-fire experiment but only 16% after the fire. If this is a true reflection of the general level of seed consumption in the area, then it shows the need for plants to store 4 or 5 years supply of seed on the plant, as a single year's crop could be consumed in its entirety. These results confirm the importance of serotiny as a mechanism for escaping granivory in fire-prone systems (Gill 1981c; Bond 1984; O'Dowd and Gill 1984; Bond and Breytenbach 1985). The numbers of seeds removed by granivores did not differ between lines at distances up to 100 m from the nearest unburnt refuge, reflecting the ability of small mammals to move freely over these distances. These results differ from those in Californian chaparral where harvester ants, rather than small mammals, removed seeds after fire (Mills and Kummerow 1989).

Seed tray experiments in Swartboskloof (Fraser 1990) indicated that the level of rodent granivory was higher in areas that had been more open before the fire. These areas were possibly open because of granivory after the preceding fire, and future granivory could perpetuate the situation. Seeds of alien *Acacia* spp. in fynbos are eaten by *R. pumilio* (Holmes 1990); up to 2300 seeds $m^{-2} yr^{-1}$ may be consumed. Soil seed banks only became large at very high levels of canopy closure (>72%). At lower levels of seed production, rodents can presumably consume most of the annual crop. Holmes concludes that in the early stages of invasion by *Acacia* spp. it is only the burial of seeds by ants that prevents rodents eating the entire seed supply. In Australian heathlands rodents were responsible for removing a small fraction (3% when ants excluded) of seeds in seed dish experiments, whereas ants removed 37% when rodents were excluded (Hughes and Westoby 1990). They conclude that (unlike the situation in South Africa) myrmecochory in Australia is a mechanism to prevent granivorous ants, rather than granivorous small mammals, from destroying the seed.

We have confirmed that rodent granivory can be an important cause of seed mortality in mature fynbos, and that it also occurs to a lesser extent after fire. In this experiment the density of seeds with embryos was about $10 m^{-2}$ one week after the fire. Given the levels of seed consumption reported by Holmes (1990), it seems unlikely that granivore satiation on its own is a sufficient explanation for the evolution of serotiny in fynbos. Large-scale seed release from cone stores, combined with the low numbers of granivores occurring immediately after the fire (this study and Midgley and Clayton 1990), and the temporary restriction of some small mammals to refuge habitats does provide a window of opportunity for successful seed germination. An earlier release of seed following a spring fire would expose seed to the reduced but still significant number of rodent granivores for a longer period leading to a decrease in reproductive success. In addition, population recovery amongst the small mammals would increase the levels of granivory to which seeds are exposed.

Studies in chaparral have shown that small mammals were a major cause of seedling mortality (Mills 1986) in the post-fire environment and were able to alter the balance between *Ceanothus greggii* and *Adenostoma fasciculatum* by selectively browsing the former. Fuentes and Echegaray

(1983) found similar selective defoliation in Chilean matorral. This is in contrast with Bonds' (1984) study in the southern Cape, where herbivores were responsible for the death of seedlings in the pre-fire environment but not in the post-fire environment. The major cause of seedling death before the fire in Swartboskloof was wilting, presumably due to water stress, whereas the major cause of death after fire was fungal attack. Herbivory was relatively unimportant and was as likely to have been herbivory by insects as small mammals. Grey kangaroos were an important cause of seedling death in Australia of regenerating *Banksia* spp. (Cowling and Lamont 1987). Small mammals appear to be more important in causing seedling death in post-fire chaparral and matorral than in fynbos or Australian kwongan.

## 10.5 Conclusions

The studies reported here have shown that small mammal communities were little affected by fire at Swartboskloof. Species richness did not change after the fire, but numbers were reduced for about 6 months after the fire, presumably due to changes in habitat structure. Although small mammals appear to limit the establishment of serotinous shrubs after fire, their numbers are sufficiently reduced after fire to create a window of opportunity in which these shrubs can escape granivory, germinate and establish. The fire in Swartboskloof was in March, and there was a relatively short time before the cold winter conditions triggered germination. A fire in September would not only have allowed a longer exposure of seeds to rodent granivores, but would also have allowed more time for the vegetation to recover, thus providing conditions suitable for higher numbers of rodents to return to the area.

*Acknowledgements.* We thank Peter Rohloff, Gordon Hockey, Pieter Badenhorst and Jan van Niekerk for ably and reliably handling most of the small mammal trapping. Gordon Hockey, Gregory Bailey and Adrian Simmers assisted with the laying out of exclosures and planting and monitoring of *P. neriifolia* seeds and seedlings. Dr. Sharon von Broembsen kindly identified the fungus on *Protea neriifolia* seedlings.

# 11 The Influence of Fire, Vegetation Age and Argentine Ants (*Iridomyrmex humilis*) on Ant Communities in Swartboskloof

A.E. De Kock, J.H. Giliomee, K.L. Pringle, and J.D. Majer

## 11.1 Introduction

In terms of numerical abundance and species richness, ants (Hymenoptera: Formicidae) are prominent in many terrestrial ecosystems (Majer 1983; Stork 1988). The total of 8800 known species of the family Formicidae is estimated to make up about 10–15% of the world's animal biomass (Hölldobler and Wilson 1990). Some of the reasons for their success lie in the fact that they have a competitive advantage over many other invertebrates because they are omnivorous (Beattie 1985), they can store food (Petal 1978) and they have effective defence mechanisms. Furthermore, they are of considerable ecological importance. In mutual ant-plant relationships they derive food and nest sites from plants, they disperse and store seeds, they protect foliage, buds and reproductive structures from folivores and seed predators, they may fertilize plants with essential nutrients, and sometimes they function as pollinators (Beattie 1985). In the fynbos, ants play a major ecological role, particularly in dispersing seeds. About 1300 (20%) of fynbos plant species are dependent on ants for the dispersal of their seeds, a process known as myrmecochory (Milewski and Bond 1982; Bond and Slingsby 1983). In this process ants, attracted to these seeds by fatty, edible coatings or appendages (known as elaiosomes), carry seeds into their nests. The seeds then benefit in that, apart from being dispersed from the mother plant, they are protected from rodents and other seed predators (O'Dowd and Hay 1980; Bond and Slingsby 1984; Bond and Breytenbach 1985). The removal of the elaisome and handling of the seeds by ants may serve as germination stimuli, and those seeds left in the nests after the elaiosomes are removed may benefit from the nutrient rich soil (references in Beattie 1985).

Ants are sensitive to disturbance and it has been suggested that their species composition and abundance may be used as an index of the conservation status of natural systems (Fox 1981; Majer 1983; Majer et al. 1984). One of the possible factors influencing the ant fauna of the fynbos is the presence of the introduced Argentine ant, *Iridomyrmex humilis*. This

**Table 11.1.** Details of four sites in Swartboskloof at which ant studies were conducted

| | Site numbers 1 | 2 | 3 | 4 |
|---|---|---|---|---|
| Post-fire age (yrs) | 6 | 29 | 29 | 8.5 |
| Argentine ants present? | Yes | Yes | No | Yes |
| Treatment in March 1987 | Burnt | Burnt | Burnt | Not burnt |
| Soil moisture[a] (%) | 14.2 | 13.3 | 12.8 | 13.1 |
| Insolation fraction[b] | 0.62 | 0.54 | 0.51 | 0.4 |
| Litter layer[c] (g $m^{-2}$) | 373.4 | 5079.8 | 5298.2 | 810.6 |

[a] Measured for the top 50 mm (mean of five measurements in 1987).
[b] Pre-fire insolation fraction at ground level, measured using a tube solarimeter (under vegetation canopy/above vegetation canopy).
[c] Oven-dried mass of leaf and twig litter.

ant species has been reported to displace some indigenous ant species, including those involved in myrmecochory (Slingsby 1982; Bond and Slingsby 1984). The extent of the Argentine ant invasion in the southwestern Cape fynbos was surveyed by De Kock and Giliomee (1989) who showed that it is associated with disturbance by humans and positively correlated with areas which have access by road.

Another disturbance factor is fire, a common natural phenomenon in fynbos, and also used in management (Kruger 1977b). The influence of fire on ants may be either direct (when the ants are killed by fire), or indirect through its impact on the structure of the vegetation and the microenvironment. Until now there have been little data from the fynbos to assess the impact of fire on the composition of ant communities. In fact, very little is known about the distribution and composition of ant communities in different habitats in fynbos, and the available data are from a few isolated sites (Slingsby and Bond 1981; Donnelly and Giliomee 1985a,b; Koen and Breytenbach 1988).

In this study we determined the effects of various disturbance factors on indigenous ant communities in Swartboskloof. We examined whether fire, vegetation age and the presence of the Argentine ant had any effect on ant community composition and abundance.

## 11.2 Methods

### 11.2.1 Locality of Study Sites

Ants were studied on four sites in Swartboskloof (Table 11.1). Sites 2 and 3 were in 29-year-old tall proteoid fynbos (Fig. 1.3) which was partly senescing. Another site (site 1) was in a 6-year-old firebreak which consisted mainly of low, resprouting plants, and site 4 was in 8.5-year-old fynbos. Argentine ants were present in three of the sites. The first three of these sites were in an area where a prescribed burn was to be carried out a year after the sampling commenced.

### 11.2.2 Extent and Rate of Invasion by Argentine Ants in Swartboskloof

We suspected that the Argentine ant would spread faster in areas with permanent sources of water such as riverbeds (Ward 1987), but this has never been verified for the fynbos. On 7 March 1988, a year after the fire, a study was conducted to determine whether this species will expand its territory faster in riparian vegetation than in adjacent fynbos, and also to determine the extent of the invasion in Swartboskloof. A sampling grid was established across the width of Swartboskloof, either side of the known limits of the distribution of the Argentine ant. A line of trapping points, at right angles to the line across the width of Swartboskloof, was established at 25-m intervals. Trapping points were 10 m apart, with five above and five below the line, giving a total of 345 points. The sampling grid kept this pattern most of its length except where it crossed a stream in the middle of the grid. Here, the grid was extended a further 50 m up the stream, for 50 m on either side of the stream. Baiting with cubes of meat was done at all trapping points within 3 h, to ensure that a change in weather would not influence the results. All the ants collected were identified to species level.

### 11.2.3 Ant Surveys

We used pitfall trapping, a method generally used to obtain a rapid census of the epigaeic (surface-active) invertebrate fauna and, to a lesser extent, those from the soil and litter (Majer and Greenslade 1988). Small pitfall traps, as described by Majer (1978) and Samways (1981) and modified by Donnelly (1983), were used. Each trap consisted of a test tube (25 × 150 mm) sunk into the soil within a plastic pipe so that the lip was flush with the soil surface. The test tubes were removed and the pipes corked when the traps were not in use. Each tube was filled to a depth of 30–40 mm with a mixture of seven parts 70% ethyl alcohol and three parts glycerol. The traps were spaced 2 m apart in four rows of four traps (i.e. 16 traps per site). Each trapping session extended over 4 days, after which all the ants collected were identified to species level and counted.

The traps were surveyed nine times over a 2-year period, five times during the first year, in March, November, and December 1986, and January and February 1987 and four times during the second, in May, August, and October 1987 and March 1988. No trapping took place during

**Table 11.2.** Temperatures and rainfall during a pitfall trap survey in Swartboskloof. The data were measured at the Swartboskloof weather station (see Chap. 2)

| Date | Temperature (°C) | | | | | Rain (mm) |
|---|---|---|---|---|---|---|
| | Mean | Mean | Highest | Lowest | Range | |
| 22–26 March 1986 | 19.2 | 8.2 | 23.3 | 4.4 | 18.9 | 45.0 |
| 31 October–4 November 1986 | 20.1 | 7.7 | 24.3 | 5.9 | 18.4 | 34.0 |
| 15–19 December 1986 | 26.1 | 15.1 | 32.3 | 12.9 | 19.4 | 1.7 |
| 15–19 January 1987 | 29.0 | 12.4 | 30.3 | 9.2 | 21.1 | 0.0 |
| 16–20 February 1987 | 28.3 | 14.6 | 32.3 | 9.8 | 22.5 | 0.0 |
| 21–25 May 87 | 26.6 | 12.0 | 29.8 | 11.3 | 18.5 | 0.0 |
| 24–28 August 1987 | 19.5 | 5.8 | 23.5 | 4.3 | 19.2 | 0.0 |
| 15–19 October 1987 | 24.0 | 11.2 | 28.3 | 7.1 | 21.2 | 0.3 |
| 24–28 March 1988 | 28.5 | 11.9 | 30.4 | 9.8 | 20.6 | 0.0 |

mid-winter as the above-ground activity of ants decreases with decreasing temperature and few of the indigenous ant species are active at this time (De Kock 1983; Donnelly and Giliomee 1988). At the end of the first year, on 17 March 1987, three of the four sites (sites 1, 2 and 3) were burnt. The ant data were pooled by year and, because there were five surveys during the first and four during the second study year, the first year's data were standardized by multiplying by 0.8 in order to make them comparable with those of the second year. Weather conditions at the time of these surveys are shown in Table 11.2.

### 11.2.4 Data Analysis

In order to compare the similarity between sites before the burn, Sørensen's modified index of similarity was calculated. This index is given by:

$$C_N = 2jN/(aN + bN)$$

Where aN = total individuals collected in habitat a,
bN = total individuals collected in habitat b, and
jN = sum of the lesser values for the species common to both habitats.

A dendrogram showing the affinity of sites in terms of their species composition was then prepared from the matrix of similarity indices by the method described in Southwood (1978).

Correspondence analysis (Greenacre 1984) was used to compare the pre- and post-fire plot data in terms of the ant species present. This technique displays the plot and species data in a multi-dimensional space. This enables the user to elucidate the affinity between plots in terms of their species composition, and also provides a graphical means of elucidating which species are contributing to the separation of plots.

The diversity of the ant community was also calculated. Patil and Taille (1979) and Dennis et al. (1979) use the following expression to generate a series of diversity indices:

$$\Delta_\beta = \sum_{i=1}^{N} P_i(1 - P_i^\beta)/\beta, \quad i = 1, 2, \ldots\ldots N,$$

where $N$ = number of species,
$P_i$ = proportion of individuals in species $i$, and
$\beta$ = a measure of the weight that is attached to evenness.

They show that if $\beta$ is in the range $-1$ to $+1$, then for

$\beta = -1$, $\Delta_{-1} = N-1$, or the species count, and for
$\beta$ approaching 0, $\Delta_0 = -\Sigma P_i lnP_i$, or the Shannon-Weaver index, and for
$\beta = +1$, $\Delta_{+1} = 1 - \Sigma P_i^2$, or the Simpson's species evenness index.

In this study the above index was modified to standardize it to a scale between 0 and 1 as follows:

$$\Delta\beta = \frac{1}{\left[1 - \left(\frac{1}{T}\right)^\beta\right]/\beta} \sum_{i=1}^{T} P_i(1 - P_i^\beta)/\beta, \; i = 1, 2, \ldots T,$$

where T = the total number of species recorded during the study period.

The standardized indices were used to construct diversity profiles which enable direct visual comparison, between both sites and years, of species richness on the left side and species evenness on the right side of the graph. The relative contribution of dominant as well as intermediate species can also be visually compared between sites by consulting the profiles.

## 11.3 Results and Discussion

### 11.3.1 Extent and Rate of Invasion by Argentine Ants in Swartboskloof

Except for the two dominant ants, *Iridomyrmex humilis* and *Anoplolepis custodiens*, 13 other ant species were collected on the grid designed to establish the extent of Argentine ant invasion. These species are not considered further here, because the way they were interspersed did not show any pattern in relation to either of the dominant ants. The Argentine ant spread more rapidly in riparian vegetation than in adjacent fynbos (see Fig. 11.1), and it successfully displaced the pugnacious ant (*A. custodiens*) as it enlarged its territory. The extent of the Argentine ant invasion will allow for future assesment of the rate of invasion.

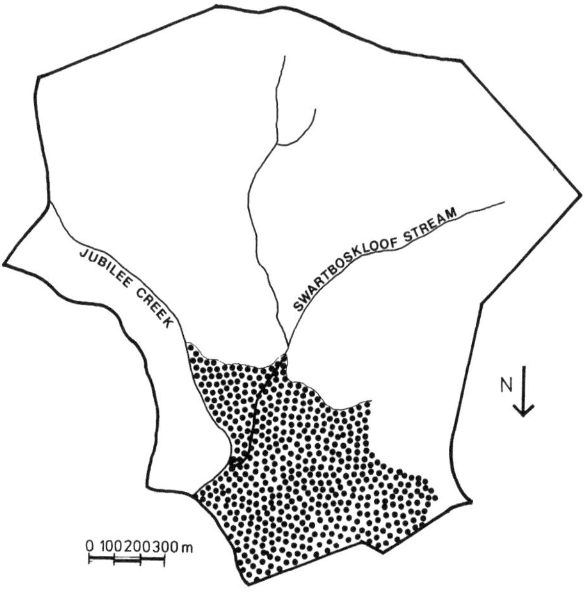


**Fig. 11.1.** Distribution of Argentine ants (*shaded*) in Swartboskloof in 1988. Note the upward extension of the distribution along the Swartboskloof stream. Areas to the *left and right of the shaded area* were not sampled, but may also be invaded by Argentine ants

## 11.3.2 Ant Species Collected

The combined catches on four sites over 2 years yielded 33 ant species. The ant species collected are listed separately in Table 11.3 for each of the four sites, both for the first and the second year. Of the species collected, 70% belonged to the Myrmicinae and 18% to the Formicinae. Other subfamilies were poorly represented. The genera *Monomorium* (eight species) and *Tetramorium* (seven species) contained the highest number of species, although some species were only represented by a single individual. Working in the same valley, Donnelly and Giliomee (1985a) also found that these were the best represented genera in terms of number of species. The most numerous species caught was the Argentine ant, *Iridomyrmex humilis*, with *Meranoplus peringueyi* and *Monomorium willowmorense* respectively being the second and third most abundant species. By contrast, Donnelly and Giliomee (1985a) found that, apart from the Argentine ant, *Anoplolepis custodiens* and *Pheidole capensis* were the most abundant species. This difference may be ascribed to the presence of *I. humilis*, as Donnelly and Giliomee only found the Argentine ant at one locality, whereas it was found at three sites in the present study.

The Argentine ant was the only abundant ant at site 2. Furthermore, at the three sites invaded by the Argentine ant (Table 11.1), the abundance of

**Table 11.3.** The abundance (numbers of individuals) of ant species collected in 2 years of pitfall trapping in Swartboskloof (1 year pre-fire and 1 year post-fire)

| Ant species | Abundance in four equivalent sampling periods | | | | | | | |
|---|---|---|---|---|---|---|---|---|
| | Pre-fire[a] | | | | Post-fire | | | |
| | Site number | | | | Site number | | | |
| | 1 | 2 | 3 | 4 | 1 | 2 | 3 | 4 |
| *Acantholepis capensis* Mayr (Formicinae) | 1 | 0 | 10 | 12 | 0 | 0 | 10 | 0 |
| *Anoplolepis custodiens* (Smith) (Formicinae) | 0 | 0 | 2 | 0 | 0 | 0 | 0 | 0 |
| *Camponotus maculatus* Fabricius (Formicinae) | 1 | 0 | 9 | 6 | 2 | 2 | 19 | 6 |
| *Camponotus niveosetosus* Mayr (Formicinae) | 2 | 0 | 2 | 10 | 1 | 0 | 7 | 2 |
| *Crematogaster liengmei* Forel (Myrmicinae) | 0 | 0 | 2 | 0 | 0 | 0 | 0 | 0 |
| *Crematogaster peringueyi* Emery (Myrmicinae) | 0 | 0 | 26 | 10 | 0 | 0 | 0 | 0 |
| *Dorylus helvolus* Linnaeus (Dorylinae) | 0 | 0 | 0 | 0 | 1 | 0 | 0 | 0 |
| *Hypoponera spei* Forel (Ponerinae) | 6 | 0 | 0 | 5 | 1 | 0 | 0 | 0 |
| *Iridomyrmex humilis* Mayr (Dolichoderinae) | 146 | 106 | 0 | 747 | 101 | 83 | 0 | 93 |
| *Leptogenys nitida* (Smith) (Ponerinae) | 0 | 0 | 1 | 0 | 0 | 0 | 0 | 0 |
| *Meranoplus peringueyi* Emery (Myrmicinae) | 0 | 0 | 0 | 242 | 0 | 0 | 0 | 128 |
| *Monomorium braunsi* Mayr (Myrmicinae) | 0 | 6 | 2 | 1 | 0 | 0 | 0 | 0 |
| *Monomorium lubricum* Arnold (Myrmicinae) | 0 | 0 | 0 | 0 | 0 | 0 | 5 | 0 |
| *Monomorium musicum* Forel (Myrmicinae) | 0 | 0 | 0 | 0 | 16 | 12 | 23 | 1 |
| *Monomorium nuptialis* Forel (Myrmicinae) | 6 | 9 | 6 | 0 | 0 | 0 | 0 | 0 |
| *Monomorium schultzei* Forel (Myrmicinae) | 0 | 0 | 0 | 0 | 1 | 9 | 3 | 1 |
| *Monomorium springvalense* Forel (Myrmicinae) | 0 | 2 | 4 | 4 | 0 | 0 | 10 | 0 |
| *Monomorium torvicte* Bolton (Myrmicinae) | 0 | 0 | 1 | 0 | 0 | 0 | 0 | 0 |
| *Monomorium willowmorense* Forel (Myrmicinae) | 122 | 0 | 7 | 76 | 51 | 1 | 3 | 55 |
| *Ocymyrmex cilliei* Prins and Roux (Myrmicinae) | 3 | 0 | 0 | 2 | 7 | 1 | 0 | 4 |
| *Pheidole foreli* Mayr (Myrmicinae) | 1 | 0 | 76 | 0 | 0 | 0 | 0 | 0 |
| *Pheidole tenuinodes* Mayr (Myrmicinae) | 0 | 0 | 8 | 0 | 0 | 0 | 0 | 0 |
| *Plagiolepis jouberti* Forel (Formicinae) | 0 | 0 | 0 | 0 | 0 | 0 | 2 | 0 |
| *Plagiolepis pygmaea* Latreille (Formicinae) | 2 | 0 | 6 | 5 | 3 | 0 | 1 | 0 |

**Table 11.3** (continued)

| Ant species | Abundance in four equivalent sampling periods | | | | | | | |
|---|---|---|---|---|---|---|---|---|
| | Pre-fire[a] | | | | Post-fire | | | |
| | Site number | | | | Site number | | | |
| | 1 | 2 | 3 | 4 | 1 | 2 | 3 | 4 |
| *Smithustroma emarginata* Mayr (Myrmicinae) | 5 | 0 | 0 | 0 | 1 | 0 | 0 | 0 |
| *Tetramorium bothae* Forel (Myrmicinae) | 10 | 0 | 0 | 4 | 18 | 5 | 1 | 3 |
| *Tetramorium capense* Mayr (Myrmicinae) | 1 | 0 | 6 | 4 | 0 | 0 | 9 | 0 |
| *Tetramorium emeryi* Mayr (Myrmicinae) | 0 | 0 | 0 | 1 | 0 | 0 | 0 | 0 |
| *Tetramorium flaviceps* Arnold (Myrmicinae) | 29 | 3 | 3 | 14 | 12 | 0 | 0 | 4 |
| *Tetramorium grassi* Emery (Myrmicinae) | 1 | 0 | 1 | 0 | 0 | 0 | 0 | 1 |
| *Tetramorium lobulicorne* Santschi (Myrmicinae) | 0 | 0 | 1 | 0 | 0 | 0 | 0 | 0 |
| *Tetramorium quadrispinosum* Emery (Myrmicinae) | 8 | 0 | 0 | 0 | 6 | 1 | 0 | 0 |
| *Tetramyrma* sp. (Myrmicinae) | 0 | 0 | 0 | 0 | 0 | 0 | 0 | 1 |
| Total numbers | 344 | 126 | 173 | 1143 | 221 | 114 | 94 | 293 |
| Total species | 16 | 5 | 19 | 16 | 14 | 8 | 13 | 12 |

[a] Five sampling periods standardized by multiplying with 0.8.

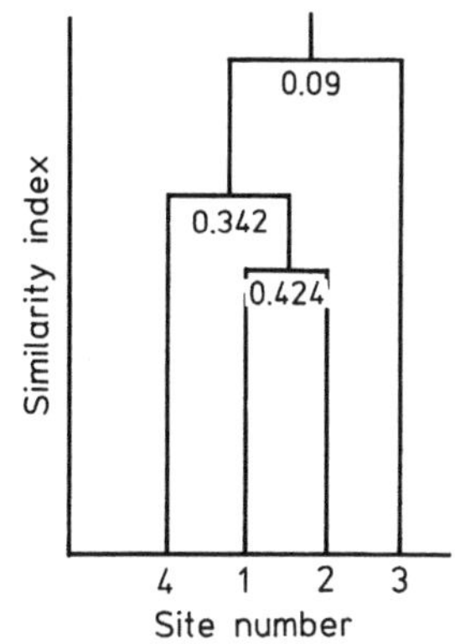


**Fig. 11.2.** Similarity grouping of pre-fire ant communities at four sites (see Table 11.1) in Swartboskloof, based on Sorensen's adjusted similarity index

indigenous ant species differed from that at the uninvaded site. In the sites with Argentine ants, the pugnacious ant, *A. custodiens*, was displaced, as has been reported in earlier studies (Slingsby 1982; Bond and Slingsby 1984). The indigenous ants, *M. willowmorense*, and to a lesser degree *Tetramorium flaviceps* at site 1, and *M. peringueyi* and *M. willowmorense* at site 4, coexisted successfully with the Argentine ant. Over the 2-year study period they contributed 31, 7, 25 and 19% respectively to total ant abundance at these two sites. The abundance pattern differed between years at uninvaded site 3. During the first year, *Pheidole foreli* and *Crematogaster peringueyi* were the most abundant species, and during the second year their position in the numerical hierarchy was replaced by *Monomorium musicum* and *Camponotus maculatus*. At site 3, these four species contributed 31, 11, 7 and 6% respectively to overall ant abundance for the years in which they were dominant.

From the above it appears that the relative abundance of the indigenous species in the sites invaded by the Argentine ant does not reflect the situation in pristine fynbos areas. This is also borne out by the fact that only two individuals of *A. custodiens*, the most numerous ant species collected by Donnelly and Giliomee (1985a,b) in the Jonkershoek Valley, were caught during this study.

### 11.3.3 Similarity of Ant Species Composition

The composition of the ant fauna at uninvaded site 3 differed substantially from the other three sites before the fire (Fig. 11.2). Although differing in vegetation age and amount of litter (Table 11.1), the other three sites had relatively similar ant faunas. This separation of site 3 from the others is supported by a plot of the first two axes of the correspondence analysis (Fig. 11.3). Regardless of whether the sites had been burnt or not, site 3 is well to the right of the centroid in Fig. 11.3, whereas all the others are to the

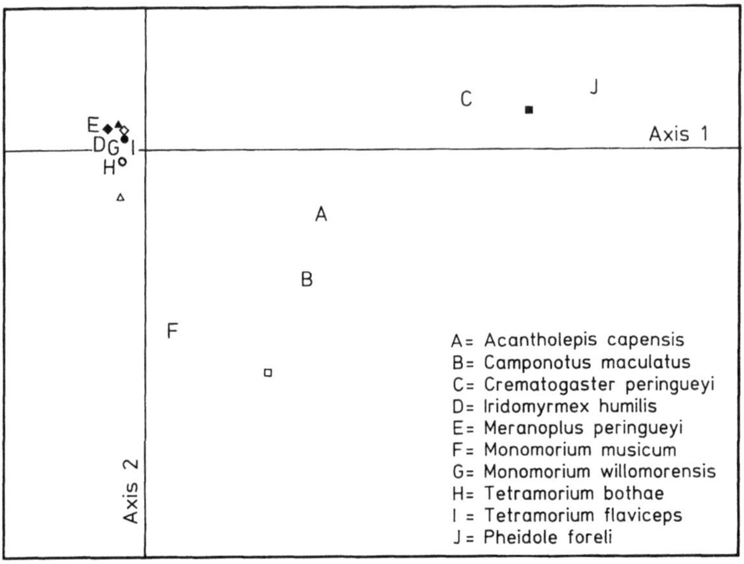


**Fig. 11.3.** Sites and ant species on the first two axes of a correspondence analysis for four sites (Table 11.1) in Swartboskloof. The ten most prevalent ant species are indicated by *letters* (see *inset*). The sites are: ● = site 1 pre-fire; ○ = site 1 post-fire; ▲ = site 2 pre-fire; △ = site 2 post-fire; ■ = site 3 pre-fire; □ = site 3 post-fire; ⧫ = site 4 pre-fire; ◇ = site 4 post-fire

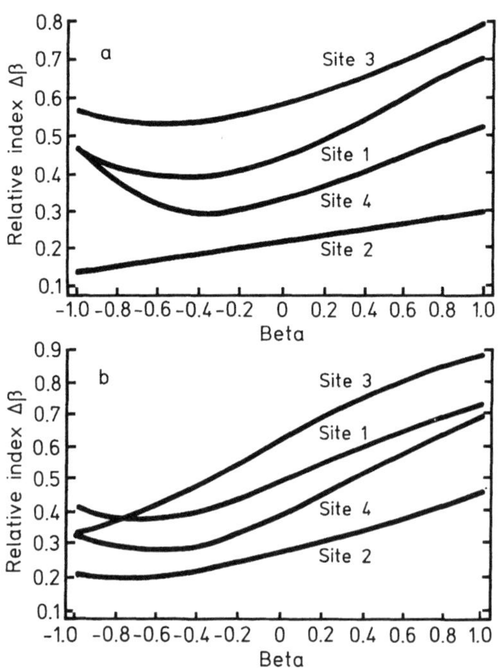


**Fig. 11.4.** Ant diversity profiles of **a** pre-fire and **b** post-fire ant communities at four sites in Swartboskloof (see text)

**Table 11.4.** A comparison between four ant study sites in Swartboskloof in terms of the relative indices for species richness and evenness (see text) 1 year before and 1 year after fire

| Site | Relative indices for species richness | | Relative indices for species evenness | |
|---|---|---|---|---|
| | Pre-fire | Post-fire | Pre-fire | Post-fire |
| 1 | 0.47 | 0.41 | 0.70 | 0.74 |
| 2 | 0.13 | 0.19 | 0.30 | 0.45 |
| 3 | 0.56 | 0.34 | 0.79 | 0.88 |
| 4 | 0.47 | 0.31 | 0.54 | 0.70 |

left. It therefore appears that presence of Argentine ants, the common disturbance factor for sites 1, 2 and 4, governed the similarity between sites, rather than vegetation age or differences in the litter layer. This is further illustrated by the fact that sites 1 and 2, both of which contained Argentine ants, were most similar in terms of their ant fauna, despite having large differences in age of vegetation (Fig. 11.2). On the other hand, site 2 (with Argentine ants) and site 3 (without Argentine ants) had least similar ant fauna, although they had vegetation of the same age. The presence at site 3 of eight ant species not found at any of the other sites indicates that these species were displaced by the Argentine ant (Table 11.3).

The displacement of native ant faunas by alien species has also been reported from other parts of the world (e.g. Haskins and Haskins 1965; Crowell 1968; Fluker and Beardsley 1970; Erickson 1971; Lieberburg et al. 1975; Jackson 1980). Similarly, the displacement of indigenous South African ant species by the Argentine ant has been observed by Hattingh (1945), Skaife (1961), Prins (1978), Bond and Slingsby (1984), and Donnelly and Giliomee (1985a).

### 11.3.4 Ant Diversity Patterns

Site 2 (with Argentine ants) and site 3 (without Argentine ants) differed markedly with respect to their indices for species richness and evenness (Fig. 11.4; Table 11.4), despite the fact that before the fire, they had vegetation of the same age and similar values for soil moisture, insolation fraction and litter mass. In all plots, the species richness and evenness values were much lower where the Argentine ant was present, suggesting that the strongly dominant Argentine ant displaced some of the indigenous ant species. Of the 11 indigenous ant species which coexisted with the Argentine ant at site 2, none were common (Table 11.3). These ant species probably coexist with the Argentine ant by avoiding interspecific competition for

**Table 11.5.** Percentage contribution of the three most abundant ant species to the total number of ants trapped at two sites before and after a fire in Swartboskloof

| Site | Ant species | Pre-fire | Post-fire |
|---|---|---|---|
| 1 | *Iridomyrmex humilis* | 43 | 46 |
| | *Monomorium willowmorense* | 35 | 23 |
| | *Tetramorium flaviceps* | 8 | 5 |
| | Total (%) | 86 | 74 |
| 3 | *Pheidole foreli* | 44 | – |
| | *Monomorium musicum* | – | 25 |
| | *Crematogaster peringueyi* | 15 | – |
| | *Camponotus maculatus* | – | 20 |
| | *Acantholepis capensis* | 6 | 11 |
| | Total (%) | 65 | 56 |

food, by spatial or temporal segregation when foraging, or by their cryptic nature. According to Greenslade and Thompson (1981), some ant species coexist with *Iridomyrmex* through differences in size of worker, time of foraging or by their submissive or evasive behaviour.

### 11.3.5 Impact of Fire

The relatively high intercept on the left side of the diversity profiles for sites 1 and 3 before the fire (Fig. 11.4a), changed to lower intercepts for both sites after the fire (Fig. 11.4b). This indicates a reduction in the number of species at these sites after the fire. Examination of the relative abundance of the species at sites 1 and 3 showed a change in percentage contribution of the three most abundant ant species at both sites (Table 11.5). The cumulative percentage contribution of the three dominant ants decreased by 12 and 9% at sites 1 and 3 respectively. There was also a dramatic change in overall species composition at site 3. On this site a total of 24 species were recorded, of which only 7 were present both before and after the fire, while 12 were only recorded before and 5 only after the fire. This appears to support Fox's (1981) statement that ant succession is a replacement of groups of species. The correspondence analysis (Fig. 11.3) indicated that *C. peringueyi* and *P. foreli* were well represented before the fire and appeared to be associated with site 3. The former was recorded at both sites 3 and 4, but in much higher numbers at site 3. It was not recorded after the fire. Since it is an arboreal species, its disappearance at site 3 might have been a result of the fire. At both site 1 and site 2, fire had no effect on the Argentine ants' dominance over other ants, with this species maintaining its dominant position on both sites.

At site 4, which did not burn, the higher value for species evenness in the second year (Fig. 11.4a,b) was probably caused by the fact that the

Argentine ant lost some of its numerical dominance (Table 11.3). Whereas it was by far the most dominant ant in the first year, contributing 66% to the total catch, it only contributed 32% in the second year. It lost its numerical dominance to the second most abundant species, *M. peringueyi*. The corresponding percentage contribution of this species to the catch in the first and second years were 21 and 43% respectively. This change in dominance at site 4 may be due to the ability of *M. peringueyi* to defend its food source against the Argentine ant by using some kind of chemical repellent (De Kock 1990a). This may result in an unstable situation where dominance fluctuates between the two species.

It is not clear how much the post-fire vegetation age influenced ant species composition before the fire of 1987. Fire occurring in tall proteoid fynbos with a relatively thick litter layer, as was the case at sites 2 and 3, could be expected to produce more drastic environmental changes than in graminoid firebreak vegetation, such as at site 1. The reason for this is that the proteoid fynbos consists predominantly of reseeders which take longer to recover after a fire. By contrast, when the low graminoid vegetation of a firebreak is burnt, the environmental change is not so drastic and the predominantly sprouter vegetation takes less time to recover. However, this was not reflected in the composition of the ant communities, as those of sites 1 and 2 (firebreak and mature, both invaded) were more similar than those of sites 2 and 3 (invaded and uninvaded) (Fig. 11.2). The presence of the Argentine ant at sites 1 and 2 might have masked both the effect of vegetation age and the fire. It can therefore be concluded that, contrary to the situation in Australia where fire has an immediate and dramatic effect on the composition of the ant community (Andersen and Yen 1985), the presence of the Argentine ant had a more pronounced effect on the indigenous ant community than either fire or vegetation age.

*Acknowledgements.* We thank Dr. Henk Geertsema, Ms. Julie McDonald and Ms. Rosemary van Zyl of the Department of Entomology and Nematology at the University of Stellenbosch for their assistance, and Dr. André Prins, previously of the South African Museum, for identifications. Adrian Simmers and Gordon Hockey assisted with fieldwork. D.P. Bands, S.A. Botha, D.C. Le Maitre, P.T. Manders, D.M. Richardson and B.W. van Wilgen provided helpful advice and constructive criticism.

# 12 The Effects of Fire on Soil Water Repellency, Catchment Sediment Yields and Streamflow

D.F. Scott and D.B. van Wyk

## 12.1 Introduction

A wide range of hydrological responses to fire has been recorded (Anderson et al. 1976; Tiedemann et al. 1979). This is due to the varied fire and site characteristics encountered: season of burn, rate and amount of energy released, geological and soil factors, basin morphology, vegetation type and climate may all be important. Thus, a uniform response to fire should not be expected. Some of the findings of fire's effects on streamflow and sediment yields, at catchment and plot levels, are summarized in Appendices 12.1 and 12.2

The hydrological consequences of fire in mediterranean-type ecosystems have often been dramatic. An early assessment of the effects of fire in Californian chaparral showed that total water yield, peak discharge and stormflows increased after fire, and were associated with accelerated soil erosion and sedimentation (Hoyt and Troxell 1934). This fire-flood-erosion sequence in mountainous chaparral is now an established phenomenon (Anderson et al. 1976). Similar flooding and erosion following summer wildfires in Australian eucalypt forests have been recorded (Brown 1972; Leitch et al. 1983). Fire in fynbos in the Abdolskloof catchment (Fig. 1.1) caused sharp (though short-lived) increases in stormflow, peak discharge and total streamflow in the first 3 months of the wet season following fire (Rycroft 1947; Banks 1964).

Such hydrological response to fire has been linked to fire-induced water repellency in the soil. Soils are usually completely wettable; water repellency is an abnormality in soils resulting from the coating of soil particles with hydrophobic organic substances which reduce the attraction between soil particles and water (DeBano et al. 1967). The source of these substances is undecomposed and decomposed plant materials (Savage et al. 1972). The time of contact between plant litter and soil is related positively to the degree of repellency shown by the soil (Teramura 1980). Water-repellent soils inhibit water infiltration and/or percolation so that overland flow may result (DeBano 1971).

Fire-induced water repellency in soils has been observed mostly in chaparral in southern California (Osborn et al. 1964; Krammes and DeBano 1965; DeBano and Krammes 1966; DeBano et al. 1970, 1979b), but also in Oregon (Dyrness 1976), Arizona (Campbell et al. 1977), Australia (Leitch et al. 1983) and South Africa (Scott and van Wyk 1990).

Concern existed that fires in the over-mature fynbos at Swartboskloof could lead to water repellency in the soils, as fuel loads were quite high (see Chap. 3), soils were fairly coarse textured, dry, and supported a large amount of organic litter and humus. The litter had been in contact with the soil for over 29 years, the time since the last fire. All these factors predispose a soil to hydrophobic behaviour (DeBano 1981).

In this chapter, we examine the hypothesis that fire would induce the development of water repellency in the soil, which would increase the probability of overland flow, causing increased stormflows and water yield, and would be associated with increased sediment yields. Investigations were conducted at three different scales: (1) streamflow, stormflow and sediment yield were analyzed to assess the response of the whole catchment to fire; (2) rain-generated overland flow was measured on small plots as a realistic measure of infiltration rate; and (3) at the finest scale, soils were tested for water repellency before and after the fire.

## 12.2 Methods and Results

### 12.2.1 Studies at the Whole Catchment Level

#### 12.2.1.1 Effects of Fire on Streamflow

The paired catchment approach was used to test for the effect of fire on streamflow. This requires the pre-treatment calibration of the catchment to be treated (in this case Swartboskloof) against a similar, but untreated, control catchment. The calibration relationship is then used as a baseline against which possible treatment effects can be measured (Wicht 1967b). Streamflow in Swartboskloof did not correlate well with that in potential control catchments (Fig. 12.1), but Tierkloof was selected as the most suitable control. Relative to Tierkloof, the dry season streamflow in Swartboskloof is much lower and the wet season flows are much greater.

Streamflow was monitored continuously using Belfort water-level recorders on compound V-notch weirs. Streamflow monitoring of Swartboskloof started in April 1985. Streamflow was totalled over fortnightly periods which smoothed the differences between streamflow in the two catchments seen over shorter periods.

The effects of fire on stormflow were assessed from storm hydrographs, which were separated using the standard method of Hewlett and Hibbert (1967). The variables stormflow (the total volume of runoff over

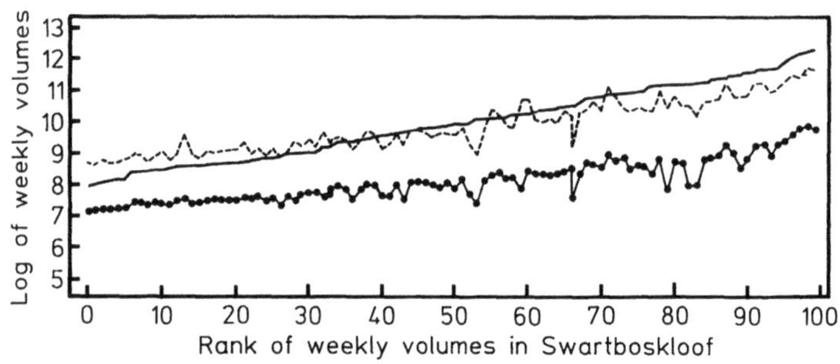


**Fig. 12.1.** Weekly streamflow volumes (log values) in Swartboskloof (—), Tierkloof (----) and Lambrechtsbos-B (●) against the rank of the Swartboskloof volumes. This shows the seasonally changing relationship between Swartboskloof and the control catchments. See Fig. 1.1 for localities of control catchments

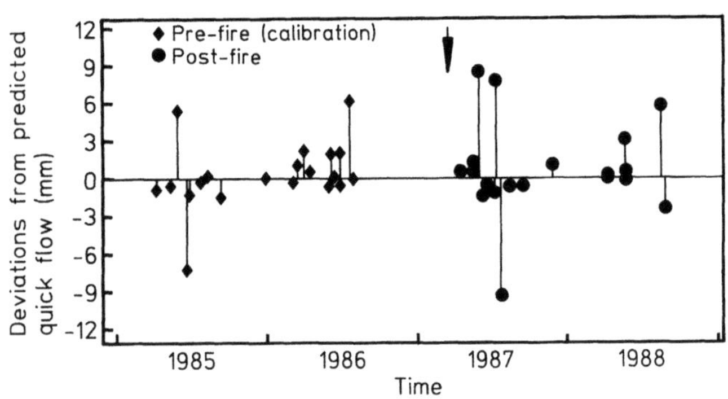


**Fig. 12.2.** Differences between observed quickflow volumes from the Swartboskloof catchment, and predictions from the calibration equation (see text). *Arrow* shows the date of fire

the period of stormflow), quickflow [that portion of stormflow above the fixed-slope separation line of Hewlett and Hibbert (1967)], peak discharge (the highest flow rate during the storm), initial discharge and stormflow duration were measured from each storm hydrograph. The associated storm rainfall depth, duration and maximum 60-min intensity were derived from recording Casella rain gauges at the foot of each catchment. Stormflow is dependent on both rainfall and the amount of water in the soil profile. Soil water can be approximated using an antecedent precipitation index (API). This was calculated for Swartboskloof as follows: the index for 1 day (i) is equal to the index for the previous day reduced by a decay factor (K) plus that previous day's rainfall (P) ($API_i = API_{i-1}.K + P$; Dunne and Leopold 1978). A fairly high decay factor of 0.92 was used to simulate the drying of the surface soil. Initial discharge gave a good measure of general catchment wetness. Response ratio for individual storms was derived by

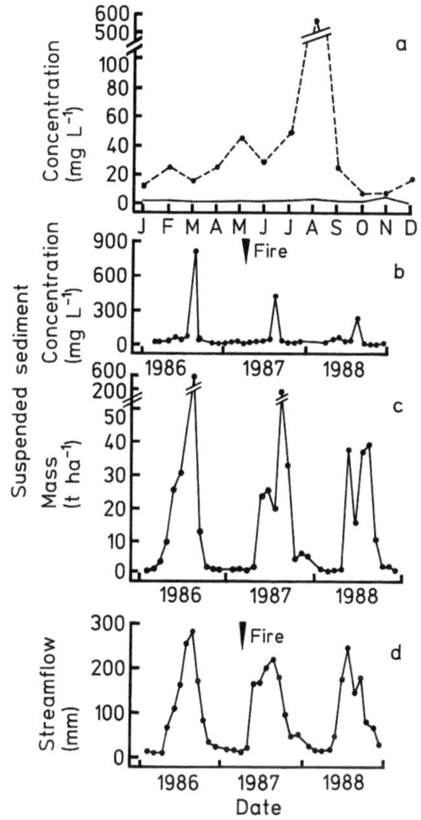


**Fig. 12.3.** Suspended sediment export from Swartboskloof in streamflow: **a** mean monthly concentrations of sediment in stormflow (----) and baseflow (—); **b** monthly mean concentrations in stormflow only from January 1986 to December 1988; **c** monthly totals of sediment exports; and **d** the total monthly streamflow from Swartboskloof. *Arrow* shows the date of fire

dividing quickflow by storm precipitation. Fifty-one post-fire storm events were compared to 52 pre-fire storms. The mean pre-fire values of the variables generated for the stormflow analysis are shown in Table 12.1

In order to test for the effect of fire on streamflow and stormflow, the data were analyzed by a multiple regression method (STEPWISE and REG procedures in SAS; SAS Institute Inc. 1985b) employing a dummy variable to account for the fire treatment (Kleinbaum and Kupper 1978). Total streamflow showed an increase of 16% above the expected volume in the first 2 years after fire; 15% (79 mm) in the first year and 17% (77 mm) in the second. Yields in the two post-fire years were not different from each other. None of the stormflow variables changed after the fire. Deviations of, for example, the larger quickflows in Swartboskloof about the calibration regression (Fig. 12.2) show, firstly, a rather weak calibration model and, secondly, that though there are fairly large quickflows above and below those expected, the pattern and size of these is not changed by the fire.

The general fit of the calibration models is not particularly good (adjusted $R^2$ values from 0.80 to 0.95; Table 12.2). Consequently, small

**Table 12.1.** Mean pre-fire (calibration period) values of variables used in the stormflow analysis in Swartboskloof and the control catchment, Tierkloof

| Catchment | n[a] | Storm rainfall[b], P (mm) | Storm duration D (h) | Maximum 1-h rainfall[b], $I_{60}$ (mm) | Initial discharge (mm $day^{-1}$) | Peak discharge (mm $day^{-1}$) | Storm flow (mm) | Quick flow, Q (mm) | Response ratio, (Q/P) 100% |
|---|---|---|---|---|---|---|---|---|---|
| Swartboskloof | 51 | 44.9 | 39.4 | 9.4 | 3.5 | 19.0 | 15.8 | 7.4 | 12.6 |
| Tierkloof | 50 | 44.7 | 35.2 | 9.1 | 2.2 | 27.5 | 16.4 | 11.5 | 18.6 |

[a] n = Number of storms; for storm rainfall and response n = 45 and 49 for Swartboskloof and Tierkloof respectively, due to missing rain data.
[b] Storm rainfall and rainfall intensity are derived from single recording rain gauges at the foot of each catchment, and as such are indices and not accurate areal assessments of these variables. Hence, the response ratio is exaggeratedly high, as P has a low bias.

**Table 12.2.** Details of the calibration (i.e. pre-fire), reduced and full multiple regression models for analysis of the streamflow data

| Dependent variable (treatment catchment) | df | Calibration model error $R^2$ | Significant predictors[a] ($P < 0.05$) |
|---|---|---|---|
| Fortnightly streamflow volume | 48 | 0.94 | C, S |
| Stormflow | 34 | 0.95 | D, P, $Q_i$, $I_{60}$, S |
| Quickflow | 34 | 0.91 | D, P, $Q_i$, $I_{60}$, S |
| Peak discharge | 35 | 0.85 | $I_{60}$, D, P, S |
| Response ratio | 37 | 0.80 | D, $I_{60}$, API |

[a] API is the antecedent precipitation index (see text); D is the storm duration (h); C is the corresponding control catchment variable; $I_{60}$ is the maximum 1-h rainfall intensity (mm $h^{-1}$); P is storm precipitation (mm); $Q_i$ is the initial discharge (mm $day^{-1}$); S is the dummy variable to account for seasonal differences between catchments.

changes in the stormflow variables as a result of fire would not have been detected. The control catchment provided relatively little explanation of stormflows in Swartboskloof. The best calibration regression models all included storm duration, storm rainfall, rainfall intensity and the dummy variable for season (Table 12.2). Antecedent wetness as initial discharge or API, was a significant "predictor" of storm- and quickflow and reponse ratio both before and after the fire.

#### 12.2.1.2 Effects of Fire on Catchment Sediment Yields

The effects of fire on catchment sediment yields were assessed from estimates of suspended sediment as well as from measurements of bedload. Baseflow grab samples were taken weekly, upstream of the gauging weir in a turbulent section of the stream where water was well mixed. Stormflow was sampled at the weir with an automatic sampler (van Wyk 1983). Suspended sediment concentrations were determined by vacuum filtering of the water samples through a pre-dried and weighed Whatman glass fibre filter with a 0.45 μ pore size. Baseflow sediment yield was obtained by multiplying the baseflow for a week by the corresponding grab sample concentration. Storm hydrographs were divided into segments to correspond with the stormflow samples, and the volume of each multiplied by the associated sediment concentration to give stormflow sediment yields.

Monthly sediment yields for 21 months post-fire were compared with those for 15 months pre-fire by the non-parametric Wilcoxon test. There was no difference in total or stormflow sediment yield due to burning (Fig. 12.3, Table 12.3), though there was an increase in baseflow sediment yield. Since sediment export occurs mainly during stormflow (95% in the 3 years of this study), the concentrations in stormflows pre- and post-fire were compared by means of an ANOVA, but no increase was detected after the fire (Fig. 12.3). Most sediment was exported from the catchment in single storm events or clusters of closely spaced storm events; for example, the sediment

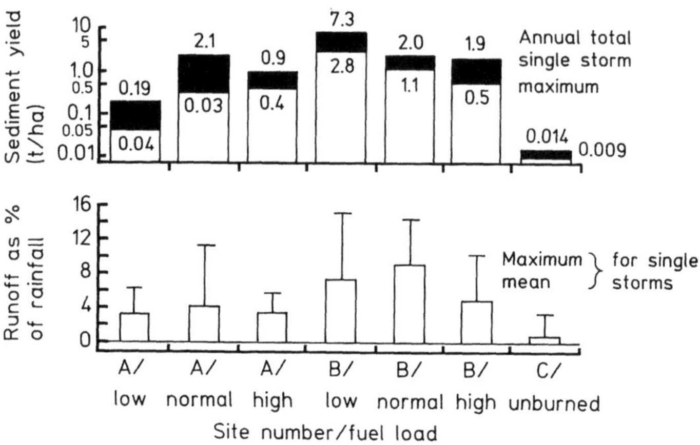


**Fig. 12.4.** Maximum (*open bar*) and total (*open plus solid bar*) sediment yield, and mean overland flow per storm (expressed as a proportion of the rainfall), from seven plots at three sites in Swartboskloof (see text) in the first year of measurement. *Whiskers* indicate maximum values for single storms. Note the log scale for the sediment yields

**Table 12.3.** Monthly mean sediment yields (kg $ha^{-1}$) before and after fire in Swartboskloof and results of the non-parametric Wilcoxon test

| | Pre-fire mean | Post-fire mean | Z | P > /Z/ |
|---|---|---|---|---|
| Baseflow | 1.04 | 2.16 | −2.278 | 0.023 |
| Stormflow | 45.57 | 24.18 | −0.706 | 0.480 |
| Total flow | 46.34 | 26.34 | −1.187 | 0.235 |

yield in the particularly wet August 1986 (0.57 t $ha^{-1}$) contributed 81% of the annual total and was greater than subsequent annual totals (0.41 and 0.15 t $ha^{-1}$).

Bedload was measured once a year by determining the volume and bulk density of material accumulated in the stilling pond above the weir. From 1986 to 1988 annual yields were estimated as 0.0118, 0.0116 and 0.0151 t $ha^{-1}$ respectively, which is an order of magnitude smaller than the annual suspended sediment yields.

### 12.2.2 Surface Runoff Plots

At two hill slope sites in Swartboskloof (sites A and B on the left and centre of Fig. 1.3 respectively), three adjacent plots (3 × 22 m) were laid lengthwise down the slope (53%). The sides and tops of each plot were formed by

1-mm flat sheet iron hammered 50 to 100 mm into the soil. At the bottom of each plot, overland flow was collected in a gutter and led to storage drums. Both sites were in tall proteoid shrubland. Site A was on sandstone-derived soils, while site B was on granite-derived soils (vegetation-soils types 3 and 1 in Table 12.7). Fuel loads were determined on the two sites using the methods described by van Wilgen (1982); fuel loads were lower (1470 $g\,m^{-2}$) on site A than on site B (3777 $g\,m^{-2}$).

To test for the influence of soil heating on infiltration in the plots three fuel load treatments were applied to each set of three plots (experiment 1). All fuel except fine ground fuels on one plot was placed on to a second plot. The third plot was left with its normal fuel load. Allocation of treatments was random. The total dry fuel load on the double fuel load treatment at site A was about 3000 $g\,m^{-2}$, whereas at site B it was about 7500 $g\,m^{-2}$. The fuel loading treatments were completed a week before the fire, and in retrospect it appears that the fuel bed had not settled much before burning, so that the transfer of heat to the soil was very inefficient. All plots, though, were burned clean of all but the largest live fuels and some unincorporated organic matter on the soil surface.

Overland flow measurements commenced before the first major storms of the wet season. After each such event the depth of rain and volume of overland flow were measured, and from a sample the suspended sediment was determined. The mass of heavier sediments which collected in the gutter was determined and added to the sediment yield on a monthly basis. Maximum rainfall intensity over a 30-min period was measured with a single day Casella recording rain gauge at the Swartboskloof weather station (Fig. 1.3).

After 1 year a second experiment (experiment 2) to examine the defferences between overland flow on the burnt sites (A and B) and an unburnt site was established in 12-year-old fynobs (site C, at the right in Fig. 1.3). The overland flow plot data were analyzed by a multiple regression method in SAS (SAS Institute Inc. 1985b). The test of treatment and site differences was done by means of the dummy variable technique (Kleinbaum and Kupper 1978).

Experiment 1 showed that fuel loading had no effect on overland flow (Fig. 12.4). Runoff and sediment response from the small plots was related to characteristics of the specific plot and independent of the fuel loading treatments. Water and sediment yields from the plots were reduced in the second year after fire (Fig. 12.5). Taken over the six plots jointly, the means were reduced from 5.6 to 1.2% runoff (down 78%), and from 2.4 to 0.06 $t\,ha^{-1}$ of sediment yield (down 97%).

Experiment 2 showed that the fire caused marked increases in overland flow and sediment yield from sites A and B. In the first year after the fire, mean runoff was 3.6 and 7.1% of rainfall, and total sediment yield was 3.3 and 11.2 $t\,ha^{-1}$ for sites A and B respectively (Fig. 12.4). In the single year of measurement at site C the corresponding figures were 0.8% runoff and

**Table 12.4.** Overall mean values, on a per storm basis, of the variables measured at overland flow plots at three sites in Swartboskloof. Sites A and B (3 plots each) were burnt, site C (1 plot) was unburnt

| Site | n | Rainfall (mm) | Maximum rainfall in 30 min (mm) | Depth of overland flow (mm) | Overland flow as percentage of rainfall (mm 100 mm$^{-1}$) | Sediment yield ($10^{-3}$t ha$^{-1}$) | Sediment yield per unit of rainfall (g 100 mm$^{-1}$) |
|---|---|---|---|---|---|---|---|
| Ste A | 204 | 42.5 | 6.0 | 0.9 | 2.2 | 17 | 2.8 |
| Site B | 201 | 43.2 | 5.9 | 1.8 | 4.5 | 57 | 8.8 |
| Site C | 35 | 43 | 6.0 | 0.5 | 0.8 | 0.4 | 0.04 |

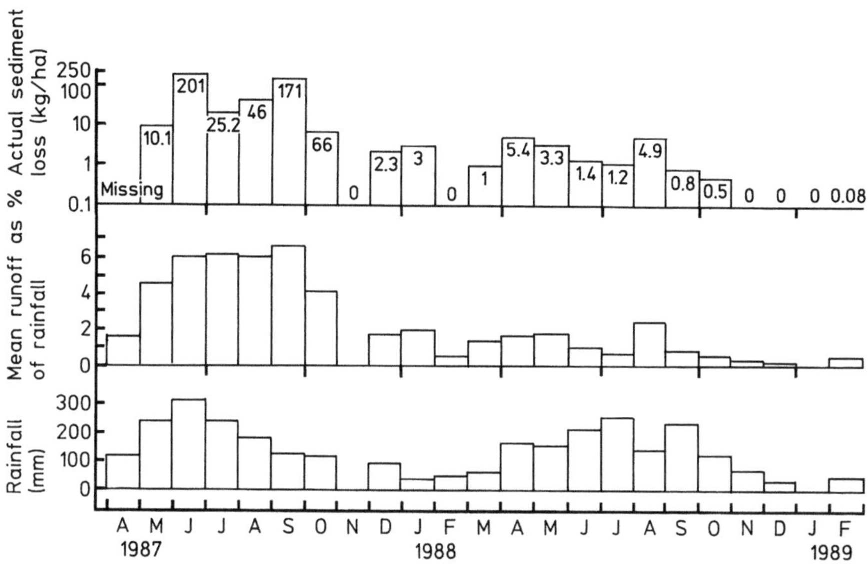


**Fig. 12.5.** Mean monthly sediment yield and runoff depth from six burned plots in Swartboskloof from the first measurements after fire in March 1987, showing marked reductions in the second year. Note the log scale for sediment yields

**Table 12.5.** Best regression models for explanation of surface runoff and soil loss from overland flow plots in Swartboskloof. An unburnt control plot was added in the second year

| Dependent variable | | Error df | Adjusted $R^2$ | Significant ($P < 0.05$) independent variables |
|---|---|---|---|---|
| First year | Surface runoff | 161 | 0.77 | Rainfall, rainfall intensity, antecedent wetness, plot |
| | Soil loss | 162 | 0.60 | Rainfall, rainfall intensity, antecedent wetness, plot |
| Second year | Surface runoff | 205 | 0.72 | Rainfall, rainfall intensity, antecedent wetness, plot, season |
| | Soil loss | 196 | 0.47 | Rainfall, rainfall intensity, antecedent wetness, plot |

$0.014\,t\,ha^{-1}$. Increases due to the fire were therefore 3.42:1 and 7.82:1 for runoff, and 238.0:1 and 819.5:1 for sediment yield at sites A and B respectively (Table 12.4).

Low $R^2$ values resulted from the regression analysis, and the most useful predictors of overland flow and soil loss were (in order of importance): rainfall amount, rainfall intensity and antecedent wetness (Table 12.5). The positive relationships found between rainfall depth and rainfall intensity, on the one hand, and runoff and sediment yield from the plots, on the other, is

**Table 12.6.** Pearson correlation coefficients between dependent and independent variables for the single storms on an unburnt plot in Swartboskloof. Number of storm events in the sample was 35 over 1 year

| Independent variables | Dependent variables | | |
|---|---|---|---|
| | Depth of surface runoff (mm) | Depth of runoff as % of rainfall | Sediment yield (g) |
| Rainfall (mm) | 0.71*** | 0.49*** | 0.48*** |
| Maximum 30-min rainfall (mm $h^{-1}$) | 0.64*** | 0.70*** | 0.60*** |
| Antecedent wetness | −0.31* | −0.48** | −0.33* |

*, **, *** indicate significance at the 10, 1 and 0.1% levels respectively.

exemplified by the correlations for the unburnt plot (Table 12.6). The unusual feature of the unburned plot was the negative relationship between antecedent wetness and runoff (Table 12.6). On the burned plots this relationship was positive, as would be expected from infiltration theory. The explanation for this unexpected behaviour is that under dry conditions (low API) the soils of the unburned plot are most water repellent, and this repellency causes a higher proportion of rainfall to be lost as surface runoff. This phenomenon has been recorded in scleropyllous eucalypt forest in Australia (Burch et al. 1989) where relatively more runoff was produced from rain events during dry periods.

### 12.2.3 Water-Repellent Soils

Soil testing was done at five sites (Table 12.7), selected by superimposing the Swartboskloof vegetation map (McDonald 1985) on the soil map (Lambrechts et al. 1986). The five sites were representative of the central and lower catchment, had high fuel loads and were in reasonable proximity to a perennial stream. These factors incorporate the hydrological importance of the units as serious repellency was expected to ensure delivery of overland flow to the stream channel.

At each site, 15 points were sampled during the 2 months before, and the 2 months after the fire. At each point soil wettability was measured at four depths: 5, 15, 75 and 150 mm (a total of 600 determinations). Two tests, the water drop penetration time (WDPT) and critical surface tension (CST) (DeBano 1981) were used to test for water repellency. WDPT is the time taken for a drop of water to be absorbed by the soil. The CST is the highest surface tension (N $m^{-1}$), determined from a range of aqueous alcohol solutions of different surface tensions, which readily wets the soil (taken as a 1s entry time). Each test score was a mean for about ten drops on a surface of approximately 0.04 $m^2$. The values obtained from these tests were combined

**Table 12.7.** Salient features of five sites sampled for water repellency in Swartboskloof

| Site | Vegetation[a] | Parent material[b] | Soil texture[b] | Gravel in top 100 mm[b] (%) | Organic carbon in top 100 mm (%) |
|---|---|---|---|---|---|
| 1 | Proteoid shrublands | Mixed granite and sandstone | Sandy loam | <10 | 5.0 |
| 2 | Proteoid shrublands | Mixed granite and sandstone | Sandy loam | <5 | 10.0 |
| 3 | Proteoid shrublands | Sandstone (quartzite) | Loamy sand | <50 | 4.2 |
| 4 | Proteoid shrublands | Sandstone (quartzite) | Sand | <10 | 3.6 |
| 5 | Tall and short riparian forests | Mixed alluvial material | Sandy loam | >50 | 8.8 |

[a] See Chapter 1 for description of vegetation types.
[b] Data from Lambrechts et al. (1986).

into a simple water repellency index, RI = WDPT/CST (DeBano 1981). The soils were classed as repellent (RI > 2000), somewhat repellent (2000 > RI > 250) or wettable (RI < 250).

The results of the water repellency testing were analyzed with an ANOVA for a split plot design (GENSTAT 1987). This allowed for testing of differences due to depth, vegetation-soil type, fire and their interactions, using the WDPT and CST measures of repellency for separate analyses.

Vegetation-soil type, fire, depth in the soil, and their interactions were significant effects. There was considerable water repellency in the soils prior to the fire (Fig. 12.6). This was concentrated in the upper two layers where a majority of soils were at least somewhat repellent. In the riparian forest (site 5) repellency was more evenly distributed with depth, probably because of even mixing of organic matter throughout the topsoil. Repellency in site 2, with its granite-influenced soils and high organic matter content, was greater than on the other sampling units.

Fire reduced repellency in the upper 15 mm of soil, in all but site 4 which was unchanged (Fig. 12.6). Repellency increased shallowly below the surface at sites 1 and 4, but was not affected at this level in the other types. Below 50 mm in the soil, there was less water repellency after fire at sites 2, 4 and 5. At these levels reductions in repellency were minor.

The extent of soil heating by the fire, assessed using guidelines of Wells et al. (1979), indicated relatively little serious heating of the soil during the fire, presumably because of high soil and litter moisture contents. In a majority of sites the insulating ground litter had been only partially consumed. However, localized patches of soil had been burned and the incorporated organic matter consumed. After the first low intensity rains these points showed the typical fire-induced repellency phenomenon where wet-

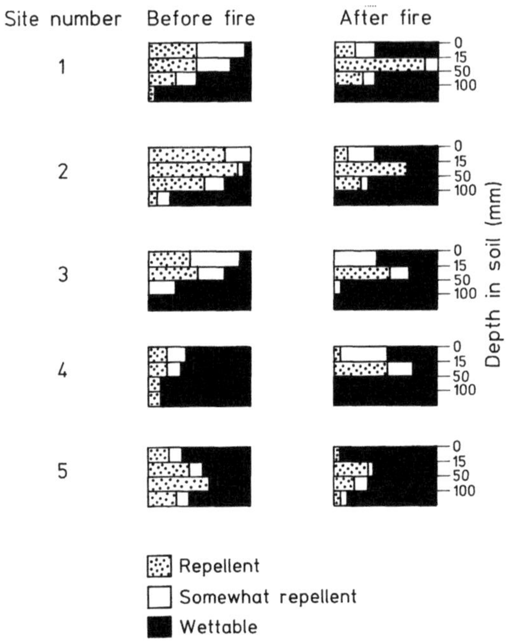


**Fig. 12.6.** The proportion of samples in three soil wettability classes at different depths at five sites in Swartboskloof, before and after fire

table surface soils lie on top of soils which are now strongly repellent (DeBano 1981). Where this occurs, water saturates the upper soil but runs off when impeded percolation cannot keep pace with rainfall inputs.

## 12.3 Discussion

### 12.3.1 Studies at the Whole Catchment Level

#### 12.3.1.1 Effects of Fire on Streamflow

The increase in streamflow at Swartboskloof is due to reductions in rainfall interception and transpiration following the removal of vegetation. Increases of this magnitude are expected and confirm the findings in the southwestern Cape and elsewhere regarding water use by shrub vegetation (Appendix 12.1; Bosch and Hewlett 1982). Bosch et al. (1986) proposed a model for predicting increases in streamflow following fire, based on pre-fire biomass and the expected rate of recovery of the vegetation. The model predicts increases in streamflow of around 120 and 93 mm in the first and second years after fire respectively; these estimates are slightly higher than those recorded (79 and 77 mm) at Swartboskloof.

The fact that fire effects on stormflows remained undetected in Swartboskloof may be partly due to the weak pre-fire calibrations, but it is clear that there was no marked response to fire of the order seen in the nearby Abdolskloof and Bosboukloof following late dry season wildfires (Appendix 12.1).

**12.3.1.2 Effects of Fire on Catchment Sediment Yields**

Stormflows and sediment yields were unaffected by the Swartboskloof fire. This is an important finding though not altogether unusual. For example, a spring wildfire in the nearby Langrivier catchment resulted in brief and minor sediment yield increases of $0.148\,t\,ha^{-1}$, which had returned to normal by the first wet season (Appendix 12.1). Sediment yields from Swartboskloof are greater than the $0.008\,t\,ha^{-1}$ yielded in the first year after fire from the Zachariashoek catchment which has a purely sandstone geology (Appendix 12.1). No flooding and little erosion resulted from a fire in southern California (Colman 1951), while a wildfire in eucalypt forest in Australia (Ronan 1986) similarly produced no significant sediment concentration increases (Appendix 12.1). Absence of hydrological effects of fire may not have been recorded in other cases simply because of their unremarkabilty.

Elsewhere, recorded increases in streamflows following fire range from low total flow increases (11%) to relatively higher stormflow increases, and peak discharges which generally show the greatest increases (up to 4400%; Appendix 12.1). Sediment discharges, being dependent on available energy and exponentially related to streamflow, show relatively greater increases, particularly at peak discharges, and a more rapid return to pre-fire levels. For example, a 1000 fold increase in peak sediment load was estimated by Brown (1972) in the first post-fire season (Appendix 12.1).

## 12.3.2 Surface Runoff Plots

There was greater overland flow on steep slopes at Swartboskloof than would have been expected from soil properties (texture, porosity, bulk density), slope and rainfall intensities. This was, in agreement with theory, positively related to the rainfall amount, intensity, and antecedent wetness of the plot. Sediment yields were similarly related to these factors, but were exponentially related to the amount of runoff produced on the plot. The high levels of runoff indicate that infiltration and/or percolation on steep slopes was inhibited, most likely by water-repellent soils.

The high runoffs measured in Swartboskloof, even off the control plot, are in sharp contrast to the almost negligible runoff (0.05%) recorded from more gently sloping (25%) plots elsewhere in the Jonkershoek Valley (Versfeld 1981), where none of the treatments of hoe, slash and burn, or thinning of pine plantation had a significant effect. Soil losses off burned plots in this study were similar to the 0.4 to $7.8\,t\,ha^{-1}\,yr^{-1}$ measured from untreated plots in the San Gabriel Mountains of California, the 2.8 to

7.3 t $ha^{-1}$ $yr^{-1}$ reported from burned plots at Santa Maria, California, and the 2.5 to 8.2 t $ha^{-1}$ measured from gentler slopes in Australia (Blong et al. 1982) after a moderate intensity wildfire (Appendix 12.2). A high intensity wildfire in a pine plantation in Bosboukloof resulted in considerably higher first year soil losses (10 to 26 t $ha^{-1}$) from mid-slope plots, at slopes of 27 to 62% (Appendix 12.1). Greater soil losses (30 to 48 t $ha^{-1}$) occurred in Australia (Atkinson 1984), largely from two single extreme rain events in the first 3 months after a wildfire, and after wildfire in the San Gabriel Mountains, California (346 t $ha^{-1}yr^{-1}$) (Appendix 12.2).

The rapid recovery to pre-fire levels of soil loss and, to a lesser extent, overland flow are typical of results reported elsewhere (Appendix 12.2). Similar short recovery periods are reported for burned chaparral in California and eucalypt forest in Australia (Rowe et al. 1951; Appendix 12.2). A definite picture emerges of high soil losses immediately following disturbance. A linear reduction of surface flow would be expected to cause an exponential reduction in soil loss, so that soil loss recovers sooner than surface runoff. A somewhat different picture emerges from some studies in California and Oregon where steep and dry (equatorial aspect) slopes with poorly aggregated surface soils may be eroded as much by dry ravel in the dry season as by rain events (Appendix 12.2; Mersereau and Dyrness 1972). In these situations erosion may continue for longer, and recovery from disturbance is slower.

Single extreme events make a disproportionate contribution to total yields in Swartboskloof (Fig. 12.4), as is the case elsewhere. Large storms following wildfire have caused catastrophic flooding and sedimentation in California, Oregon and Australia (Appendices 12.1, 12.2).

The failure of the fuel load treatments to produce consistent effects from the plots is probably due to minor heating resulting from inefficient transfer of heat on the double fuel load plots (as fuels were elevated), and the insulating effects of moist litter and soil. Plot response was therefore dominated by inherent inter-plot differences.

Soil losses off burned slopes in Swartboskloof over one fire cycle (>10 yrs) will be well within the 12.5 t $ha^{-1}$ limit used as a guideline for arable land in South Africa (Russell 1983), though higher than the natural erosion rates measured on undisturbed plots in the USA (Smith and Stamey 1965). Once the burned site is stabilized, which seems to occur largely in the first year, soil losses drop to orders of magnitude less than tolerable levels for cultivated land. This ensures that fire can be used as a viable management practice in catchment areas, where erosion control measures would not be feasible.

### 12.3.3 Water-Repellent Soils

Although soils in Swartboskloof showed inherent repellency in the upper layers prior to the fire, this is unlikely to influence hydrological behaviour on a large scale, as other factors (high ground cover and coarse, friable soils) allow for sufficient infiltration.

Plant litter is the source of hydrophobic substances which transfer into soil by incorporation of organic matter, by leaching downward over time, or by vapour transport during burning (DeBano and Krammes 1966; Savage et al. 1972). Thus soils with a higher humus content (sites 1 and 2, Table 12.7) would show more repellency than those with less humus (site 3, Table 12.7). Water repellency decreases as organic matter content decreases at greater depths, causing the differences due to depth in the profile. The absence of fire in the riparian forest and the incorporation of partially decomposed plant matter to greater depths in the soil profile, explain the high repellency ratings to greater depths at this site.

The heating of the soils is reflected most in the increased wettability of the surface layer. Soil at Swartboskloof which was ashed (blackened in the fire) was especially wettable. The organic compounds responsible for hydrophobic behaviour of soil are destroyed or denatured by excessive heating. Both temperature and the duration of heating are important factors; for example, temperatures in excess of 300°C for 30 min would remove a hydrophobic effect (Krammes and DeBano 1965).

Where less soil heating occurs incipient repellency can be intensified (DeBano 1966). Vapourized hydrophobic substances from upper soil layers may condense at lower levels in response to a temperature gradient (DeBano 1966), thereby broadening the band of repellency in the soil. Thus where soil heating is sufficient, repellency should increase at lower levels in the soil (e.g. Dyrness 1976). Our results show that typical fire-induced repellency occurred more obviously in the soils on sites 1, 3 and 4 (Table 12.7) although this pattern, though never continuous, was present in most of the cleanly burned sites.

Soil wetness is critical in determining the amount and spread of soil heating during a fire. Dry soils are heated more than wet soils because water greatly increases their heat capacity and conductivity (Hillel 1980). The unseasonally high rainfalls in February and March 1987 greatly reduced the potential for soil heating during the fire. The mean gravimetric wetness of soil in the upper 50 mm of the profile in Swartboskloof was 13.8% in early March.

As water-repellent soils gain moisture so they become more wettable (DeBano et al. 1967; Gilmour 1968). The lower post-fire repellency ratings below 50 mm (Fig. 12.6) may be an artefact of the moister soil conditions during the post-fire soil survey. Regardless of the degree of repellency present in any vegetation-soil type, there were frequent patches of soil which showed no repellency, and these could have accommodated water from areas where percolation was impeded.

### 12.3.4 Integration of Catchment, Plot and Profile Level Studies

Studies in Swartboskloof were conducted at three levels (whole catchment, plot and soil profile). The plot level studies showed that overland flow and sediment levels increased after fire, but were short-lived. Studies at the soil profile level indicated that such increases may be due to, at least in part, increases in soil water repellency after fire. The increases were not reflected at the whole catchment level, however, where stormflow and sediment yields showed no change after fire.

Several features of the Swartboskloof catchment may explain the lack of response to fire at the catchment level. In the first place, riparian forests remained intact, and these would have buffered any effects observed at the plot level. Secondly, the weak and patchy nature of the repellency would have allowed water to infiltrate the soils at the base of steep slopes before such sediment loads could reach the drainage channels. Swartboskloof also does not have the network of logging roads found in many other catchments. For example, flooding and soil losses after fire were severe in the Bosboukloof catchment (Fig. 1.1; Scott and van Wyk 1990), where roads and skid paths aided the delivery of surface water to the drainage channels.

Soil water repellency has been cited as contributing to increases in stormflow in many studies. A "severe" burn in a pine forest in Arizona induced repellency and reduced the infiltration rate by 66% (Campbell et al. 1977), while streamflow and "runoff efficiency" increased by 800 and 450% respectively in the first wet season following the fire. High soil losses ($22\,t\,ha^{-1}$) that occurred after a wildfire in an eucalypt forest in Australia were also associated with fire-induced repellency (Leitch et al. 1983). Similarly, the large response to fire in Bosboukloof (Scott and van Wyk 1990) was linked to the inducement of water repellency in the soil. The absence of a marked stormflow response to fire in Swartboskloof may have been due to the a patchy development of a water-repellent layer in the soil.

Geology and soils are important controlling variables of erosion rates. Some mountains are inherently less stable; for example, the steep slopes of the San Gabriel Mountains of California have naturally high erosion rates (Anderson et al. 1959). Soil parent materials may produce less cohesive and more erodible soils, such as the granites of the southern Highlands of Victoria (Brown 1972) and the non-cohesive Wilson Diorite of the San Gabriel Mountains (Anderson et al. 1959). Granite-derived soils in fynbos catchment areas, while not especially erodible, seem to produce higher sediment yields than the sandstones. Thus the erosion rates of the Jonkershoek catchments, with some exposure of granite, are higher than comparable catchments on sandstone alone (van Wyk 1981a).

Although soils of fynbos catchment areas have a low erodibility, fire may increase this rating. Adequate heating can consume organic matter incorporated in the soil, thereby breaking down soil aggregates and increasing erodibility (DeBano et al. 1977; Giovannini and Lucchesi 1983). This fire effect is important in determining soil loss after fire in areas such as Swartboskloof where sediment supply limits erosion.

## 12.4 Conclusions

We set out to examine the hypothesis that fire would induce water repellency in the soil (thereby increasing the probability of overland flow, the magnitude of stormflows and water yield) and would increase sediment yields. The hypothesis was only partially supported. Some repellency was noted, and overland flow and sediment yields at the plot level increased for a short period after the fire. However, no effects were evident at the catchment level, and we conclude that the fire had little effect on the catchment's hydrological behaviour.

It is important to note that the fire of March 1987 was relatively mild (see Chap. 1), and different effects may result from wildfires that burn under relatively hot and dry conditions that result in higher fire intensities. This was the case in the nearby Bosboukloof catchment (Scott and van Wyk 1990), where fuel loads were also much higher due to afforestation.

The implication for management is that fires can be applied in late summer or early autumn without having a marked hydrological effect. Key provisions for successful burning were high soil and fuel wetness, and protection of the riparian zone. Critical levels of soil and fuel wetness would depend on the energy available for soil heating, as well as insulation of soil by litter. Judging from this study, a mean soil wetness of 15% in the top 50 mm and fuel wetness of around 30% was adequate to restrict fire effects to acceptable limits.

*Acknowledgements.* We thank H. Dicks and G.P.Y. Clarke of the Department of Mathematics and Statistics at the University of Natal, Pietermaritzburg, and H.J. van Hensbergen for advice and assistance with statistical analyses. H.J. Barnard built and installed the sediment and bedload samplers. G. Moses coordinated most of the fieldwork. H.J. Barnard, L. Carelse, E. Theron, R. Robyntjies, M. September, D. Willemse and H. Julius assisted in the field and laboratory. We thank the Computing Centre for Water Research, University of Natal, for the use of computing facilities. G.G. Forsyth kindly provided rainfall summaries. The editors and reviewers are thanked for considerable help and encouragement.

**Appendix 12.1.** Summary of catchment scale assessments of fire effects on hydrology in mediterranean-type ecosystems

| Catchment and location | Area (ha) | MAP[a] (MAR) mm | Description | Treatment | Streamflow[b] | Sediment yield[b] ($t\ ha^{-1}\ yr^{-1}$ unless specified) | Reference |
|---|---|---|---|---|---|---|---|
| *South Africa* | | | | | | | |
| 1. Abdolskoof, Jonkershoek, SW Cape | 12 | 1275 (385) | Deep coarse loams; mountain fynbos | High intensity wildfire (April 1945) | *TF and PF*: 133 and 400% increases in 1st wet season respectively | | Rycroft (1947); van Wyk (1987) |
| 2. Kasteelskloof and Zachariashoek, SW Cape | 324<br>287 | 1443 (694)<br>1003 (379) | Quartzitic sandstones; shallow to deep sandy loams; mountain fynbos | Prescribed burns at 6- and 12-year cycles | *TF*: increases of up to 85 mm (15%) in first year reducing to zero by 4th year | *SS*: up to $0.97 \times 10^{-3}$ in 1st year<br>*B*: $6–7 \times 10^{-3}$; sediment yields normal within 10 months | van Wyk (1981a); Lindley et al. (1988) |
| 3. Langrivier, Jonkershoek, SW Cape | 246 | 2261 (1603) | Deep gravelly loams; overmature mountain fynbos | Wildfire at the end of the wet season (October 1987) | | *SS*: 0.148; sediment yield normal by the first wet season | van Wyk and Lesch (1989) |
| 4. Bosboukloof, Jonkershoek, SW Cape | 201 | 1296 (593) | Deep gravelly loams; 2nd rotation pine plantation | High intensity wildfire (February 1986) | *TF*: 70 mm (12.4%) increase in 1st year<br>*SF*: 62% increase in 1st year<br>*PK*: 290% increase | *SS*: 6<br>*B*: 1.8 | Scott and van Wyk (1990) |
| 5. Swartboskloof, Jonkershoek | 191 | 2300 | Deep gravelly loams mature fynbos | Prescribed burn; moderate conditions | *TF*: 16% increase in 1st 2 years. Stormflows unchanged | *SS*: 0.416 (3-year mean)<br>*B*: $13.8 \times 10^{-3}$ (3-year mean) | This study |

*Australia*

| | | | | | | | |
|---|---|---|---|---|---|---|---|
| 5. Yarrangobilly River and Wallaces Creek, SE NSW | 22450<br>4310 | (ca. 520)<br>(ca. 550) | All-year rainfall; wet to dry eucalypt forests | High intensity wildfire (March 1965) | *TF and PF*: large increases. A sharp additional peak in storm hydrographs | *SS*: 25000 and 100000% increases in peak sediment loads; recovery within 2 years | Brown (1972) |
| 6. Kosciusko N.P., NSW | | ca. 500 | Alpine and sub-alpine forest and shrubland | Mid-summer wildfires after long dry periods | | *SS*: 250–350% increases at peak discharge | Good (1973) |
| 7. 7 large catcahments, southern Highlands, Victoria | 1430–12800 | ca. 1450<br>450 | Deep clay loams; high (70 m) *Eucalyptus regnans* forest | High intensity wildfire (January 1939) | *TF*: reductions in flow averaging 214 mm (24%) for 21 years | | Langford (1976) |
| 8. Eden, NSW | | ca. 900 | Grey sandy loams; dry sclerophyllous forest | Mid-summer wildfire, after long dry period | *SF*: increased by 100–300%.<br>*PF*: increased by 100–900%.<br>Effects dependent on % area burned and severity of burn | | MacKay and Cornish (1982) |
| 9. Bushrangers, ACT | 97.5 | 1230 | Deep soils; all-year rainfall; sub-alpine forest | High intensity prescribed fire | Summer baseflow increased by 32 mm (106%) in 1st year only | | O'Loughlin et al. (1982) |
| 10. Warburton, Victorian Central Highlands | 35 | | Shallow, gravelly loams; mixed eucalypt forest; moderately steep slopes | High intensity wildfire in late summer | *PF*: estimated maximum of 51–103 mm $h^{-1}$ from single, 17 mm storm | *SY*: estimated net loss of 22 t $ha^{-1}$ from single storm | Leitch et al. (1983) |

**Appendix 12.1** (continued)

| Catchment and location | Area (ha) | MAP[a] (MAR) mm | Description | Treatment | Streamflow[b] | Sediment yield[b] ($t\ ha^{-1}\ yr^{-1}$ unless specified) | Reference |
|---|---|---|---|---|---|---|---|
| *USA* | | | | | | | |
| 11. Montrose – La Crescenta, California | | | Mountain chaparral | Wildfire followed by severe rainstorm | | *SY*: estimated 306 $t\ ha^{-1}$ from the single event | Colman (1951) |
| 12. Yucaipa, California | 253 | | Mountain chaparral | Wildfire in July; high intensity thunderstorm | | *SY*: estimated 25 t $ha^{-1}$ from the single event | Colman (1951) |
| 13. Wheeler Springs Fire, California | 10625 | | Chaparral | Wildfire (1948) | No floods | Little erosion | Colman (1951) |
| 14. San Gabriel Mountains, California | | | Steep slopes; chaparral | Fire | *PF*: increases between 100–4400% dependent on storm size, in 1st year | | Rowe et al. (1954) |
| 15. Wilson River, Tillamook Burns, Oregon | 40800 | | Temperate; coastal rain-forest | 3 Wildfires (1933–1945) | *TF*: increase of 229 $mm\ yr^{-1}$ (ca. 11%) over 15 years; baseflows increased by 16%. *PF*: 45% higher immediately after fire, declining over 8 years | *SY*: 400–700% increase, 3 $t\ ha^{-1}\ yr^{-1}$ 17 years after first fire | Anderson (1976) |

| | | | | | | | |
|---|---|---|---|---|---|---|---|
| 16. Burns Watershed, Washington State | | 579 (112–175) | | Wildfire in summer (August 1970) | *TF*: increase of about 200 mm $yr^{-1}$ (120%) for 1st 3 years | *SY*: massive debris, rock and soil flows | Helvey et al. (1976) |
| 17. Rattle Burn, Coconino N.F., Arizona | 8.1 | ca. 737 | Stoney soils; <20% slopes; open pine forest | High intensity wildfire (May 1972) | *TF*: 700 and 280% increases in 1st and 2nd years. Response increased 357%, to 3.8% of rainfall | *SY*: 1.7 t $ha^{-1}$ (ca. 40 000% increase) in first 2 years | Campbell et al. (1977) |
| 18. Three Bar Watersheds, Arizona | | 620–750 | Deep, coarse soils; sparse deep-rooted chaparral | Wildfire (June 1959) | *TF*: 276 mm (900%) increase in 1st year. Baseflows increased | | Pase and Ingebo (1965) |

[a] MAP, Mean annual precipitation.
[b] TF, Total runoff; SF, stormflow; QF, quickflow; PF, peak discharge; SY, sediment yield; SS, suspended sediment; B, bedload.

**Appendix 12.2.** Summary of plot measurements of overland flow and sediment movement in (mainly) mediterranean-type ecosystems

| Location | Plot size | MAP (mm) | Description | Treatment | Overland flow (as % of rainfall) | Sediment yield ($t\ ha^{-1}\ yr^{-1}$) | Notes | Reference |
|---|---|---|---|---|---|---|---|---|
| *South Africa* | | | | | | | | |
| 1. Jonkershoek, SW Cape | 0.08 ha | 1156 | Deep sandy loams; 25% slope; fynbos and pine | Various, including hoe and burn | Mean 0.01–0.07%; max. 0.65% | Negligible | No treatment effects | Versfeld (1981) |
| 2. Langrivier, Jonkershoek, SW Cape | 3 × 22 m | 2260 | Deep sandy loams; 53% slope; Over-mature fynbos | Wildfire at end of wet season (October 1987) | 1.2–2.2% | 0.4 (in 1st year) | | Scott (1989) |
| 3. Swartboskloof, Jonkershoek | 3 × 22 m | 2270 | A. Deep sandy loams | Prescribed burn | 3.1–4% | 0.2–2.1 | 1st year figures | This study |
| | | | B. Deep gravelly loams | | 5.1–8.8% | 1.9–7.3 | | |
| *Australia* | | | | | | | | |
| 4. Canberra, ACT | 0.0006 ha | Simulated rain | 7% slope; *Pinus radiata* plantation | Prescribed fire; rain of 20 mm at 80 mm $h^{-1}$ | | None | Non-significant decrease in infiltration | Gilmour and Cheney (1968) |
| 5. Narrabeen Lagoon, NSW | 2 × 4 m | ca. 1200 | Deep sandy topsoil; 21% slope; dry forest | Moderate intensity wildfire | Mean 3–5%; max 9.6% | 2.5–8.2 | 40% Soil loss from 1 event | Blong et al. (1982) |
| 6. Royal National Park, Sydney, NSW | | 340 in 3 months | Shallow sandy soils; 6–13% slopes; dry bushland | High intensity wildfire (Jan. 1983) | | 30–48 in 3 months | 35% Organics in 1st debris | Atkinson (1984) |

| | | | | | | | | |
|---|---|---|---|---|---|---|---|---|
| 7. Wallaby Creek, NE Victoria | 0.002 ha | 1220 | Stoney clay loams; 36% slope; dry forest | Hot prescribed fuel-reduction burns | Doubled to 2.5% after high intensity burn | Doubled to 0.1–0.35 | Recovery: runoff 3–5 yrs; erosion 1–5 yrs | Ronan (1986) |
| *USA* | | | | | | | | |
| 8. North Fork, Madera Co., California | 3 × 33 m | 840 | Deep sandy clay loams; 36% slope; chaparral-woodland | 1. Undisturbed<br>2. Twice burned<br>3. Annually burned | 0.04%<br>1.87%<br>14.2% | 0.014<br>9.1<br>257 | 9 yr averages, majority of yields in 1st year | Rowe (in prep.) cited by Adams et al. (1947) |
| 9. San Gabriel Mountains, California | Various | | Non-cohesive soils; 55–95% slopes; unburned chaparral | No treatment: baseline measurements | | 0.4–7.8 | Majority as dry ravel | Anderson et al. (1959) |
| 10. 12 contrasting areas in USA | 0.004–2.2 ha | 500–1750 | Humid sites suitable for agriculture; 2–30% slopes | No treatment, plots under good cover | 0.03–13.8 | 0.22–1.4 | Considered to be natural erosion rates | Smith and Stamey (1965) |
| 11. W San Gabriel and Santa Ana Mountains, California | | 773<br>949 | Gravelly<br>Granitic loams | Wildfire<br>Wildfire | 9.7<br>9.9 | 346<br>52 | | Krammes and Osborn (1969) |
| 12. Los Padres NF, Santa Maria, California | 3 × 12 m | | Gravelly loams; 20 and 50% slopes; mature chaparral | Hot prescribed burn, 66% of fuel consumed | | 2.8 Gentle slopes, 7.3 steep slopes | | DeBano and Conrad (1976) |
| 13. El Oso Creek, Arizona | Various | | Chaparral | Wildfire; reduced canopy to almost zero | 3–90% | 0.005–1.47 | | Heede et al. (1988) |

# 13 Fire and Catchment Chemical Budgets

D.B. van Wyk, W. Lesch, and W.D. Stock

## 13.1 Introduction

Natural and man-induced disturbances are increasingly being recognized as essential parts of the functioning of ecosystems. Disturbances can be highly variable as they occur at different spatial and temporal scales and with different degrees of intensity. This variability can have significant effects upon nutrient cycling patterns including causing changes in pathways and processes of nutrient transfer within ecosystems. Studies of budgets of elements gained or lost from a catchment have been used to provide a conceptual and empirical basis for understanding ecosystem responses to disturbance. Disturbance effects upon chemical budgets of catchments and in particular stream water draining such catchments have been extensively studied (Tiedemann et al. 1979; Raison 1980; Swank and Waide 1988). Conflicting results and conclusions from these studies show the complexities involved in separating disturbance effects from geographical, seasonal and annual variability in ecosystem behaviour. An important disturbance in the southwestern Cape Province is the burning of the fire-prone fynbos at intervals of 12–20 years. Fynbos is burned to increase water yields of mountain catchments, to reduce fire hazard, to maintain fynbos species diversity and to control invasive plants (van Wilgen et al. 1991). Fires are highly variable events and to a large extent depend on factors such as fuel load and vegetation characteristics (see Chap. 3) and weather patterns (see Chap. 2). Few catchment studies have been undertaken in mediterranean-climate regions, and little is known about the fate of mineral nutrients released by fires. It is also not known whether regular burning may lead to any significant depletion of catchment nutrient capital.

In this chapter we evaluate the impact of fire on fynbos ecosystem functioning by quantifying input-output budgets of mineral elements in the Swartboskloof catchment (Fig. 13.1). To do this we determined: (1) the input of chemical constituents via wet and dry deposition; (2) the loss of chemical constituents in stream water and stream bedload leaving the catchment; and (3) volatilization losses. We used the input-output data to

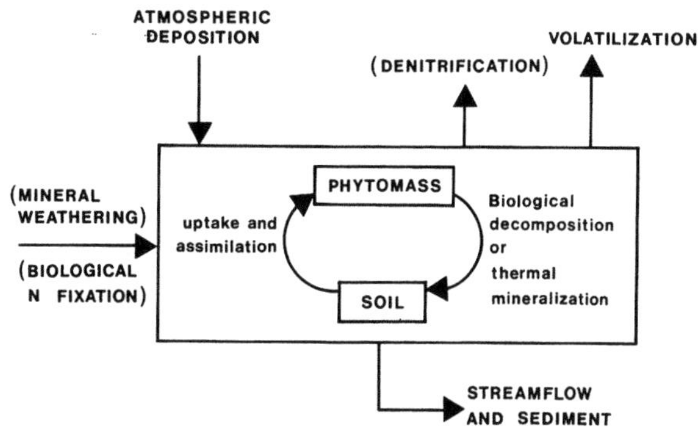


**Fig. 13.1.** Diagrammatic representation of the main processes involved in the nutrient budget of an ecosystem subjected to fire. Processes in parentheses were not quantified in this study

construct budgets of 14 chemical constituents. The diversity of chemical elements followed in the study allows a complete evaluation of fire effects as elements susceptible and resistant to pyric volatilization were included in the study.

## 13.2 Approach and Methods

### 13.2.1 Data Collection and Analysis

Streamflow at Swartboskloof was quantified using a weir with a 90° sharp-crested compound V-notch, which was monitored continuously with a Belfort water level recorder. Stream water samples were taken at the stream-gauging station (see Fig. 1.3). Weekly grab samples were taken from a well-mixed, free-flowing spot where the stream entered the stilling pond of the weir. Samples were also collected at the weir immediately after every spate had subsided using a stage sampler (van Wyk 1983). Samples were stored at 4°C for a maximum of 3 days until analyzed. The specific conductivity (25°C) of all samples and the pH of the grab samples were determined. Full chemical analyses were performed on the first grab sample of each month, and one selected spate for each month. Samples were preserved with mercuric chloride before chemical analyses were carried out by the Hydrological Research Institute in Pretoria.

Rainfall was measured in the Jonkershoek valley with a SA standard rain gauge with a Nipher shield. Atmospheric deposition samples were collected either in bulk deposition samplers (van Wyk 1981b) or in an Aerochem-metrics wet/dry atmospheric deposition sampler (Anon 1979). Bulk deposition samples were stored in 20-l containers in underground chambers to reduce bacterial activity and algal growth between weekly collections. Wet deposition samples from the wet/dry sampler were collected at 08:00 only on days after rain had fallen. Dry deposition was collected weekly as the 500 ml deionized water rinsate with which the sampler was spray-rinsed. Input values were corrected for impurities in the deionized water.

Rainfall, stream water and rinsate samples were analyzed according to the set standard water analytical procedures (van Vliet et al. 1988). Samples were analyzed for 14 chemical constituents [$Na^+$, $K^+$, $Ca^{2+}$, $Mg^{2+}$, $NH_4^+$, $Cl^-$, $SO_4^{2-}$, $NO_3^-$, $PO_4^{3-}$, $HCO_3^-$, TP (total phosphate), KN (Kjeldahl/ total nitrogen), Si and F], covering a full spectrum, rather than a selected few plant macro-nutrients, so as to detect any possible effects of fire on stream water quality.

Nutrient content of soils and vegetation of the catchment were determined to provide baseline data. Phytomass was estimated prior to the fire in 1987 by clipping vegetation in five plots (1 × 1 m) at each of three sites in proteoid shrublands. Samples were divided into dead (necromass) and live (biomass) components. Nutrient contents of the vegetation were determined, and multiplied by phytomass to obtain estimates of the nutrient contents of the vegetation (Table 13.1). Volatilization losses were determined by combusting ground (0.5 mm sieve) material in a muffle furnace. From each site six replicate samples of necromass and biomass were heated in crucibles for 15 min (estimated exposure time to fire) at 100, 200, 300, 400 and 500°C respectively. Paired samples were bulked at each temperature setting to give three samples for chemical analysis. In total, 740 samples were heat-treated and of these 320 pooled samples chemically analyzed. Results of analyses in each category were pooled to give final volatilization loss values. Erosion losses were estimated from bedload collected annually in the stilling pond behind the weir. Bedload samples were taken in 1986, 1987 and 1988.

## 13.2.2 Statistical Treatment of Data

Concentrations of chemical constituents in atmospheric deposition and stream water were compared by weighting each respectively with rainfall depth and streamflow volume. Mass chemical imports and exports were expressed as $kg\,ha^{-1}\,year^{-1}$ and were calculated by multiplying concentration values with either rainfall or streamflow volumes. Pre-fire data (January 1986 to February 1987) were compared with post-fire data (April 1987 to December 1988) to determine the effects of fire on the concentrations of chemical constituents in the stream water chemistry of Swartboskloof. A non-parametric Wilcoxon-Mann-Whitney test (Zar 1984, p. 139) was applied to pre- and post-fire mass exports to determine the impact of fire on the chemical constituents in the stream. Time trend (regression) analyses (Zar

**Table 13.1.** Estimated nutrient contents (kg $ha^{-1}$) of fynbos phytomass (necromass + biomass) of fynbos of three ages

| Chemical constituent | 29-year-old Fynbos (this study) | | | 21-year-old Fynbos (van Wilgen and Le Maitre 1981) | | | 11-year-old Fynbos (Low 1983) | | |
|---|---|---|---|---|---|---|---|---|---|
| | Biomass | Necromass | Phytomass | Biomass | Necromass | Phytomass | Biomass | Necromass | Phytomass |
| $K^+$ | 111.70 | <0.01 | 111.70 | 86.47 | 5.70 | 92.17 | 36.7 | 1.8 | 38.5 |
| $Ca^{2+}$ | 31.96 | 86.71 | 118.67 | 69.92 | 29.94 | 99.86 | 174.7 | 19.9 | 194.6 |
| $Mg^{2+}$ | 13.87 | 20.30 | 34.17 | 15.57 | 7.13 | 22.70 | 10.9 | 2.5 | 13.4 |
| N | 107.50 | 125.11 | 259.61 | 105.17 | 92.68 | 197.85 | 76.7 | 12.6 | 89.3 |
| P | 6.67 | 7.27 | 13.94 | 6.35 | 2.85 | 9.20 | 7.2 | 0.8 | 8.0 |

**Table 13.2.** Rainfall and streamflow-weighted mean concentrations (mg $l^{-1}$) of chemical constituents in bulk atmospheric deposition (inputs) and in stream water outputs (spates and baseflow) from the Swartboskloof catchment for the period January 1986 to December 1988

| Chemical constituent | Input | Output |
|---|---|---|
| $Na^+$ | 2.578 | 4.300 |
| $K^+$ | 0.490 | 0.828 |
| $Ca^{2+}$ | 0.770 | 0.755 |
| $Mg^{2+}$ | 0.290 | 0.669 |
| $NH_4^+$ | 0.061 | 0.0614 |
| $Cl^-$ | 3.968 | 8.643 |
| $SO_4^{2-}$ | 3.370 | 2.975 |
| $NO_3^-$ | 0.062 | 0.035 |
| $PO_4^{3-}$ | 0.003 | 0.0031 |
| $HCO_3^-$ | 5.407 | 5.043 |
| TP | 0.011 | 0.013 |
| KN | 0.204 | 0.307 |
| Si | 0.799 | 3.215 |
| F | 0.077 | 0.046 |
| pH | 4.85 | 5.88 |

1984, p. 268) for each variable were undertaken over the whole study period to detect possible trends in the chemical composition of the stream water related to fire. Cosine trends, corrected for seasonality (Cryer 1986), were fitted for selected elements, and the residuals tested for differences between years. SAS (SAS Institute Inc. 1985a,b) was used for all statistical tests.

The use of short-term data sets in water quality studies has often been criticized because of year-to-year variability in the hydrological regime. Hill (1986) found that six consecutive years of recording were sufficient to ensure that errors of greater than 20% would not occur in more than 5% of the recording period. On the other hand, Keller et al. (1986) found high coefficients of variation for water quality variables with a long-term study (15 years), but they were element-specific, ranging from 21% for $Ca^{2+}$ to 113% for $NH_4^+$. We therefore present our data graphically in addition to the statistical analysis, as many important small changes in the concentrations of chemical constituents in atmospheric deposition and stream water could be masked by using summarized and averaged data.

## 13.3 Atmospheric Deposition

The chemical composition of the atmosphere is highly variable and its constituents may originate from many sources including the oceans, the lithosphere, volcanic activity, organic matter decay and anthropogenic sources (Likens et al. 1977; Bowen 1979). Man-made pollutants appear to have a negligible effect on atmospheric chemistry at Swartboskloof (van Wyk 1990) and there is no volcanic activity in the area. High concentrations of $Na^+$, $Mg^{2+}$, $Cl^-$ and $SO_4{}^{2-}$ in atmospheric deposition at Swartboskloof (Table 13.2) relative to those found at inland sampling sites (van Wyk 1990) suggest a marine origin for atmospheric aerosols in the Jonkershoek valley. This contention is also supported by the $Cl^-:Na^+$ and $SO_4{}^{2-}:Cl^-$ ratios of 1.5 and 0.9 respectively which more closely resemble those found in seawater, namely 1.0 and 0.2 (Gambell and Fisher 1966), than the inland ratios for $Cl^-:Na^+$ of 2.3 and $SO_4{}^{2-}:Cl^-$ of 1.2 and 1.4 (van Wyk 1990). Agricultural activity around Stellenbosch could also have a pronounced effect on chemical concentration of atmospheric deposition, as is indicated by the $Ca^{2+}:Mg^{2+}$ ratio of 2.7 which is closer to the 3.7 determined by Eaton et al. (1973) for continental air masses, than the 0.2 for seawater. Except for the marine influence, Swartboskloof is not unusual in terms of atmospheric inputs, and is similar to other undisturbed catchment ecosystems around the world (Tables 13.2, 13.3).

Values of all non-marine-derived chemical constituents in atmospheric deposition at Swartboskloof are low. For example, concentrations of total N and P were less than half those recorded in other regions of South Africa [KN at 3.87, 10.20 and 15.17; TP at 0.50, 4.05 and 1.01 kg $ha^{-1}$ $year^{-1}$ for Jonkershoek, Sabie (Eastern Transvaal) and Winterton (Natal) respectively (van Wyk 1990)]. Mass values of these chemical constituents at Swartboskloof are also low in relation to atmospheric inputs in the industrialized countries of the northern hemisphere, i.e. 15.81 kg $ha^{-1}$ $year^{-1}$ $NO_3{}^-$, in the USA (a mean for 17 stations; Pack 1980) and 6–24 kg $ha^{-1}$ $year^{-1}$ $NO_3$-N in western Germany (Rehfuess 1989).

The acidity of atmospheric deposition recorded at Swartboskloof (pH 4.85) reflects the lack of industrial activity in surrounding areas. This is in contrast to other more polluted areas of South Africa such as the Eastern Transvaal Highveld that have much lower pH values ranging from pH 3.9–4.6 (Tyson et al. 1988).

## 13.4 Stream Water Chemistry

Concentrations of all chemical constituents in stream water, spates and baseflow were very similar (Fig. 13.2). Concentrations of $NH_4{}^+$, $NO_3{}^-$, $PO_4{}^{3-}$, and TP were exceptionally low in both baseflow and spates (Fig. 13.2). Studies of catchment mineral balances in North America show that in catchments with stable substrata such as granites and quartzite (as in Swartboskloof), most ions in streamwater are derived from atmospheric precipitation (Peters 1984). This is the probable case in Swartboskloof, with the higher output to input concentration values for $Na^+$, $K^+$, $Mg^{2+}$ and $Cl^-$

**Table 13.3** Summary of data on atmospheric deposition (imput) and streamwater chemistry (output) of undisturbed catchments. All values are means, in mg $l^{-1}$ (±SD)

| Area: | Hubbard Brook, USA | | Coweeta, USA | | Switzerland | Oregon, USA | | British Columbia, Canada | | |
|---|---|---|---|---|---|---|---|---|---|---|
| Vegetation: | Northern hardwoods | | Mixed hardwood | | Coniferous: spruce and fir | Douglas fir–130 yr old | | Douglas fir and Red alder: 1 = 20; 2 = 90 yr old | | |
| Sample period (yr): | 11 | | 11 | | 15 | 7–13 | | 0.75 | | |
| Source: | Likens et al. (1977) | | Swank and Waide (1988) | | Keller et al. (1986) | Martin and Harr (1988) | | Binkley et al. (1982) | | |
| Variable | Input | Output | Input[a] | Output[b] | Output[d] | Input | Output | Input | Output 1 | Output 2 |
| $Na^+$ | 0.12 (0.01) | 0.88 (0.02) | 0.172 (0.009) | 0.80 (0.23) | 2.1 (0.9) | 0.25 (0.03) | 2.49 (0.06) | | | |
| $K^+$ | 0.07 (0.02) | 0.23 (0.01) | 0.124 (0.006) | 0.35 (0.09) | 0.7 (0.3) | 0.09 (0.02) | 0.43 (0.03) | 0.02 (0.002) | 0.11 (0.02) | 0.05 (0.02) |
| $Ca^{2+}$ | 0.17 (0.02) | 1.65 (0.06) | 0.236 (0.008) | 0.53 (0.11) | 45.1 (9.1) | 0.18 (0.03) | 2.91 (0.02) | 0.20 (0.008) | 1.66 (0.08) | 1.50 (0.04) |
| $Mg^{2+}$ | 0.05 (0.01) | 0.38 (0.01) | 0.047 (0.002) | 0.28 (0.04) | 3.5 (2.9) | 0.04 (0.01) | 0.58 (0.02) | 0.06 (0.001) | 0.40 (0.02) | 0.32 (0.005) |
| $NH_4^+$ | 0.22 (0.01) | 0.04 (0.01) | 0.117 (0.003) | 0.003 (0.001) | 0.04 (0.04) | | | 0.009 (0.001) | 0.001 (0.0001) | 0.002 (0.0002) |
| $Cl^-$ | 0.51 (0.09) | 0.54 (0.02) | 0.283 (0.007) | 0.55 (0.06) | 0.8 (0.4) | | | 0.60 (0.04) | 0.82 (0.07) | 0.82 (0.04) |
| $SO_4^{2-}$ | 2.87 (0.11) | 6.23 (0.11) | 0.675 (0.008) | 0.20 (0.10) | 10.1 (5.6) | | | 1.34 (0.10) | 1.78 (0.19) | 2.11 (0.10) |
| $NO_3^-$ | 1.43 (0.11) | 1.93 (0.31) | 0.174 (0.002) | 0.006 (0.006) | 0.24 (0.11) | 0.031 (0.002) | 0.003 (0.001) | 0.68 (0.06) | 3.91 (0.56) | 0.25 (0.06) |
| $PO_4^{3-}$ | 0.008 – | 0.002 – | 0.0056 (0.0004) | 0.001 (0.0005) | | 0.004 (0.001) | 0.022 (0.001) | | | |
| $HCO_3^-$ | – – | 1.62 (0.12) | | | | 2.0 (0.1) | 19.9 (0.6) | 0.31 (0.06) | 7.81 (0.49) | 7.81 (0.18) |
| TP | | | | | | | | | | |
| KN | | | | | | 0.069 (0.005) | 0.042 (0.003) | | | |
| Si | – – | 4.43 (0.11) | – – | 5.79 (1.57) | | 0.26 (0.07) | 12.35 (1.01) | 0.06 (0.006) | 3.12 (0.28) | 2.19 (0.06) |
| F | | | | | | | | | | |
| pH | 4.13 (0.00) | 4.90 (0.05) | 4.68 – | 6.7[c] | | 5.4 (0.1) | 7.3 (0.04) | 4.5 (<0.1) | 6.51 (0.06) | 6.8 (<0.1) |

[a] Mean weekly concentrations in bulk precipitation within the greater catchment.
[b] Mean concentrations for 7 control catchments in the greater catchment area, averaged to give single values with standard deviations of each set of 7 values.
[c] Calculated from $H^+$ concentration in stream water flowing from a single catchment in the greater catchment area (Coweeta **WS** 2).
[d] Mean concentrations for 6 catchments, averaged to give mean values and standard deviation for the greater catchment area.

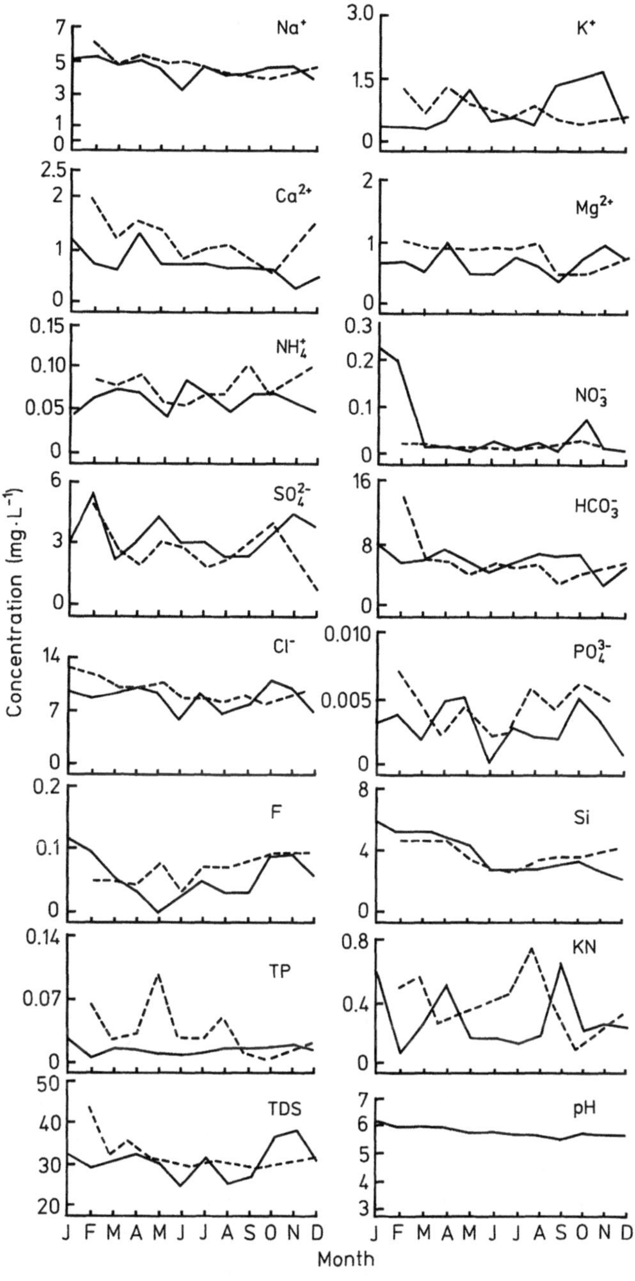


**Fig. 13.2.** Monthly mean concentrations (mg $l^{-1}$) of chemical constituents in stream water at Swartboskloof, from January 1986 to December 1988, for baseflow (____) and spates (----)

**Table 13.4.** Chemical analyses of soil samples at four sites in Swartboskloof before the fire

| Variable | Soil samples | | | |
|---|---|---|---|---|
| | Site 1 | Site 2 | Site 3 | Site 4 |
| Soil grain size | | | | |
| Clay (%) | 18.0 | 8.0 | 8.0 | 8.0 |
| Silt (%) | 32.0 | 12.0 | 10.0 | 10.0 |
| Gravel (%) | 13.2 | 14.8 | 11.7 | 11.7 |
| Sand-medium (%) | 29.5 | 41.6 | 42.0 | 41.3 |
| Sand-fine (%) | 7.4 | 23.6 | 28.1 | 28.9 |
| Conductivity (ohms) | 6200 | 4050 | 4700 | 4500 |
| Cation exchange capacity | 1.92 | 2.95 | 2.63 | 1.93 |
| pH | 4.1 | 3.8 | 3.7 | 3.9 |
| Na (mEq) | 0.07 | 0.04 | 0.07 | 0.07 |
| K (mEq) | 0.20 | 0.13 | 0.09 | 0.09 |
| Ca (mEq) | 0.54 | 1.12 | 0.67 | 0.34 |
| Mg (mEq) | 0.13 | 0.43 | 0.38 | 0.17 |
| Na (ppm) | 55 | 21 | 14 | 33 |
| K (ppm) | 72 | 53 | 38 | 36 |
| Ca (ppm) | 0.80 | 0.88 | 0.61 | 0.42 |
| Mg (ppm) | 0.36 | 0.36 | 0.33 | 0.19 |
| P (ppm) | 7 | 3 | 2 | 1 |
| Carbon (%) | 1.26 | 0.89 | 1.21 | 0.96 |
| Nitrogen (%) | – | 0.093 | 0.087 | 0.075 |

(Table 13.2) being the result of evapotranspiration. Weathering of bedrock certainly contributes to inorganic constituent outflow concentrations in most systems (Bricker and Rice 1989). Key minerals present in the bedrock have pronounced effects on ion concentrations in outflow (Bailey et al. 1987). The main bedrock materials at Swartboskloof are granite of the Cape Granite Suite and sandstone from the Peninsula Formation (Söhnge 1988). Silicon in the form of quartz (in the sandstone and granite) and feldspar (in the granite) is the main constituent. Thus weathering of the geological substrate of the catchment would explain the high Si export concentrations compared to imports of this element (Table 13.2). Weathering of the alkali-feldspar-rich granites found at Swartboskloof result in clayey soils (Söhnge 1988) and it is probable that the $K^+$ and $Na^+$ released by this process could account in part for the high outflow concentrations of these cations. This is supported by the pre-fire soil analyses (Table 13.4) which show $K^+$ and $Na^+$ to be the major cations in soils derived from both geological substrata.

Total mass outflow of most chemical constituents from the catchments in both the unburnt and burnt state (Fig. 13.3) shows sudden increases at the onset of the rainy season (April/May). The early winter losses from the catchment may be the outcome of source area expansion as described by Hewlett and Nutter (1969). During the drier summer months the source area is restricted to the lower streams. Elements made available by mineral-

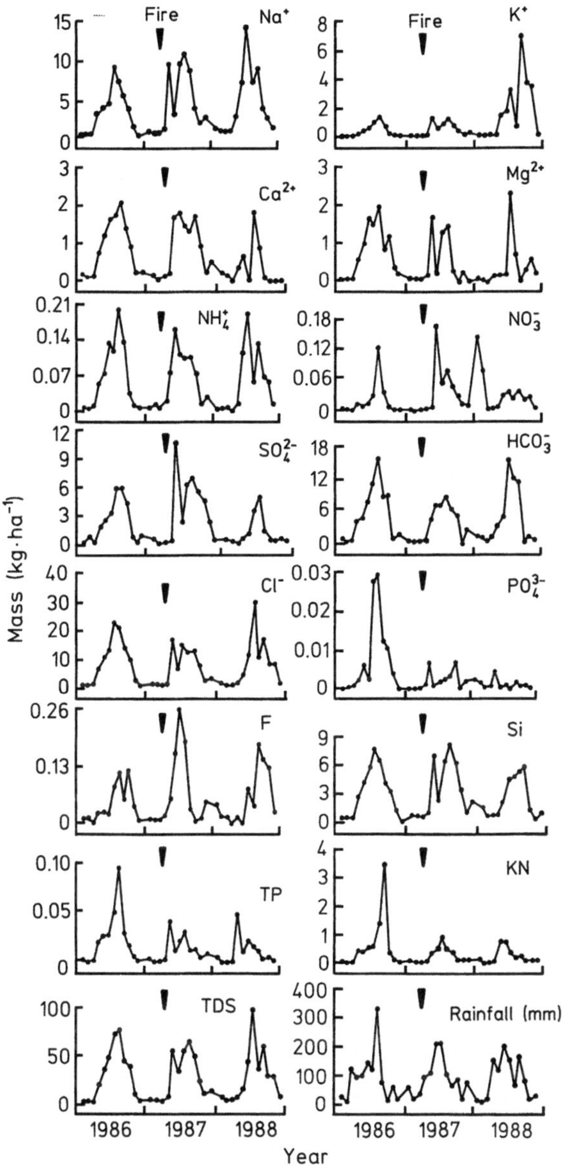


**Fig. 13.3.** Time trends of monthly total mass outflow ($kg\,ha^{-1}\,month^{-1}$) of chemical constituents in stream water at Swartboskloof, from January 1986 to December 1988. The fire of 17 March 1987 is indicated

ization processes (decomposition of organic matter and weathering of bed-rock) throughout the catchment only reach the stream with the rapid expansion of the source area with the onset of the winter rains. These first winter rains would also be responsible for leaching soluble ions from

**Table 13.5.** Annual total mass budgets and nett losses or gains (kg $ha^{-1}$ $yr^{-1}$) for the Swartboskloof catchment (1986 to 1988). Atmospheric deposition represents input while stream water chemistry (excluding the effect of fire) plus volatilization (including fire) represents output. Bedload losses were ignored as they were negligible (see Table 13.9)

| Chemical constituent | Input atmospheric deposition | Output | | Nett gain/loss without fire | Nett gain/loss with fire |
|---|---|---|---|---|---|
| | | Stream water | Volatilization | | |
| $Na^+$ | 47.891 | 46.755 | ND[a] | + 1.136 | ND |
| $K^+$ | 9.997 | 10.773 | 2.899 | − 0.776 | −3.675 |
| $Ca^{2+}$ | 12.969 | 8.158 | 2.648 | + 4.811 | +2.163 |
| $Mg^{2+}$ | 4.984 | 6.737 | 0.757 | − 1.753 | −2.510 |
| $NH_4^+$ | 2.551 | 0.663 | ND | + 1.888 | ND |
| $Cl^-$ | 96.083 | 96.478 | ND | − 0.395 | ND |
| $SO_4^{2-}$ | 60.057 | 27.316 | ND | +32.741 | ND |
| $NO_3^-$ | 1.009 | 0.355 | ND | + 0.654 | ND |
| $PO_4^{3-}$ | 0.540 | 0.043 | ND | + 0.497 | ND |
| $HCO_3^-$ | 103.044 | 57.108 | ND | +45.936 | ND |
| TP | 1.158 | 0.188 | 0.332 | + 0.970 | +0.638 |
| KN | 7.868 | 4.461 | 6.334 | + 3.407 | −2.927 |
| Si | 9.091 | 52.384 | ND | −43.293 | ND |
| F | 0.622 | 0.647 | ND | − 0.025 | ND |
| TDS | 373.829 | 337.335 | | +36.494 | ND |

[a] ND = not determined.

litter material, which falls predominately in summer (November to January, King et al. 1987; Mitchell and Coley 1987). The first major spate after summer would also remove most riparian-derived detritus from the rivers (King et al. 1987) thus removing this material as a future source of ions. Increased activity of microorganisms also accounts for some of the outflow (see Chap. 14).

## 13.5 Mineral Nutrient Budgets

Mass imports and exports were calculated from wet/dry atmospheric deposition and spates plus baseflow respectively to provide estimates of net gains and losses of each chemical constituent (Table 13.5). The budget shows a net gain of the important mineral nutrients (N and P) into the catchment from the atmosphere. The total gain in TDS is about 36 kg $ha^{-1}$ $year^{-1}$. $HCO_3^-$, $SO_4^{2-}$, $Ca^{2+}$, KN, $NH_4^+$, $NO_3^-$, $Na^+$, TP and $PO_4^{3-}$ were retained, while losses of $K^+$, $Mg^{2+}$, $Cl^-$, Si and F occurred. Exceptional cases are the high loss of Si (43 kg $ha^{-1}$) and the high gains of $HCO_3^-$ (46 kg $ha^{-1}$) and $SO_4^{2-}$ (33 kg $ha^{-1}$). The loss of Si has been attributed to leaching of weathered bedrock and of soils (see above). Other ecosystems show a similar high net loss of Si, either a gain or loss of $SO_4^{2-}$, but $HCO_3^-$ is usually also lost from the systems in substantial amounts (see Table 13.6). In fact, most chemical constituents are lost, and only essential plant nutrients such as phosphate and nitrate are retained in most ecosystems (Table 13.6).

Net gain of chemical constituents in Swartboskloof, many of which are essential plant mineral nutrients, could occur as a result of their absorption by the vegetation in the catchment and their eventual incorporation into the soil organic matter, which is the largest reserve of plant nutrients in fynbos systems (Mitchell et al. 1987). Many elements, such as N and P, are held in phytomass and are cycled slowly in fynbos ecosystems (Mitchell and Coley 1987; Stock et al. 1988) because of the slow decomposition and biological mineralization rates. These elements are therefore not susceptible to loss by leaching. Fire has been cited as the major factor responsible for nutrient cycling in fynbos (Brown and Mitchell 1986; Stock and Lewis 1986a) but it introduces the danger of post-fire nutrient leaching.

## 13.6 The Effect of Fire

Chemical changes induced by fire can increase or decrease the susceptibility of each element to loss from a catchment. Most elements contained in phytomass are physically mineralized during burning and most of this is

**Table 13.6.** Summary of data on average annual nutrient budgets for undisturbed catchments. Input (atmospheric deposition) and output (stream water) values and the net difference are all presented in kg $ha^{-1}$ $yr^{-1}$ (mean ± SD)

| Area: | Hubbard Brook, USA | | | Coweeta, USA | | |
|---|---|---|---|---|---|---|
| Vegetation: | Northern Hardwoods | | | Mixed hardwood | | |
| Sample period (yr): | 11 | | | 11 | | |
| Source: | Likens et al. (1977) | | | Swank and Waide (1988)[a] | | |
| Variable | Input (SD) | Output (SD) | Net loss/gain | Input (SD) | Output (SD) | Net loss/gain |
| $Na^+$ | 1.59 (0.12) | 7.23 (0.47) | −5.64 (0.50) | 4.24 (0.78) | 10.87 (1.71) | − 6.63 (2.15) |
| $K^+$ | 0.89 (0.19) | 1.89 (0.14) | −0.99 (0.27) | 1.96 (0.18) | 4.79 (0.48) | −2.84 (0.36) |
| $Ca^{2+}$ | 2.17 (0.21) | 13.7 (1.1) | −11.6 (1.2) | 3.90 (0.37) | 7.49 (1.96) | −3.59 (1.73) |
| $Mg^{2+}$ | 0.58 (0.07) | 3.13 (0.21) | −2.55 (0.26) | 0.93 (0.14) | 3.94 (0.72) | −3.01 (0.66) |
| $NH_4^+$ | 2.90 (0.18) | 0.34 (0.08) | +2.55 (0.21) | 2.19 (0.28) | 0.05 (0.02) | +2.15 (0.26) |
| $Cl^-$ | 7.01 (1.1) | 4.89 (0.33) | +2.02 (1.10) | 6.91 (1.52) | 7.70 (1.52) | −0.79 (1.52) |
| $SO_4^{2-}$ | 38.4 (2.5) | 53.8 (4.4) | −15.3 (2.8) | 11.69 (1.34) | 3.13 (2.24) | +8.56 (1.81) |
| $NO_3^-$ | 19.7 (2.1) | 17.1 (3.2) | +2.6 (2.7) | 3.34 (0.56) | 0.10 (0.11) | +3.25 (0.48) |
| $PO_4^{3-}$ | 0.131 | 0.023 | +0.108 | 0.10 (0.01) | 0.02 (0.004) | +0.08 (0.01) |
| $HCO_3^-$ | – | 7.64 (0.77) | −7.64 (0.77) | 1.27 | 40.44 | −39.16 |
| TP | | | | | | |
| KN | | | | | | |
| Si | – | 38.4 (3.0) | −38.4 (3.0) | 0.54 (0.10) | 79.03 (15.08) | −78.10 (16.4) |
| F | | | | | | |

[a] Mean annual from 7 control catchments were averaged to give mean input, output and net loss/gain values for all variables. Standard deviation was determined on the 7 values from the catchments

deposited at the soil surface in the ash, where loss can occur due to erosion (Zinke 1977). DeBano and Conrad (1978) found in chaparral systems that soluble cations such as $K^+$, $Ca^{2+}$ and $Mg^{2+}$ were susceptible to erosion and runoff after fire, whereas elements such as P were retained within the system. This is similar to fynbos mountain catchments with lower rainfall (less than 1300 mm $year^{-1}$, van Wyk 1981a). Baseflow concentration values of $Na^+$, $Cl^-$, $SO_4^{2-}$ and $HCO_3^-$ increase in the first 12 weeks after fire, while changes in levels of $Mg^{2+}$, $Ca^{2+}$ and $K^+$ are undetectable. Increased $Na^+$, $K^+$, $Ca^{2+}$, $Mg^{2+}$, $Cl^-$, $SO_4^{2-}$ and $HCO_3^-$ concentrations were found only in the first few spates following fire, while other chemical constituents appeared to be in low concentrations since $PO_4^{3-}$, $NH_4^+$ and $NO_3^-$ were below the levels of detection of the methods employed (van Wyk 1981a). All changes in stream water chemistry were of short duration and values all returned to pre-burn levels within 10 months after fire.

The 1987 fire in Swartboskloof had a significant ($P < 0.05$) effect only on the mass export of $K^+$ and $PO_4^{3-}$ according to trend analysis (Table 13.7), while the non-parametric Wilcoxon-Mann-Whitney test (Table 13.8) showed significant differences for $K^+$ and $NO_3^-$ ($P < 0.05$). Trends are often only evident in spates, but the results remained the same when spate flow was analyzed separately. Graphically evident changes for $K^+$, $PO_4^{3-}$, TP and KN (Fig. 13.3) were corrected for seasonality by fitting cosine trends to the data (Cryer 1986) and then testing the residuals for differences between years. Differences between years were statistically significant ($P <$

**Table 13.6** (continued)

| Oregon, USA<br>Douglas fir -130 yr old | | | Plynlimon, Wales<br>Grasslands (G1&2); Sitka spruce (A1&2) | | | | SWCape, South Africa<br>Fynbos | | |
|---|---|---|---|---|---|---|---|---|---|
| 7 – 13<br>Martin and Harr (1988) | | | 3 (4 catchments)<br>Reynolds et al. (1989) | | | | 2 – 8<br>van Wyk (1981a) | | |
| Input (SD) | Output (SD) | Net loss/ gain | Net loss/ gain (G1) | Net loss/ gain (G2) | Net loss/ gain (A1) | Net loss/ gain (A2) | Input | Output | Net loss/ gain |
| 5.4 (0.58) | 30.7 (2.83) | −25.3 | −23.6 | −25.0 | −33.1 | −39.6 | 27.01 | 23.93 | +3.08 |
| 2.0 (0.29) | 5.6 (0.85) | −3.6 | −1.1 | −1.1 | −0.8 | −0.5 | 2.34 | 0.75 | +1.59 |
| 3.7 (0.68) | 35.0 (4.46) | −31.3 | −23.6 | −10.2 | −11.8 | −18.0 | 5.76 | 3.30 | +2.46 |
| 0.9 (0.09) | 7.1 (0.63) | −6.2 | −11.1 | −7.5 | −8.7 | −9.2 | 5.24 | 3.74 | +1.50 |
| 1.0 (0.05) | 0.5 (0.07) | +0.5 | +7.0 | +7.1 | +6.9 | +7.3 | | | |
| | | | −29.6 | −31.5 | −43.6 | −54.2 | 56.51 | 42.11 | +14.40 |
| | | | −41.1 | −21.3 | −36.1 | −48.8 | | | |
| 0.7 (0.06) | 0.1 (0.01) | +0.6 | +1.4 | −1.0 | −14.1 | −8.3 | | | |
| 0.1 (0.01) | 0.3 (0.04) | −0.2 | | | | | | | |
| 8.6 (0.65) | 47.9 (4.29) | −39.3 | −19.4 | −8.8 | −5.0 | −13.6 | | | |
| 2.4 (0.63) | 72.0 (8.23) | −69.6 | −20.3 | −19.3 | −28.7 | −26.6 | | | |

**Table 13.7.** Trends in the mean monthly mass exports of 15 chemical constituents from the Swartboskloof catchment for the period 1986 to 1988 (* = significant at $P < 0.05$, NS = not significant)

| Variable | Mean monthly export (kg $ha^{-1}$) | Standard deviation | Slope[a] (b) | t | Probability >t | Significance of trend |
|---|---|---|---|---|---|---|
| $Na^+$ | 3.896 | 3.666 | 0.06571 | 1.121 | 0.270 | NS |
| $K^+$ | 0.898 | 1.422 | 0.06445 | 3.170 | 0.003 | * |
| $Ca^{2+}$ | 0.680 | 0.693 | −0.01640 | −1.502 | 0.142 | NS |
| $Mg^{2+}$ | 0.561 | 0.639 | −0.00999 | −0.974 | 0.337 | NS |
| $NH_4^+$ | 0.060 | 0.059 | 0.00013 | 0.143 | 0.887 | NS |
| $Cl^-$ | 8.040 | 7.460 | 0.05645 | 0.466 | 0.644 | NS |
| $SO_4^{2-}$ | 2.276 | 2.548 | −0.02323 | −0.563 | 0.577 | NS |
| $NO_3^-$ | 0.030 | 0.041 | 0.00046 | 0.699 | 0.489 | NS |
| $PO_4^{3-}$ | 0.004 | 0.007 | −0.00022 | −2.118 | 0.042 | * |
| $HCO_3^-$ | 4.759 | 4.752 | −0.03587 | −0.038 | 0.970 | NS |
| TP | 0.016 | 0.019 | −0.00042 | −1.370 | 0.180 | NS |
| KN | 0.372 | 0.606 | −0.01111 | −1.148 | 0.259 | NS |
| Si | 2.910 | 2.473 | −0.00905 | −0.225 | 0.823 | NS |
| F | 0.054 | 0.064 | 0.00122 | 1.193 | 0.241 | NS |
| TDS | 28.111 | 25.254 | 1.79854 | 0.362 | 0.719 | NS |

[a] Positive values indicate increasing trend, negative values indicate decreasing trend.

0.05) for all elements tested. It appears that fire has a pronounced effect upon ecosystem function. Reasons for the sudden increase in $K^+$ losses are not clear. The decreased concentrations of the other chemical constituents could be due to their retention by post-fire resprouting plants. In all three sampling years, concentrations of chemical constituents in the bedload

**Table 13.8.** Mean monthly total mass exports (kg $ha^{-1}$ $month^{-1}$) of chemical constituents from Swartboskloof before (January 1986 to February 1987) and after (April 1987 to December 1988) the fire, using a Wilcoxon-Mann-Whitney test (* = significant at $P <$ 0.05, NS = not significant)

| Variable | Mean monthly export (kg $ha^{-1}$) Before | After | Z | Probability | Significance |
|---|---|---|---|---|---|
| $Na^+$ | 2.74 | 4.72 | −1.701 | 0.089 | NS |
| $K^+$ | 0.35 | 1.29 | −2.439 | 0.015 | * |
| $Ca^{2+}$ | 0.72 | 0.65 | 0.482 | 0.619 | NS |
| $Mg^{2+}$ | 0.65 | 0.50 | 0.513 | 0.608 | NS |
| $NH_4^+$ | 0.05 | 0.06 | −0.642 | 0.521 | NS |
| $Cl^-$ | 7.20 | 8.64 | −1.251 | 0.211 | NS |
| $SO_4^{2-}$ | 1.83 | 2.59 | −0.898 | 0.369 | NS |
| $NO_3^-$ | 0.02 | 0.04 | −2.342 | 0.019 | * |
| $PO_4^{3-}$ | 0.01 | 0.00 | 0.515 | 0.606 | NS |
| $HCO_3^-$ | 4.73 | 4.78 | −0.289 | 0.773 | NS |
| TP | 0.02 | 0.01 | 0.000 | 1.000 | NS |
| KN | 0.51 | 0.28 | −0.289 | 0.773 | NS |
| Si | 2.61 | 2.13 | −1.155 | 0.248 | NS |
| F | 0.03 | 0.07 | −1.379 | 0.168 | NS |
| TDS | 24.75 | 30.51 | −1.219 | 0.223 | NS |

**Table 13.9.** Mass and chemical export (kg $ha^{-1}$ × $10^{-4}$) by bedload from the Swartboskloof catchment before and after fire. Samples were taken on 22 August 1986, 7 August 1987 and 23 October 1988

| Chemical constituent | Sample | | |
|---|---|---|---|
| | August 1986 | August 1987 | October 1988 |
| Oven-dry bedload | 147600 | 115900 | 151400 |
| Na | 2.95 | 1.97 | 2.12 |
| K | 5.90 | 0.46 | 0.91 |
| Ca | 0.34 | 12.26 | 7.27 |
| Mg | 13.29 | 2.09 | 2.73 |
| N | 183.00 | 63.49 | 15.14 |
| P | 2.95 | 1.81 | 1.39 |
| C | 7010 | 1274 | 606 |

(Table 13.9) were very low. Although there is a post-fire increase in Ca and decreases in K, Mg, N and C, export values are negligible when compared to that of stream water.

Increased nutrient export (Fig. 13.2) seemed more pronounced for most constituents at the start of the rainy season of the post-fire year (1987) than for 1986 and 1988. This was not due to any differences in rainfall (Fig. 13.2), suggesting that the minor increase in export of some constituents was due to immediate fire effects. A change resulting from long-term influences of the fire was the prominent sudden surge in mass export of $NO_3^-$ in the first summer after the fire. This is not explained by any increases in rainfall, nor is it paralleled by increases in the output of any other chemical constituent (Fig. 13.3). Of the nitrogen remaining in the ecosystem after fire, some is chemically altered by soil heating and subsequent microbial activity to the highly available $NH_4^+$ and $NO_3^-$ forms (DeBano et al. 1977; Dunn and DeBano 1977; Dunn et al. 1979; Wells et al. 1979; Stock and Lewis 1986a). Stock and Lewis (1986a) have reported increased nitrification activity and $NO_3^-$ accumulation in post-fire coastal fynbos soils and with the first rains this could lead to a surge in mass export (Fig. 13.3) as nitrate is leached rapidly from soils. The depleted N capital of an ecosystem can be replaced either by atmospheric inputs [which Stock and Lewis (1986b) estimate would take between 10–80 years in fynbos] or via $N_2$-fixation (see also Chap. 14). The period required for $N_2$ fixation to replace volatilized N in fynbos ecosystems is unknown. Though such large changes in the forms of soil N can occur, only negligible losses of N in streamflow after chaparral fires have been reported (DeBano and Conrad 1976, 1978) and similar low $NO_3^-$ and $NH_4^+$ values have been also reported in streamflow after other fynbos fires (van Wyk et al. 1984; van Wyk and Lesch 1989).

One important factor that could influence the concentrations of mineral nutrients in stream water output after fire is volatilization. From laboratory volatilization studies it is clear that for all the elements studied ($K^+$, $Ca^{2+}$, $Mg^{2+}$, N and P) there is an increase in volatilization losses with increased temperature (Table 13.10). Nitrogen is regarded as the most important nutrient volatilized during fires (DeBano et al. 1977, 1979a; DeBano and Conrad 1978, Wells et al. 1979) and losses of up to 80% have been shown for chaparral biomass combusted at 528°C during laboratory studies (Dunn and DeBano 1977). Arianoutsou and Margaris (1981) reported a 96% volatilization loss of N during laboratory combustion studies of maquis. These figures concur with the mean loss of 80% N at 500°C from phytomass in the Swartboskloof laboratory study (Table 13.10). Combustion of maquis at 540°C recorded a 54% loss of $K^+$ (Arianoutsou and Margaris 1981) and in field-fire studies, DeBano and Conrad (1978) recorded 50% $K^+$ losses for chaparral. In contrast to this, the combustion of fynbos to 500°C caused a loss of 75% $K^+$. Mass volatilization losses of mineral nutrients from phytomass at Swartboskloof (Table 13.10), calculated from the nutrient content of pre-fire fynbos (Table 13.1), show that losses due to a single fire could greatly affect the nutrient status of the catchment. If, however, fire cycles are considered (here 29 years), the budget shows that, except for N, the major output from a catchment still occurs through stream water but that these losses are balanced by similar inputs from atmospheric deposition (Table 13.5). Rate of $N_2$-fixation for fynbos ecosystems are unknown, as is the rate of weathering of the soil and bedrock. Both these factors would increase the input figures calculated for constituents in this study, which could lead to balanced budgets, showing fire to have little effect on the long-term nutrient capital in fynbos ecosystems.

**Table 13.10.** Experimentally determined mass volatilization losses at different temperatures in kg $ha^{-1}$ and as a percentage of the total amount of the element concerned in the sample (in parentheses), of chemical constituents from the vegetation of Swartboskloof

| Chemical constituent | 100°C | 200°C | 300°C | 400°C | 500°C | Mean 400–500°C |
|---|---|---|---|---|---|---|
| *Biomass* | | | | | | |
| $K^+$ | 16.72(15) | 25.20(23) | 39.48(35) | 74.77(67) | 84.08(75) | 79.43(71) |
| $Ca^{2+}$ | 0 | 0 | 0.37(1) | 18.14(57) | 23.18(73) | 20.66(65) |
| $Mg^{2+}$ | 0 | 0 | 0.25(2) | 7.73(56) | 10.03(72) | 8.88(64) |
| N | 3.79(4) | 17.58(16) | 39.87(38) | 82.95(77) | 92.28(86) | 87.62(81) |
| P | 0.76(11) | 1.24(19) | 2.44(37) | 4.56(68) | 5.19(78) | 4.88(73) |
| *Necromass* | | | | | | |
| $K^+$ | 0 | 0 | 0 | 0 | 0 | 0 |
| $Ca^{2+}$ | 6.32(7) | 14.33(17) | 28.66(33) | 42.83(49) | 53.61(62) | 48.22(56) |
| $Mg^{2+}$ | 2.80(14) | 5.43(27) | 8.38(41) | 10.37(51) | 11.91(59) | 11.14(55) |
| N | 2.79(2) | 7.92(6) | 50.17(40) | 79.62(64) | 91.40(73) | 85.51(68) |
| P | 0.52(7) | 0.43(6) | 2.45(34) | 3.97(55) | 4.45(61) | 4.21(58) |
| *Phytomass* | | | | | | |
| $K^+$ | 16.72(15) | 25.20(23) | 39.48(35) | 74.77(67) | 84.08(75) | 79.43(71) |
| $Ca^{2+}$ | 6.32(5) | 14.33(12) | 29.03(24) | 60.98(51) | 76.79(65) | 68.89(58) |
| $Mg^{2+}$ | 2.80(8) | 5.43(16) | 8.63(25) | 18.10(53) | 21.94(64) | 20.02(59) |
| N | 6.58(3) | 25.50(10) | 90.04(35) | 162.56(63) | 183.68(71) | 173.12(67) |
| P | 1.28(9) | 1.67(12) | 4.89(35) | 8.52(61) | 9.64(69) | 9.08(65) |

Losses of chemical constituents through leaching after fire depend to a large extent upon the rapidity with which the vegetation recovers, and the soil properties of the catchment (Zinke 1977). Rapid colonization in the first year of post-fire chaparral environments by herbaceous plants and seedlings of chaparral shrubs has been noted (Schlesinger and Gill 1980). This is similar to patterns of post-fire fynbos recovery (Kruger and Bigalke 1984; Hoffman et al. 1987). The resprouting elements of the vegetation in Swartboskloof (see Chap. 6) are particularly important in this respect, as they would assimilate nutrients released by the fire. Higgins et al. (1987) found some 46% of the below-ground biomass was composed of the roots of resprouters in Swartboskloof. These roots were located mainly in the surface layers of the soil profile (67% of soil profile root phytomass in the top 100 mm). Thus fynbos can exhibit rapid regrowth which, together with a permanent root network, could potentially trap most nutrients released from fires. Considering the poor nutrient retentivity (low cation exchange capacity; Lambrechts 1979) of the sandy soils of the region, this biological sink could be of major importance in preventing large-scale losses of soluble elements released by fire.

A further factor possibly reducing post-fire stream water export of chemical constituents could be the largely unaffected riparian zone of the Swartboskloof catchment. In agricultural research, riverine vegetation has proved to be an effective buffer against chemical leaching from catchments as it acts as an excellent filter for released nutrients (Karr and Schlosser 1978; Lowrance et al. 1984; Peterjohn and Correll 1984; Jacobs and Gilliam 1985; Lowrance et al. 1985). A healthy riparian zone is propagated as a general control for non-point-source pollution (Phillips 1989). The riparian zones of the six catchments used in van Wyk's (1981a) earlier studies, however, were poorly developed because of regular short interval fires (van Wilgen and Kruger 1981). This probably accounts for difference in the magnitude of post-fire nutrient losses between Swartboskloof and other catchments mentioned. A sound policy would be to administer fire so that, within the constraints of suitable burning conditions and fire cycles, a healthy riparian zone can be maintained to act as a filter preventing excessive nutrient losses from catchments.

## 13.7 Conclusions

Atmospheric inputs to the Swartboskloof ecosystem are low, and no industrial influence is evident, although marine influence can be seen. In the absence of fire, outputs are also low, and generally the budgets balance. Some outputs are greater than inputs (for example Si), and these are related to the geology of the area. It is only after the first heavy rains each year that outflows from the system are apparent.

Fire had little overall effect on the budgets, or on the nutrient capital of the catchment. It was evident that vegetation (both regrowth from burnt fynbos, and unburnt riparian zones) played an important role in acting as a nutrient filter or trap. The management of fynbos catchments using fire therefore appears to be a viable long-term option, and fears of nutrient depletion appear unfounded. The input-output budgets appear to be a suitable technique for assessing the effects of fire in this case.

*Acknowledgements.* We are indebted to H.J. Barnard and F.W. Prinsloo for constructing and maintaining the weir. H.J. Barnard built and installed the water quality and bedload samplers. H.J. Barnard, D. Willemse and H. Julius collected field data; R. Robyntjies managed the data; L. Carelse managed the local laboratory; chemical analyses were done by the Hydrological Research Institute, Pretoria; soil analyses were done at the Elsenberg Agricultural Research Centre; plant material and bedload were analyzed at the Saasveld Forestry Research Centre; D.T. Mitchell provided guidance in the volatilization studies, and the University of Cape Town provided facilities for the volatilization studies.

# 14 Soil Microorganisms and Activities in Relation to Season, Soil Factors and Fire

C.A. van Reenen, G.J. Visser, and M.A. Loos

## 14.1 Introduction

Microorganisms play major roles in cycling carbon, nitrogen and other elements in nature, in particular in the mineralization of organic compounds and transformations of inorganic nutrients (Alexander 1977). Fire can be expected to affect the nutrient status of the soil (Viro 1974; Christensen and Muller 1975; Dunn and DeBano 1977; DeBano and Conrad 1978; DeBano et al. 1979a,b; Wells et al. 1979; Gray and Schlesinger 1981; DeBano and Dunn 1982; Rundel 1983; Brown and Mitchell 1986; Mitchell and Brown 1986; Stock and Lewis 1986a; Mitchell 1987) and soil microbial populations (Ahlgren 1974; Christensen and Muller 1975; Dunn and DeBano 1977; Dunn et al. 1979, 1985; Wells et al. 1979; Warcup 1981; Arianoutsou-Faraggitaki and Margaris 1982). The nutrient budget of the Swartboskloof

**Table 14.1.** Physical and chemical properties of three soil forms in Swartboskloof. Data from Fry (1987) for soil to depth 250 mm (Glenrosa, Clovelly forms) or 200 mm (Fernwood form)

| | Glenrosa | Clovelly | Fernwood |
|---|---|---|---|
| Sand (2 to 0.02 mm) (%) | 72.2 | 89.5 | 92.5 |
| Silt (0.02 to 0.002 mm) (%) | 12.4 | 4.2 | 6.1 |
| Clay (<0.002 mm) (%) | 15.4 | 6.3 | 1.4 |
| Organic C (%) | 7.9 | 1.8 | 0.9 |
| Total N (%) | 0.34 | 0.052 | 0.043 |
| P (1.0 M HCl-extractable) (μg/g) | 310.4 | 65.5 | 33.8 |
| Cations ($Na^+$, $K^+$, $Ca^{2+}$, $Mg^{2+}$) [cmol(+)/kg][a] | 1.76 | 0.16 | 0.53 |
| Cation exchange capacity (c.e.c.) [cmol(+)/kg][b] | 15.68 | 3.33 | 2.70 |
| Base saturation = cations × 100/c.e.c. (%) | 11.0 | 5.0 | 20.0 |
| pH ($H_2O$) | 4.84 | 4.85 | 4.6 |
| pH (0.1 M KCl) | 3.77 | 3.85 | 3.3 |

[a] Leached with 0.2 M $CH_3COONH_4$.
[b] Adsorbed $NH_4^+$ (from 0.2 M $CH_3COONH_4$) leached with 0.2 M $K_2SO_4$.

catchment in relation to fire (Chap. 13) does not indicate the roles of microorganisms nor the effects of fire on microbial determinants. Microorganisms through their mineralization activity might contribute to the seasonal increase in the outflow of nutrients in water from the catchment, while their nitrogen-fixing abilities might replenish losses of nitrogen. We therefore investigated in two soil types over 2–5 years: (1) soil pH, levels of soil microbial respiratory activity (mineralizing activity), ATP as an indicator of metabolically active microbial biomass, total fungal and bacterial (actinomycete and non-actinomycete) biomass; (2) the effects of a fire in March 1987 thereon, including the destructive effects to different soil depths and recovery of the microorganisms and their respiratory activity post-fire; and (3) populations of free-living nitrogen-fixing bacteria in three soils and the rhizospheres of plants after fire.

## 14.2 Soil Chemical, Physical and Microbiological Investigations

### 14.2.1 Experimental Soils

Most studies were conducted on soils of the Clovelly and Glenrosa forms (MacVicar et al. 1977) on the central plateau in Swartboskloof (Fig. 1.3). Clovelly soil, derived from quartzite, was a typical acidic, sandy, nutrient-poor (Table 14.1), strongly leached mountain fynbos soil (cf. Specht and Moll 1983). Glenrosa soil, derived from a mixture of granite and quartzite, was a sandy loam with markedly higher levels of nutrients (Table 14.1), but a similar pH. The vegetation on both soils was tall proteoid shrubland (see Chap. 1; Sect, 1.4.2) dominated by *Protea neriifolia*. A single study of nitrogen-fixing bacteria in the rhizosphere of *Protea repens* involved sampling from quartzite-derived Fernwood soil (MacVicar et al. 1977) at an altitude of about 450 m. This soil was similar to the Clovelly soil (Table 14.1).

### 14.2.2 Soil pH

The pH of sieved (<2 mm) soil (measured in water, 0.01 M $CaCl_2$ and 1.0 M KCl; Jackson 1958) showed pre- and post-fire seasonal fluctuations often close to 1 pH unit (Fig. 14.1). The fire caused immediate increases of 0.3–0.5 pH units, followed by further rises, especially in the upper 0–50 mm of the soils, in which the pH (in water) eventually reached 5.7 (Glenrosa) and 5.9 (Clovelly). The mean pH values for the entire data set showed no significant differences due to fire, soil depth or soil type.

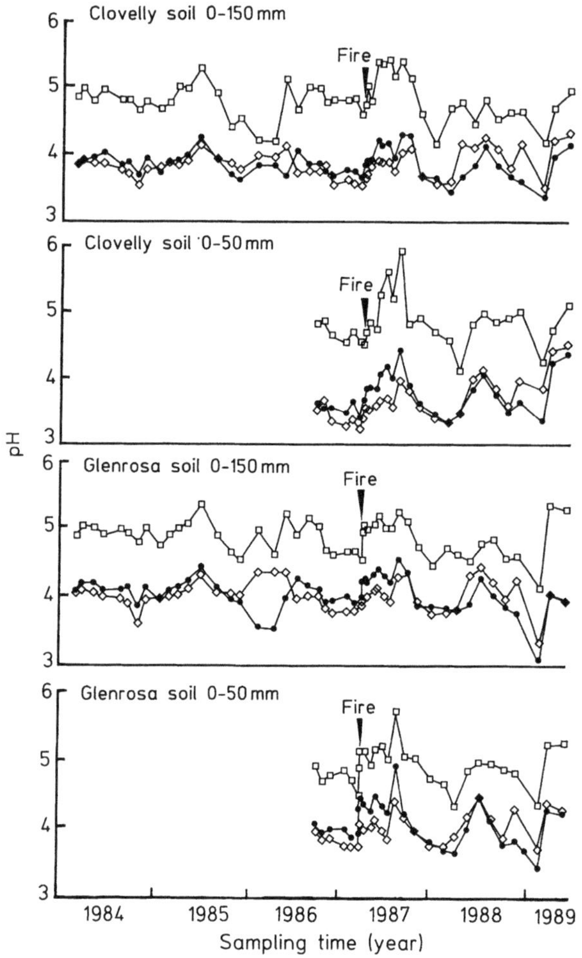


**Fig. 14.1.** Mean pH (means for two plots) of 1:2.5 suspensions of Clovelly and Glenrosa soils in Swartboskloof in water (□), 0.01 M $CaCl_2$ (●) and 1 M KCl (◇) from April 1984 to May 1989

### 14.2.3 Soil Moisture

Mean moisture contents of sieved soils from two depths (Fig. 14.2A,D) reflected the dominant winter rainfall. Rain in summer was accompanied by increases in soil moisture during the summers of 1984–85, 1986–87 and 1987–88 (see also Chap. 2). Glenrosa soils had a water-holding capacity, under field conditions, 2.5–3 times greater than Clovelly soil.

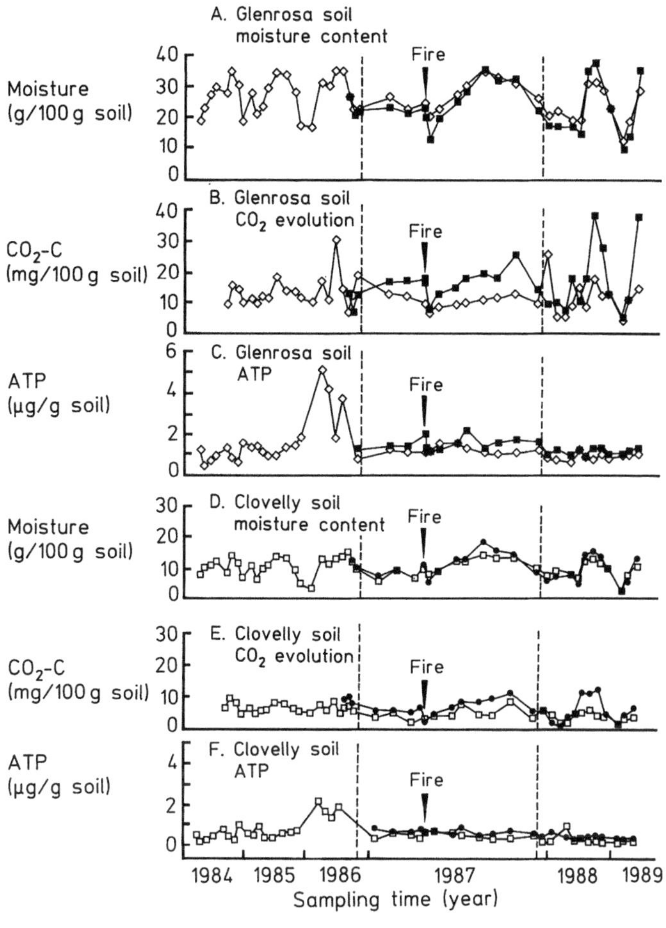


**Fig. 14.2A–F.** Mean moisture content, ATP content and evolution of $CO_2$ (means for two plots) for two soils in Swartboskloof from depths of 0–150 mm (Glenrosa soil = ◇, Clovelly soil = □) and 0–50 mm (Glenrosa soil = ■, Clovelly soil = ●) from April 1984 to May 1989. Note the change in scale on the x axis for 1987

## 14.2.4 Soil Microbial Respiratory Activity

Respiratory activity in sieved soils was determined in the laboratory at 26°C, using Jenkinson and Powlson's (1976) method for non-fumigated control soils, but with no manipulation of the soil moisture, to determine the effect of soil moisture on respiratory $CO_2$ evolution. The $CO_2$ evolution from both soils (Fig. 14.2B,E) reflected the trends in soil moisture. Declines in $CO_2$ evolution immediately after the fire were evident. The respiratory activity was consistently higher in the Glenrosa soil than the Clovelly soil; with both soils the 0–50 mm samples usually had higher values than the 0–150 mm samples.

### 14.2.5 Soil Microbial ATP

Microbial ATP was extracted using the chloroform-$NaHCO_3$ procedure (Paul and Johnson 1977) and determined luminometrically following a 1 in 9 dilution of the extract with 20 mM Tris (pH 7.75)-2 mM EDTA and a 1:1 addition to this dilution of nucleotide-releasing agent NRB (Lumac Systems AG 1979). This technique shows trends but not absolute microbial ATP values, as the chloroform and $NaHCO_3$ extract only about 40% of the maximum extractable ATP (Jenkinson et al. 1979; Martens 1985). The ATP levels in both soils (Fig. 14.2C,F) fluctuated, apparently with season, both pre- and post-fire, but not in agreement with soil moisture and the measured respiratory activity of the same samples. ATP levels tended to show peaks during late autumn/early winter and late winter/spring. Particularly high ATP levels were recorded in the last autumn and spring before the fire, following the dry summer of 1985–86. The ATP was higher in the Glenrosa soil than in the Clovelly soil; with both soils the values for 0–50 mm were usually higher than for 0–150 mm. The fire caused a sharp but short-lived reduction in ATP in the Glenrosa soil at 0–50 mm, but had no effect at 0–150 mm. The immediate effects of the fire on ATP in Clovelly soil appeared small.

### 14.2.6 Fungal, Actinomycete and Non-Actinomycete Bacterial Biomass

Fungal, actinomycete and non-actinomycete bacterial populations were monitored using a direct microscopic procedure (Jones and Mollison 1948) from shortly before the fire (0–150 mm samples) or immediately thereafter (0–50 mm samples) to May 1989. Fungal and actinomycete hyphae were measured and total lengths in five different diameter ranges determined. Bacteria in six different diameter/length ranges were counted. Biovolumes of the three groups were calculated and converted to mg dry biomass/g dry soil using the conversion factors of Schnürer et al. (1985).

Fungi were the dominant microorganisms in the soils, at both depths (Fig. 14.3). In 0–150 mm soil sampled the day following the fire, the biomasses of the three groups in the Glenrosa soil were similar to those 30 days before the fire, whereas in the Clovelly soil, the fungal, actinomycete and bacterial biomasses were 51, 40 and 171%, respectively, of pre-fire values. As the staining method did not distinguish living from dead cells, mortality due to fire cannot be deduced. The declines in biomass recorded 8 days later may reflect disappearance of hyphae and cells killed by the fire, or a reduction resulting from changed soil conditions, or both. Recovery of most biomasses was evident 51 days post-fire, following rain. Thereafter the biomasses, particularly of the fungi, were moderate to high through the winter, declining or showing reduced increases at the end of October as the soils dried (Fig. 14.2A,D). Following rain in March 1988, the fungal

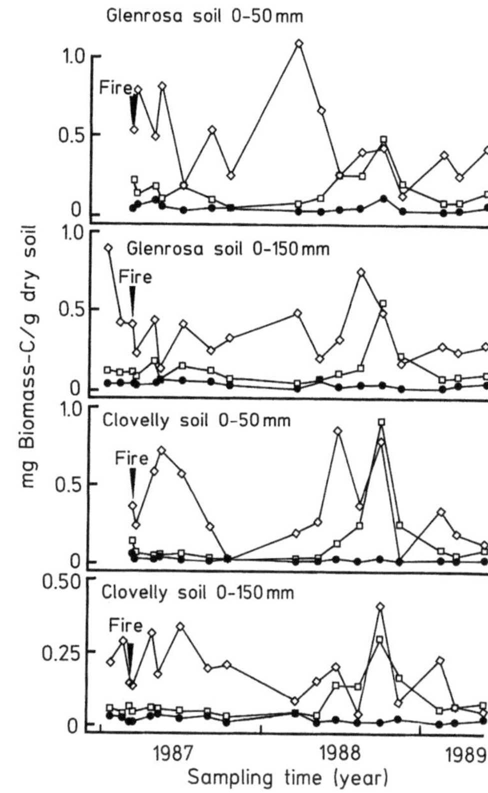


**Fig. 14.3.** Fungal (◇), actinomycete (●) and non-actinomycete bacterial (□) biomass-C concentrations (for pooled samples from two plots) in two soil types in Swartboskloof from January 1987 to May 1989. Biomass-C = 0.5 × biomass

biomass in the Glenrosa 0–50 mm soil showed a massive increase before declining to more moderate winter levels and increasing again in spring. The Glenrosa 0–150 mm soil also showed the March increase in fungal biomass, but the second increase was in winter (July) rather than spring. The fungal biomass in the Clovelly soil (both depths) showed a later first increase to a high level in response to autumn rain only in May and June 1988 (late autumn/early winter), then decreased before showing a spring maximum in September 1988. The bacterial biomasses in all soils increased sharply to exceptionally high levels during September 1988. The increases in fungal biomass during February 1989 when soil moisture was low (Fig. 14.2A,D) were unexpected, but were probably caused by 10 mm rain 2 weeks previously, causing a flush of fungal growth.

### 14.2.7 Post-Fire Soil Populations of Fungi and Bacteria by Plate Count

Survival and recovery trends of fungi and bacteria following the fire were investigated by dilution plate counts of viable fungal and bacterial pro-

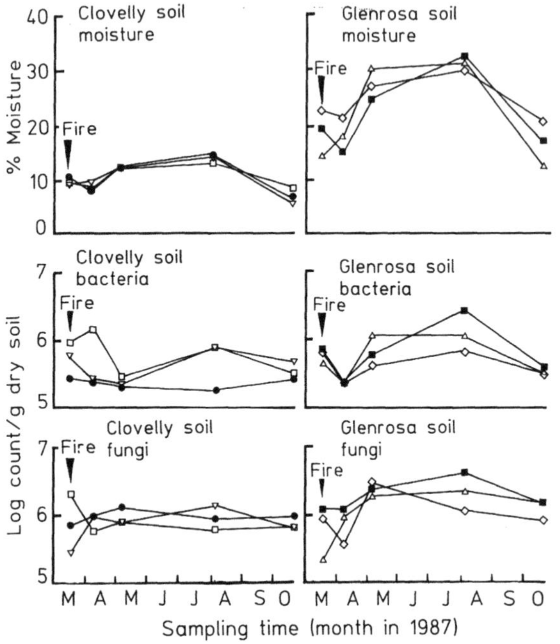


**Fig. 14.4.** Mean populations by plate count (means for two plots) of bacteria and fungi in two soils in Swartboskloof at 0–25 mm (▽ = Clovelly soil, △ = Glenrosa soil), 0–50 mm (● = Clovelly soil, ■ = Glenrosa soil) and 0–150 mm (□ = Clovelly soil, ◇ = Glenrosa soil) from March 1987 (immediately after the fire) to October 1987

pagules able to produce colonies on a non-selective glucose-yeast extract-soil extract agar (Loos et al. 1979) incubated aerobically at 26°C. Soil samples were from 0–25, 0–50 and 0–150 mm. The first counts 1 day after the fire showed large surviving populations of fungi and bacteria, even in the 0–25 mm samples, which would have been most affected by the fire (Fig. 14.4). Thereafter, most populations decreased, then recovered. However, the fungi in the 0–25 mm samples of both soils showed sharp initial increases. The viable fungal populations at all depths seemed to have recovered within 51 days post-fire. Bacterial numbers showed similar rapid recoveries in some samples, but declined in others. After 8 May, bacterial counts, except those for the Clovelly 0–50 mm and Glenrosa 0–25 mm soils, increased to August then declined.

### 14.2.8 Post-Fire Soil and Rhizosphere Populations of Bacteria and Apparent Nitrogen-Fixing Bacteria by Plate Count

The anticipated rhizosphere effect of regenerating plants in stimulating bacterial populations, in particular nitrogen-fixing bacteria, was investigated

Chan HT (1980) Reproductive biology of some Malaysian dipterocarps. II. Fruiting biology and seedling studies. Malaysian For 43: 438–51
Charnov EL, Shaffer WM (1973) Life-history consequences of natural selection: Cole's result revisited. Am Nat 109: 791–93
Chesson PL, Warner RR (1981) Environmental variability promotes coexistence in lottery competitive systems. Am Nat 117: 923–943
Christensen NL, Muller CH (1975) Effects of fire on factors controlling plant growth in *Adenostoma* chaparral. Ecol Monog 45: 29–55
Clarholm M, Rosswall T (1980) Biomass and turnover of bacteria in a forest soil and a peat. Soil Biol Biochem 12: 49–57
Coetzee JH, Giliomee JH (1985) Insects in association with the infloresence of *Protea repens* (L) (Proteaceae) and their role in pollination. J Entomol Soc South Afr 48: 303–314
Coetzee JH, Giliomee JH (1987) Seed predation and survival in the infructescences of *Protea repens* (Proteaceae). S Afr J Bot 53: 61–4
Collins BG, Rebelo AG (1987) Pollination biology of the Proteaceae in Australia and southern Africa. Aust J Ecol 12: 387–421
Colman EA (1951) Fire and water in southern California's mountains. USDA For Serv Calif For Range Exp Stn, Misc Pap 3
Comins HN, Noble IR (1985) Dispersal, variability and transient niches: species coexistence in a uniformly variable environment. Am Nat 126: 706–723
Cone GB (1973) The floras of the south western part of South Africa. Tuatara 20: 160–164
Connell JH (1978) Diversity in tropical rain forests and coral reefs. Science 199: 1302–1310
Connell JH, Slatyer RO (1977) Mechanisms of succession in natural communities and their role in community stability and organization. Am Nat 111: 1119–1144
Conrad CE, Oechel WC (eds) (1982) Dynamics and management of mediterranean-type ecosystems. USDA For Sev Gen Tech Rep PSW-58
Countryman CM, Philpot CW (1970) Physical characteristics of chemise as a wildland fuel. USDA For Serv Res Pap PSW-66
Cowling RM (1983a) Diversity relations in Cape shrublands and other vegetation in the south-eastern Cape, South Africa. Vegetatio 54: 103–127
Cowling RM (1983b) The occurrence of $C_3$ and $C_4$ grasses in fynbos and allied shrublands in the south eastern Cape. Oecologia (Berlin) 58: 121–127
Cowling RM (1984) A syntaxonomic and synecological study in the Humansdorp region of the fynbos biome. Bothalia 15: 175–227
Cowling RM (1987) Fire and its role in coexistence and speciation in Gondwanan shrublands. S Afr J Sci 83: 106–112
Cowling RM (ed) (1991) The ecology of fynbos: Nutrients, Fire and Diversity. Oxford University Press, Cape Town
Cowling RM, Campbell BM (1980) Convergence in vegetation structure in the mediterranean communities of California, Chile and South Africa. Vegetatio 43: 191–197
Cowling RM, Gxaba T (1990) Effects of a fynbos overstorey shrub on understorey community structure: implications for the maintenance of community-wide species richness. S Afr J Ecol 1: 1–7
Cowling RM, Lamont BB (1987) Post-fire recruitment of four co-occurring *Banksia* species. J Appl Ecol 24: 645–658
Cowling RM, Moll EJ, Campbell BM (1976) The ecological status of the understorey communities of pine forests on Table Mountain. S Afr For J 99: 13–23
Cowling RM, Campbell BM, Mustart P, McDonald DJ, Jarman ML, Moll EJ (1988) Vegetation classification in a floristically complex area: the Agulhas Plain. S Afr J Bot 54: 290–300.
Crawley MJ (1986) The population biology of invaders. Philos Trans R Soc Lond B 314: 711–731

**Table 14.2.** Populations of aerobic bacteria (non-specific) and apparent nitrogen-fixing bacteria by dilution plate count in non-rhizosphere (S) and rhizosphere soils (R) associated with fynbos plants in Swartboskloof

| Plant species | Soil form | Sampling date | Aerobic bacteria[a] (millions $g^{-1}$) | | | $N_2$-fixing bacteria[a] (millions $g^{-1}$) | | |
|---|---|---|---|---|---|---|---|---|
| | | | S | R | (R:S ratio) | S | R | (R:S ratio) |
| *Aristea major* | Glenrosa | 22 June 1988 | 3.0 | 36.9 | (12.6) | 2.9 | 33.9 | (12.9) |
| | | 29 September 1988 | 3.8 | 56.2 | (14.4) | 18.4 | 219.2 | (11.2) |
| | | 14 December 1989 | 0.62 | 2.8 | (4.3) | 1.0 | 21.3 | (21.9) |
| | Clovelly | 26 July 1988 | 1.7 | 12.1 | (8.1) | 5.1 | 22.5 | (4.6) |
| | | 29 September 1988 | 1.8 | 11.0 | (6.2) | 4.1 | 40.1 | (10.3) |
| | | 14 December 1989 | 0.49 | 1.0 | (1.9) | 0.51 | 2.1 | (4.2) |
| *Watsonia borbonica* | Glenrosa | 11 October 1988 | 1.2 | 8.3 | (7.1) | 2.7 | 19.1 | (7.2) |
| | | 15 December 1989 | 1.7 | 3.1 | (1.8) | 1.7 | 9.9 | (6.1) |
| | Clovelly | 18 October 1988 | 1.8 | 7.6 | (4.2) | 0.84 | 20.2 | (24.1) |
| | | 15 December 1989 | 0.56 | 1.5 | (3.1) | 0.65 | 2.7 | (4.1) |
| *Protea repens* | Fernwood | 1 August 1989 | 0.73 | 3.0 | (4.1) | 0.63 | 2.5 | (4.0) |
| | | 24 August 1989 | 0.54 | 1.5 | (2.8) | 0.56 | 1.4 | (2.5) |
| | | 14 September 1989 | 0.40 | 1.6 | (4.0) | 0.51 | 1.5 | (2.9) |

[a] The counts and R:S ratios of 11 and 18 October 1988 for *Watsonia borbonica* and all counts and R:S ratios for *Protea repens* are determinations for single plots. All other values are means of separate determinations for duplicate plots.

(Brookes et al. 1983; Martens 1985), both of which are linked to respiratory activity (Schlegel 1985). ATP peaked under the warm, moist conditions of autumn/early winter and spring, and declined during winters and dry summers. High ATP levels before the fire, following a very dry summer, were possibly caused by litter production from the summer leaf fall (Kruger 1981) and increased movement of nutrients such as C, N and P into the soils with the first rains, stimulating soil microbial populations to greater activity (Nannipieri et al. 1978; Fairbanks et al. 1984). Immediate increases in the ATP and adenylate energy charge of the microbial populations may have followed moistening of the dry soils (Brookes et al. 1983). By comparison, the seasonal ATP fluctuations after the fire, when the young vegetation would have produced much less litter, were of modest amplitude.

Dominant fungal biomass showed the trends expected from the ATP results in the second post-fire year (Fig. 14.3). Both soils showed autumn/early winter and early spring peaks of fungal development, with lower levels during the winter, and the preceding and succeeding late spring/summer. The fungal biomass during the first post-fire wet season showed peaks during late autumn and winter, followed by declines in spring as the soils dried (Fig. 14.2A,D). Non-filamentous bacteria and actinomycetes were minor components of the microbial biomass except during the second spring post-fire.

The pre- and post-fire trends of soil microbial respiratory activity and biomass partly explain the increase in the outflow of nutrients, both before and after the fire (Chap. 13), at the onset of the rainy season. The rapid expansion of the source area from which nutrients are leached with the onset of the winter rains was postulated as an important factor, but the concomitant increase of metabolically active microbial biomass associated with increased respiration and mineralization of organic residues may also be significant.

## 14.4 Effects of Soil Type and Depth

The Glenrosa soil, with three to four times as much clay plus silt and organic C as the Clovelly soil, contained much higher concentrations of cations, P and N than the Clovelly soil and had a much higher water-holding capacity, though pH was similar (Table 14.1). These differences provide explanations for the consistently higher levels of respiratory activity and ATP in the Glenrosa soil.

In both soil types, the respiratory activity and ATP in the 0–50 mm samples were usually higher than in the corresponding 0–150 mm samples (Fig. 14.2B,C,E,F). The differences cannot be ascribed to moisture status (Fig. 14.2A,D), but may reflect nutrient and/or oxygen gradients or effects of the fire, which were not investigated.

## 14.5 Effects of Fire

Fynbos fires are light, reaching maximum temperatures of about 300°C at the soil surface and 50°C at 50 mm (Rundel 1983). The Swartboskloof fire was a low intensity fire. Post-fire increases of soil pH, incorporating both fire and seasonal effects, did not exceed 1 pH unit (Fig. 14.1). The probably low soil temperature rise and modest pH increase can be expected to have caused only limited disturbance of the soil microbial populations, mainly in the upper 50 mm. Soil moisture declined sharply in the upper 50 mm during the fire and immediately thereafter to the sampling of 27 March (Fig. 14.2A,D), which preceded the first post-fire rain of 6–7 April.

Mean $CO_2$ evolution and ATP values for the pre- and post-fire periods (2–3 year means) indicated no significant changes in the soil microbial respiratory activity and ATP as a result of the fire in the longer term. The short-lived declines in soil microbial respiratory activity and ATP at the time of and immediately following the fire, seemed restricted mainly or entirely to the upper 50 mm of the soils (Fig. 14.2B,C,E,F). After the first post-fire rain, both respiratory activity and ATP recovered, reaching maximum values for the first post-fire rainy season within 2 to 5 months.

Fungal, actinomycete and non-actinomycete bacterial biomass indicate small short-lived declines immediately after the fire (Fig. 14.3). Thereafter they recovered rapidly, although with fluctuation, during the first post-fire rainy season. The dominance of the fungal biomass suggests that much of the soil respiratory activity and ATP, and the effects of the fire thereon, would have been due to fungi, although bacteria were important in the spring of 1988.

Plate counts, which started on the day following the fire, indicate high survival of fungi and bacteria, even in the upper 25 mm (Fig. 14.4). As fungi are not highly heat-resistant, the soils appeared to have been only partially sterilized. This supports results of the respiratory activity, ATP and biomass determinations on the 0–50 mm soil samples. Also in agreement with these other results, most fungal and bacterial plate counts showed declines to about the time of the first rains and subsequent increases in the wet season to peaks within 2–5 months post-fire.

## 14.6 Rhizosphere Stimulation of Nitrogen-Fixing Bacteria in Post-Fire Soils in Relation to General Rhizosphere Effect

The stimulation of populations of microorganisms in the rhizosphere is a normal phenomenon associated with root development (Alexander 1977). R:S ratios of 1.8–14.4 were recorded for non-specific aerobic bacteria in the rhizospheres of the selected species after fire (Table 14.2). Surprisingly, high levels of apparent nitrogen-fixing bacteria occurred in all rhizosphere and non-rhizosphere samples, often at higher levels than the non-specific aerobic bacteria. This does not imply that the latter bacteria were the same organisms as the apparent nitrogen fixers, but only that two small segments of the total bacterial populations (412–2600 millions/g dry non-rhizosphere soil according to direct microscopic counts) were counted by the different plating procedures. Many of the R:S values for the apparent nitrogen-fixing bacteria were similar to those for non-specific aerobic bacteria (although their range at 2.5–24.1 was somewhat higher than that of the non-specific aerobic bacteria), suggesting little specific stimulation of nitrogen-fixing bacteria in the rhzospheres, but rather their participation in the general rhizosphere stimulation of bacteria. The nitrogen-fixing abilities of the bacteria are not known, but many colonies on the plates were large, slimy colonies of which a selection has been tentatively identified as *Azotobacter* species. *Azotobacter* species can fix nitrogen (Postgate 1982), and could be significant in replenishing nitrogen in the ecosystem following volatilization losses (Chap. 13).

## 14.7 Intercontinental Comparisons

### 14.7.1 Microbial Activity and Biomass in Relation to Seasonal Factors

Our results in respect of seasonal effects show trends in agreement with those found elsewhere in mediterranean- and non-mediterranean-type regions where general principles have been established. The mediterranean climate seems to produce most favourable conditions for microbial development and activity in autumn and spring. Declines of $CO_2$ evolution in winter, when temperatures are low, and during summer drought, leave peaks of activity in spring and autumn, as has been reported by Lossaint (1973) in France and by Arianoutsou-Faraggitaki and Margaris (1982) in Greece. In both studies temperature effects were dominant over moisture effects, resulting in only slight declines in $CO_2$ evolution in summer, but marked declines during winter. Soil microbial dehydrogenase activity and nitrifying capacity showed autumn peaks consistently but spring peaks inconsistently in the Greek study.

We suggest (Sect. 14.3) that the regular pH increases in autumn (Fig. 14.1) may have resulted from a flush of microbial mineralization of organic material following wetting of the soils at the end of summer with attendant leaf fall. The pH of unburnt Oregon forest soil (Wright and Bollen 1961) also increased with the main flush of microbial development in the soil. This occurred during late winter/spring, notwithstanding the wet winter-dry summer rainfall regime, but low temperature may have kept microbial activity low in winter as in France (Lossaint 1973). In forest soils in Washington State (Vogt et al. 1980), ATP showed a prominent peak in spring, lowest levels in summer/autumn, and surprisingly high levels during the cold winter. During the driest period (summer/autumn), ATP and $CO_2$ evolution appeared inversely related.

The dominance of fungi in soils is possibly the usual situation. Fungal respiration was responsible for 65–90 and 60–80% of the total microbial respiration of agricultural and acid forest soils in Germany (Anderson and Domsch 1975). Furthermore, fungal biomass trends in acid coniferous forest soils in western Oregon (Hunt and Fogel 1983) and central Sweden (Söderström 1979; Bääth and Söderström 1982), where dry summer conditions prevailed, resembled those in Swartboskloof with low summer levels and peaks in spring and/or autumn. Bacterial populations, which showed no clear relation to season in our study, showed similar trends to fungal biomass in the Swedish soil (Clarholm and Rosswall 1980; Lundgren and Söderström 1983). The bacterial peak in autumn was particularly prominent.

### 14.7.2 Microbial Activity and Biomass in Relation to Soil Factors

Rates of $CO_2$ evolution in nutrient-poor fynbos soils, even the more favourable Glenrosa soil, were low compared with those elsewhere. Jenkinson and Powlson (1976), Jenkinson et al. (1979) and Oades and Jenkinson (1979), using the same technique for studying $CO_2$ evolution except that soil moisture was first adjusted to 50 or 55% of water-holding capacity, recorded the evolution from a range of unfumigated Australian and English soils of 2.0–26.4 and 4.3–38.9 mg $CO_2$-C/100 g soil, respectively. $CO_2$ evolution in soils at Swartboskloof is thus comparable to the lower values for the Australian and English soils, suggesting that the mineralization of organic C to $CO_2$ proceeded relatively slowly in Swartboskloof both before and after fire. Decomposition of litter in Swartboskloof was also extremely slow (Mitchell and Coley 1987), and appeared to have no seasonal pattern. The calculated turnover time was 14.94 years, which is slower than the times (2.4–7.9 years) in low altitude chaparral (760–1060 m; Gray and Schlesinger 1981) but comparable to the values (11.4–23.0 years) for high altitude chaparral (1370–1650 m).

### 14.7.3 Effects of Fire on Soil Microorganisms and Their Activities

The fire on our plots was probably equivalent to a light burn in chaparral, with soil temperatures reaching at most about 100°C at 25 mm and 50°C at 50 mm (DeBano et al. 1977). A proportion of the fungi (particularly heat-shock fungi) and of the bacteria were found to survive such temperatures in heated chaparral soil (Dunn and DeBano 1977; Dunn et al. 1979, 1985), in agreement with our first day post-fire plate counts, supported by the levels of ATP and respiratory activity. After watering of chaparral soil, heated to simulate a moderate burn (DeBano et al. 1977, 1979b; Dunn et al. 1979), fungi and bacteria multiplied steadily as long as the soil remained moist, then declined when it dried (Dunn et al. 1979). In field studies of a northeastern Minnesota soil under burned jack pine, fungi and bacteria multiplied rapidly to high levels following the first post-fire rain (Ahlgren 1974). Our studies showed similar increases of soil microbial respiratory activity, ATP, fungal and bacterial numbers and biomass during the first post-fire rainy season after initial declines (most determinations) prior to the rain.

The evident lack of long-term effects of fire on soil microbial respiration, ATP, fungal, actinomycete and bacterial biomass supports results elsewhere. A Greek soil showed for burned and unburned plots similar seasonally fluctuating levels of $CO_2$ evolution and dehydrogenase activity (Arianoutsou-Faraggitaki and Margaris 1982), as well as similar rates of cellulose decomposition (Margaris 1977); likewise, fire in Douglas fir in Oregon had no effect on soil fungi, actinomycetes and other bacteria (Wright and Bollen 1961). However, soil pH was higher by as much as 2–3 pH units after fire in Oregon, while the burned Greek soil showed higher nitrification rates. In chaparral soil, fire had no effect on rates of ammonification and nitrification, which appeared to be heterotrophic in the burned soil (Dunn et al. 1979). Nitrifying microorganisms were not investigated in our study, but the export of nitrate from Swartboskloof during the summer following the fire (Chap. 13) and the post-fire accumulation of nitrate in a coastal fynbos soil (Stock and Lewis 1986a) suggest that they would have been active.

Different microorganisms may become dominant in soils after fires (Ahlgren 1974; Warcup 1981), particularly the pyrophilous fungi. Dunn et al. (1982) list the principal pyrophilous or heat-shock fungi of southern California chaparral soils and consider their possible beneficial role in the stabilization of ash beds. This role remains to be investigated in fynbos soils.

The generally slow microbial decomposition of litter and organic residues in fynbos and chaparral soils (Sect. 14.7.2) indicates that microbial mineralization of organic material in these soils is inadequate and that periodic fires have an important function as a mineralizing agent (DeBano and Dunn 1982; Stock and Lewis 1986a; Mitchell and Coley 1987).

### 14.7.4 Nitrogen-Fixing Bacteria in Mediterranean Regions in Relation to Fire

Replenishment of nitrogen compounds from the atmosphere, by bacterial nitrogen fixation, is needed to replace losses by volatilization or by post-fire runoff in fire-prone ecosystems (DeBano et al. 1979a; DeBano and Dunn 1982; see also Chap. 13). Nitrogen fixation by symbiotic bacteria-plant systems is considered important in mediterranean-type ecosystems; examples are the symbioses of *Frankia* with *Ceanothus* or *Cercocarpus* and *Rhizobium* with various legumes in California (Wells et al. 1979; DeBano and Dunn 1982; Rundel 1983), as well as numerous symbioses of *Rhizobium* with legumes, *Frankia* with *Casuarina* or *Myrica* and blue-green bacteria with cycads in mediterranean Western Australia and South Africa (Groves 1983; Lamont 1983). Free-living nitrogen-fixing bacteria have been poorly investigated in mediterranean regions, but they apppear to make a contribution to nitrogen gains in chaparral (DeBano and Dunn 1982) and nitrogen-fixing aerobic bacterial populations occur in soils of Italian beech and beech-fir forests (Virzo De Santo et al. 1978). Becking (1959) found *Azotobacter* and *Beijerinckia* in soil of an arboretum and *Beijerinckia* in a pine plantation soil in the fynbos area of South Africa, but neither genus in other soils under pine plantations, natural forest or fynbos. We investigated free-living nitrogen-fixing bacteria in fynbos rhizospheres and soils in view of the scarcity of legumes in many fynbos communities, the significance of associative nitrogen fixation for the nutrition of a range of different plants (Vose and Ruschel 1981) and the report by Jorgensen and Wells (1971) of marked increases in apparent nitrogen fixation associated with annual burning in loblolly pine in South Carolina.

## 14.8 Conclusions

This study has clarified several aspects of the role of microorganisms in the ecology of a fire-prone ecosystem. Autumn/early winter and spring seemed to be the seasons most favourable for microbial development and metabolic activity in Swartboskloof. Mineralization of organic matter during the flush of microbial development after the dry summer may partly explain the flush of inorganic nutrients in water leaving the catchment at the start of the wet season. Large populations of fungi, bacteria and actinomycetes survived the fire and re-established populations responding approximately as expected to the seasons. The large populations of free-living aerobic apparent nitrogen-fixing bacteria found in the fynbos soils and the rhizospheres of fynbos plants may be important in replenishing nitrogen after fires. However, much remains to be done before the importance of these organisms in contributing to the stability of the ecosystem can be clearly understood.

*Acknowledgements.* We thank Anthony M. Crossley, Paul Kempton-Jones, Karin S. Dreyer, Susara M. le Roux and Melané A. Mulder for assistance with the plate counts. The Foundation for Research Development through the Co-operative Scientific Programmes and the University of Stellenbosch are thanked for financial assistance.

# 15 Ecosystem, Community and Species Response to Fire in Mountain Fynbos: Conclusions from the Swartboskloof Experiment

D.M. Richardson and B.W. van Wilgen

## 15.1 Introduction

Full appreciation of the functioning of an ecosystem requires perspectives gained at different levels. Studies of individual plants or animals, populations of single species, interactions between species and the environment within homogeneous areas, or between the biota and the environment in a landscape are all useful in this regard. The management of fynbos, however, is currently based on an understanding of the population biology of a few important species. For example, a much-quoted "rule of thumb" for determining fire frequencies in fynbos is that prescribed burns should only take place once 50% of the population of the slowest-maturing species in an area has flowered for at least three successive seasons (Kruger and Lamb 1979; Kruger 1982). This rule is usually applied to shrubs in the family Proteaceae, since these are generally the slowest to mature. Although studies have been conducted at all levels of organization in mountain fynbos (Table 15.1), community- and ecosystem-level perspectives have not been used to formulate fire management policies, nor are they incorporated in management prescriptions.

The research at Swartboskloof examined the effects of fire at different scales, ranging from an assessment of responses of populations of single species (level 1), through community-level responses (level 2) to catchment- or ecosystem-level influences (level 3). At all levels, attention was given to describing the processes that cause the observed effects. Important objectives of this volume were to compare the ecological effects of fire in mountain fynbos with effects in other mediterranean-type ecosystems, and to explore the implications of the results for management. For convenience, the studies were divided into those that addressed the regeneration and recolonization strategies of the biota (essentially level 1), the stability of boundaries between vegetation communities (essentially level 2), and those that examined the effects of fire on resource availability (essentially level 3). Some studies do not fit neatly into this framework, and sought perspectives at more than one level of organization. Nevertheless, the population-community-ecosystem hierarchy provides a useful framework for syn-

**Table 15.1.** Major spheres of fire-related studies undertaken to date in mountain fynbos, showing the various levels addressed by these studies

| Sphere of study | Level Addressed | | | | Key references |
|---|---|---|---|---|---|
| | Individual | Population | Community | Ecosystem | |
| *Widdringtonia cedarbergensis* | X | X | | | Manders (1987) |
| *Watsonia borbonica* | X | X | | | Kruger and Bigalke (1984); Le Maitre (1984) |
| Serotinous Proteaceae | X | X | X | | Bond et al. (1984) |
| Myrmecochorous Proteaceae | X | X | X | | Bond and Slingsby (1984); Yeaton and Bond (1991) |
| Small mammals | | X | X | | Bond (1984); Kruger and Bigalke (1984) |
| Ants | | X | X | | Bond and Slingsby (1984) |
| Vegetation studies | | | X | | McKenzie et al. (1977); van Wilgen (1981); van Wilgen and Kruger (1981); Manders (1990a,b) |
| Invasive plants | X | X | X | (X) | Kruger (1977c); Macdonald and Richardson (1986); Richardson and Brown (1986); Richardson et al. (1987) |
| Nutrient cycling | | | X | | van Wilgen and Le Maitre (1981); Mitchell and Coley (1987) |
| Catchment hydrology | | | | X | van der Zel and Kruger (1975); Bosch et al. (1984); Bosch et al. (1986) |

thesizing the results of the experiment. In this chapter we identify major components of resilience (here defined as the ability of an organism, community or ecosystem to return to its initial structure and function after disturbance) at each level of organization, and formulate the most important findings in terms that are relevant to managers.

## 15.2 Population-Level Effects of Fire

Mountain fynbos plants exhibit a wide range of regeneration strategies and fire-survival mechanisms. Most species can resprout after fire, and appear to be resilient under a wide range of fire regimes. Few species (14% of a sample of 210 species; Chap. 4) rely exclusively on seeds for regeneration. Most obligate re-seeders regenerate from seeds stored in the soil, and very few indigenous species (e.g. *Protea neriifolia* and *P. repens*) maintain seed stores in the canopy. Despite the poor representation of serotiny in the Swartboskloof flora, these species make up the bulk of phytomass over most of the area, and therefore exercise considerable influence on community- and ecosystem-level processes. The rarity of sprouting trees and shrubs in the overstorey of mountain fynbos is unusual for mediterranean-type shrublands (Le Maitre and Midgely 1991). Regeneration of the serotinous overstorey shrubs is highly variable over space and time; populations are susceptible to decimation or local extinction if fire cycles are shorter than the juvenile period, or if post-fire conditions are unfavourable for seedling recruitment. Large fluctuations in population sizes of these components clearly influence community dynamics (see later.)

Inter-fire recruitment of serotinous overstorey shrubs is more common than was previously recognized. Evidence from Swartboskloof supports the hypothesis (see Bond 1984; Breytenbach 1984) that seed predation and herbivory, rather than seed biology or fire per se, regulate recruitment and account for the even-agedness of mountain fynbos communities in most areas (Chap. 7).

Serotiny is well represented in the alien flora of mountain fynbos (Chap. 9). Introduced serotinous species such as *Hakea sericea*, *Pinus halepensis*, *P. pinaster* and *P. radiata* are often dominant components of mountain fynbos. The extreme resilience of populations of these taxa under any fire regime (relative to the indigenous taxa) is essentially because they mature quickly, and accumulate large stores of highly mobile seeds. The first property is inherent to the species, but the second is influenced by the absence of natural enemies. In combination, these attributes confer resilience in two ways: in situ resilience, and resilience by migration (sensu Grubb and Hopkins 1986).

A comparative study of the life histories of *Protea neriifolia* (a serotinous reseeder) and *P. nitida* (a facultative resprouter) sheds new light on the evolution of the reseeding strategy in fire-prone shrublands (Chap. 7). The principle of limited resource availability in life history theory predicts that there is a trade-off between allocation to reproduction and to growth and longevity. In accordance with this principle, net population growth of *Protea neriifolia* after fires is greater than that of *P. nitida*. Le Maitre (Chap. 7) argues that increased seedling recruitment increases fitness, and may have favoured the evolution of the reseeding strategy in fire-prone systems. When recruitment of a seeder is successful, it is very successful, but when recruitment fails it may become locally extinct. A sprouter, even under optimum conditions, only produces a few seedlings, but local extinction is rare. Contrary to arguments based on life history theory, mortality rates of seedlings of the two species were not markedly different. However, Le Maitre suggests that the increasing mortality with age in *P. neriifolia*, compared with the constant mortality rate in *P. nitida*, provides evidence for the link between reproductive investment and longevity.

Most (86%) fynbos plants in Swartboskloof flower within 1 year of a fire, and only 2% of species takes longer than 3 years to flower (Chap. 4). Since fires at intervals of less than 4 years are uncommon, populations of most fynbos plants are resilient when fire cycles exceed 4 years. Some species have short life spans, and resilience declines if intervals are longer than about 40 years (see later). Of special interest is the phenomenon of fire-stimulated flowering, which is a common trait in herbaceous plants in fire-prone systems. The detailed studies of the life cycles of the geophytes *Watsonia borbonica* and *Cyrtanthus ventricosus* in Swartboskloof (Chap. 8) showed that the stimulus for the flowering of these species is apparently not a direct effect of fire (e.g. through heat-induced damage to leaves or apical buds), but is an indirect effect linked to changes in the environment such as altered soil temperature regimes. Fire-stimulated flowering in *W. borbonica* is an efficient means of satiating seed predators and ensuring successful seedling recruitment. Although geophytic species are considered to be highly resilient to changes in the fire regime (because they flourish in frequently burnt firebreaks), results from Swartboskloof showed that frequent fires reduce ramet size, increase mortality rates and reduce seedling recruitment. Similar patterns are likely for other resprouting plants. Fortunately, the results presented in this volume suggest that the current management strategy, which was implemented to maximize regeneration of reseeding shrubs, of burning in late summer or early autumn every 12–15 years, also favours geophytes like *Watsonia borbonica* and resprouting shrubs like *Protea nitida*.

Changes in the fire regime in mountain fynbos affect different groups of plants in different ways. For example, fire frequency is important in determining the success of serotinous Proteaceae (sensitive to short and long intervals), and species with soil-stored seeds that are stimulated to germinate by fire (sensitive to long intervals; Table 15.2). Fire intensity affects the balance between sprouters (more sensitive to high intensity fires) and reseeders. Fire intensity also influences the response of myrmeco-

**Table 15.2.** Components of resilience at the population level in groups of plants with different regeneration strategies in mountain fynbos

| Regeneration strategy | Populations resilient under following conditions: | Populations not resilient under following conditions: |
|---|---|---|
| Sprouting | – Frequent fires<br>– Fire in any season<br>– Long fire cycles | – Intense fires |
| Fire-stimulated flowering | – Frequent fires<br>– Fire in any season<br>– Long fire cycles | – Intense fires |
| Fire-stimulated germination | – Intense fires | – Long fire cycles |
| Serotiny | – Intense fires | – Short fire cycles<br>– Fires in winter and spring<br>– High densities of seed predators post-fire |
| Myrmecochory | – Most intense fires<br>– High densities of seed predators post-fire | – Mild fires (spring, winter)<br>– Invasion by Argentine ant |

chorous species. The season of burns is an important determinant of resilience for serotinous species. Research into the management of fynbos ecosystems by prescribed burning was initially focused on determining the most appropriate interval between fires. It later became apparent that seasonal variation in fire could have effects that were as marked as frequency effects (Bond et al. 1984). The effects of variation in fire intensity are less well understood, but may be as important as seasonal and frequency effects (Bond et al. 1990). Burning to attain desired levels of fire intensity is a major challenge in the fynbos. In addition, further research into the effects of intensity is needed before we can predict the long-term implications of substituting relatively mild prescribed fires for intense wildfires.

Direct mortality due to fire is rare among the faunal components, and the changes in habitat due to the fire appear to be more important in deciding the fate of animals. Species dominance in ant and small mammal communities changed after the fire. Fire-induced changes in population sizes of small mammals also have considerable community-level effects (see later). The effects of fire intensity and season on faunal communities are poorly understood.

## 15.3 Community-Level Effects of Fire

A major objective of the Swartboskloof experiment was to explore the determinants of stability of boundaries between plant communities, since it is at this level that management is usually focused. There was no evidence that fire played a role in determining the boundary between communities within the fynbos vegetation groups (Chap. 4), but fire is a major factor in determining the boundary between fynbos and forest (Chap. 5). Fire also creates conditions that allow for invasion of fynbos by alien trees and shrubs. A re-occurring theme in this volume is the concept of invasibility – the "openness" of communities to invasion or recolonization by indigenous and alien taxa. The conditions under which the "invasion game" is played are determined by the fire, especially by the frequency, season and intensity of fires. The variation in fire regime that prevails over much of the fynbos biome has allowed a myriad of species to coexist. When this process is disrupted, either through excluding fire or by introducing alien species, new stable states may result.

Two community-level questions are particularly important for management of mountain fynbos: (1) Why are trees restricted to ravines and riparian habitats?; and (2) Why is fynbos open to invasion by alien trees and shrubs? Both questions were addressed in Swartboskloof and provide complementary insights concerning the determinants of plant community structure in fynbos. Fire is the integral factor determining the forest-fynbos boundary. Differences in the arrangement and nature of fuels in shrublands and forests result in significantly different fire behaviour in the two vegetation types (see Chap. 3). Whereas the shrublands have porous, fine fuels, forests accumulate compact layers of fine fuels close to the ground. Live foliage in the forest canopy is sparsely packed, has high moisture contents and low energy contents relative to live foliage in fynbos. Fires frequently burn through the shrublands but carry into the forest canopy only under exceptionally severe weather conditions. Consequently, the forests form islands of fire-proof habitat in fynbos.

Seeds of forest species are dispersed into the shrublands, and seedlings sometimes establish in mature fynbos communities, especially in sheltered sites. Colonizing forest species are usually killed by fire before forest patches have developed sufficiently to exclude fire. Where fire-free intervals are long (>50 years), colonizing forest elements form dense thickets which eventually coalesce to form continuous forests that exclude fire, thus perpetuating the forest. Much of the fynbos biome, and certainly all of Swartboskloof, is bioclimatically suitable for trees. The studies of the determinants of the forest-fynbos boundary showed that it is the frequency of fire, in relation to the rate of forest development, that determines the extent of forest development. Soil moisture and nutrients per se do not limit forest distribution, but affect the rate of forest development (Chap. 5).

Each fire has unique effects on recruitment levels of the serotinous proteoid component. Important determinants of recruitment success include pre-fire factors (e.g. the size of seed banks), the fire itself (e.g. heat pulse for releasing seed), and post-fire conditions (e.g. drought-induced seedling mortality or the density of granivores). Although small mammal numbers decline after fire, granivory by small mammals does limit seedling recruitment of serotinous species, especially after fires in spring. Over several fire cycles each species encounters at least one set of conditions favourable for recruitment: competitive hierarchies fluctuate with the changing relative abundances of species, and coexistence is maintained. This fire-induced variation in the floristic structure of fynbos communities implies that local extinction of community components and invasion (or recolonization) is a normal and regular process. Fire introduces recurrent fluctuations in community connectance (sensu Pimm 1989), creating transient invasion windows. This non-equilibrium model of fynbos community dynamics can explain the "drifting clouds of species abundance" (sensu Grubb 1986) reported for species such as *Leucadendron laureolum* (Cowling and Gxaba 1990; see Chap. 9). Furthermore, it provides a framework for understanding the invasion of fynbos by pines and hakeas.

The conditions suitable for establishment, growth and proliferation (the invasion window sensu Johnstone 1986) of indigenous forest trees in fynbos are very different to those required for alien taxa such as *Hakea sericea* and the serotinous *Pinus* species. The alien trees and shrubs that invade mountain fynbos are fire-adapted species. Whereas fire eliminates indigenous forest elements from the shrublands, its regular occurrence is essential for invasion and population growth of alien trees and shrubs. The aliens initially behave much like the indigenous shrubs but they are well buffered against fire-induced population crashes because of their short juvenile periods and large reserves of viable seeds. Invasion thus disrupts the non-equilibrium status of natural communities and results in new depauperate steady states, which have further impacts at community and ecosystem levels.

Variations in fire frequency affect the structure and composition of fynbos communities in different ways, depending on whether alien trees such as pines are present (Fig. 15.1). In uninvaded fynbos, fire maintains a mosaic of shrublands and forest patches. Significant changes in vegetation structure occur when fires occur at sufficiently short intervals to prevent the encroachment of indigenous forest elements, or eliminate reseeding shrubs. Invasion of the fynbos by alien trees and shrubs disrupts the normal situation and results in dense stands of aliens under moderate fire frequencies. Under high fire frequencies, an open savanna, with a mixture of sprouting fynbos plants and young pines and hakeas is likely to exist (Fig. 15.1). Exclusion of fire will invalidate some advantages enjoyed by fire-adapted alien trees, and a mixture of alien and indigenous trees will result.

The long inter-fire intervals needed for the establishment of forests in fynbos catchments are not practically attainable, and they would possibly reduce species diversity in the species-rich shrublands. Management directed towards encouragement of forest along streams is perhaps appropriate, because of the importance of riparian forests as filters (see Chaps. 12, 13 and 14). However, of more pressing concern in most areas is manage-

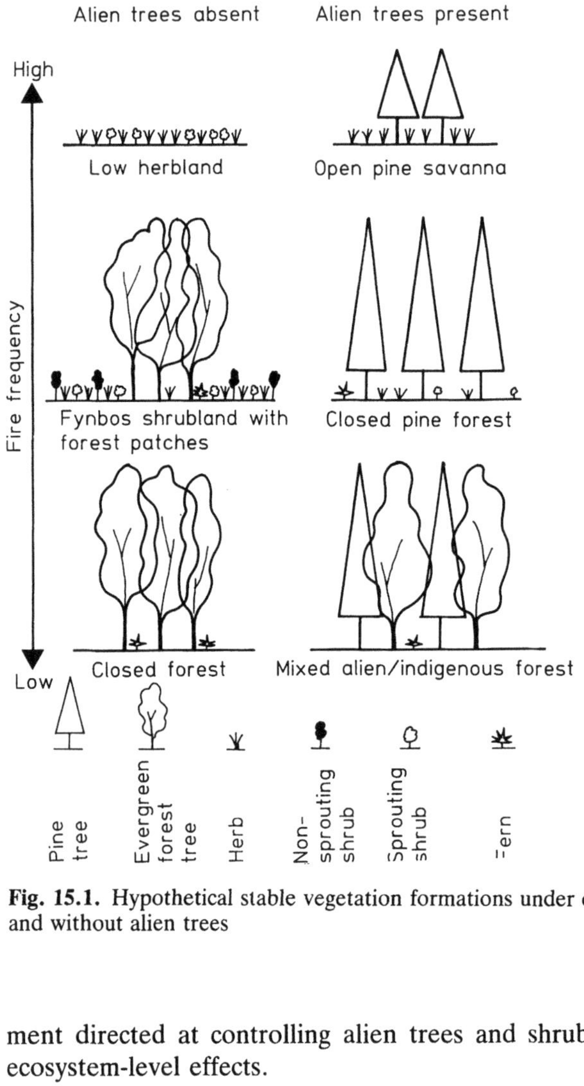


**Fig. 15.1.** Hypothetical stable vegetation formations under different fire frequencies with and without alien trees

ment directed at controlling alien trees and shrubs which have important ecosystem-level effects.

## 15.4 Ecosystem-Level Effects of Fire

Studies at the ecosystem level in Swartboskloof concentrated on nutrient budgets, streamflow and erosion. The fire in Swartboskloof resulted in increased overland flow and sediment yield at the scale of small plots. These increases, which were of short duration, did not alter stormflow or sediment yield at the catchment level. The lack of ecosystem-level response was ascribed to the riparian forests which acted as a buffer (Chap. 12). The importance of riparian zones in catchment management is often emphasized (see e.g. DeBano and Schmidt 1990), and has been reinforced by studies in Swartboskloof. Additional factors contributing to low levels of sediment yield at Swartboskloof are the inherently low erodibility of soils, and the pristine nature of the catchment (e.g. the absence of logging roads).

Bosch et al. (1986) proposed a model for predicting the magnitude of streamflow increases that could be expected from fynbos catchments after fire. In this model, the magnitude of increases is dependent on pre-fire biomass, as this determines levels of interception and transpiration; the rate of return to pre-fire levels is dependent on the rate of recovery of the vegetation, which in turn depends on the ratio of resprouting to non-sprouting plants. In systems such as Swartboskloof, with relatively high biomass and dominance by non-sprouting, slow-growing shrubs, regular burning could increase streamflow levels. The measured increases at Swartboskloof support the Bosch model (see Chapter 12; Sect. 12.3.1.1).

Chemical budgets also appear to be little influenced by fire. Considerable amounts of nitrogen are lost through volatilization but these losses appear to be balanced by inputs from atmospheric deposition and nitrogen fixation (Chap. 13). The studies at Swartboskloof directed at quantifying the role of microorganisms in the nutrient budgets have shed light on previously unexplained increases in outflow from the catchments. For example, it appears that the flush of nutrients in streamflow after the first winter rains could be due to increased microbial activity in the soils (Chap. 14). The importance of riparian vegetation as a filter to nutrient loss was also emphasized. The budgets show that regular burning (at intervals of 15–30 years) will not affect the nutrient status of catchments. However, frequent burning, as applied in firebreaks, leads to the elimination of dominant non-sprouting shrubs and causes large reductions in biomass (van Wilgen 1982; Fig. 15.1). Invasion by alien trees, on the other hand, may lead to five-fold increases in biomass (Versfeld and van Wilgen 1986; Fig. 15.1). The resilience of altered systems under the fire regimes that would then prevail is not understood.

Fire also plays a role in maintaining species diversity at the ecosystem level. A very small percentage of fynbos plant species, notably the serotinous shrubs, are vulnerable to *local* extinction (i.e. at the community scale) following fire. Richardson and Cowling (Chap. 9) argue that the sequence of local extinction and recolonization is an inherent feature of mountain fynbos dynamics. Indeed, it appears that vagile and marauding populations of these overstorey species impose on the landscape a dynamic array of habitats which promote floristic patchiness at the ecosystem level (Cowling 1987). Disruption of the non-equilibrium system occurs when alien trees and shrubs are added to the system, resulting in new depauperate steady states (Chap. 9; Fig. 9.4). Plant invasions have caused significant reductions in the richness of indigenous plant species at the scale of small quadrats and at the community level. In this respect, the impacts of plant

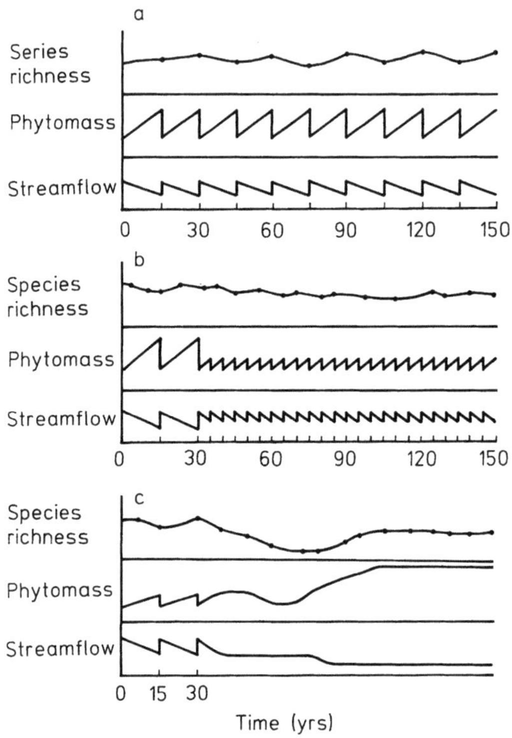


**Fig. 15.2.** Hypothetical changes in three parameters (plant species richness, phytomass and streamflow) at the ecosystem level for three fire frequencies at Swartboskloof. The fire frequencies depicted are: **a** fire every 15 years; **b** fire every five years; and **c** fire excluded. *Tick marks* on the x-axis indicate the occurrence of fire

invasions in mountain fynbos have been exceptionally severe (cf. Vitousek 1986 for North America and Crawley 1987, p. 441 for Britain). As is the case in Britain (Crawley 1987), no plant extinctions in the fynbos can *yet* be directly attributed to invasions (Richardson et al. 1989). However, many endemic plant species in the fynbos biome are currently threatened with extinction due to the effects of invasions by trees and shrubs, and extinctions seem inevitable. Indeed, there is likely to be a cascade of extinctions in the next few decades as more communities are disrupted by invasions, and as the time since the establishment of thickets of these species increases (with concomitant attrition of seed banks of suppressed endemic species).

The results of the Swartboskloof experiment allow for the examination of resilience of selected ecosystem parameters to fire (Fig. 15.2). Ecosystem properties such as plant species richness, phytomass and streamflow appear to be resilient under a regime of fires at intervals of 12–30 years, with increases in biomass and decreases in streamflow with increasing post-fire age. Frequent fires (e.g. every 5 years) will reduce species richness slightly by eliminating seed-reproducing shrubs, and phytomass will also decrease. Streamflow, on the other hand, will increase. Fire exclusion will cause a net reduction in plant species richness as the vegetation becomes moribund. This may recover to some extent as forest tree species invade, but overall diversity should be markedly lower than for frequently burnt fynbos. Phytomass will follow similar trends, but final phytomass in forest vegetation will be much higher than that in frequently burnt fynbos. Streamflow will be reduced in proportion to the increases in biomass.

## 15.5 Managing Fire in Mountain Fynbos

Management in mountain fynbos is usually aimed ultimately at conserving species and representative ecosystems, reducing fire hazard, and ensuring that catchments continue to deliver high quality water. Fire is the most important management practice for achieving these aims and is applied at intervals of 12–15 years during late summer or autumn (see van Wilgen et al. 1990a, 1991 for reviews of the evolution of this policy). All results from the Swartboskloof experiment indicate that this is the most appropriate regime for achieving the objectives mentioned above (Fig. 15.2). The application of a rigid fire regime (every 12–15 years in March) removes much of the stochasticity that is necessary for maintaining coexisting species. However, in practice mountain fynbos is subjected to a variety of fires other than prescribed burns, notably firebreak burns in spring that escape, and wild-fires in summer. These fires account for more than 70% of the total area burnt (van Wilgen et al. 1990a). Also, some areas are not burnt on schedule because of management constraints, thus creating longer intervals between fires. As long as this situation prevails, fire regimes should be sufficiently variable to safeguard the stochasticity necessary for species coexistence.

The biggest problem currently facing managers in mountain fynbos is the control of invasive trees and shrubs. The currently accepted method of control involves the felling of the trees and shrubs, leaving them for between 12 and 24 months to allow for the depletion of seeds, and then burning to kill any regenerating seedlings. This method is usually effective in suppressing the aliens but is extremely expensive and sometimes causes further damage to the invaded systems (Richardson and van Wilgen 1986b; Breytenbach 1989). Biological control measures against *Hakea sericea* and the Australian *Acacia* species are likely to reduce population levels in the long term (review in van Wilgen et al. 1991). However, managers are still faced with the problem of clearing existing dense stands, and need to coordinate felling and burning operations. Invasive pines are a special case since authorities are reluctant to sanction research on biological control agents, because of the economic importance of pines for timber. Pine plantations adjoin many mountain catchments and provide a constant source of seeds for re-invasion following control (Richardson and Brown 1986).

Reduced funding for alien plant control (see Preface) creates major logistic problems for managers. Van Wilgen et al. (1991) examined four scenarios of funding and their implications. They showed that current levels of funding will probably allow prescribed burning to take place once every 15 years, but the cost of weed control programmes would mean that each management unit could only be cleared every second fire cycle (Scenario 1). The known rates of proliferation for *Pinus radiata* (Richardson and Brown 1986) show that canopy cover of aliens would exceed 30% in the second fire cycle. It would return to low levels again following clearing. To keep areas clear of aliens, funding would have to be increased to a level where clearing would be possible in conjunction with every prescribed fire (Scenario 2). Under this scenario, it would be possible to keep the cover of aliens down to <1%. A more probably scenario is that funding will decrease in real terms, so that clearing will only be possible every third fire cycle (Scenario 3). The cover of aliens would increase to over 30% in the second cycle, and to 90% in the third cycle. Clearing of these dense stands would then be extremely costly. Under a worst-case scenario (Scenario 4), funding for control of aliens is terminated. Aliens would increase to almost 100% cover after about four fire cycles, and would remain at this level. The implications of these scenarios for species diversity, water yield and fire hazard are described by van Wilgen et al. (1991) who conclude that a major effort and expenditure of funds is required to prevent important ecosystem-level changes.

# References


Abele LG (1976) Comparative species richness in fluctuating and constant environments: coral-associated decapod crustaceans. Science 192: 461–463

Acherar M, Lepart J, Debussche M (1984) La colonisation des friches par le pin d'Alep (*Pinus halepensis* Miller) en Languedoc méditerranéen. Acta Oecol Oecol Plant 5: 179–189

Adams F, Ewing PA, Huberty MR (1947) Hydrologic aspects of burning brush and woodland-grass ranges in California. California Division of Forestry

Agren GI, Fagerstrom T (1984) Limiting dissimilarity in plants: randomness prevents exclusion of species with similar competitive abilities. Oikos 43: 369–375

Ahlgren IS (1974) The effect of fire on soil organisms. In: Kozlowski TT, Ahlgren CE (eds) Fire and ecosystems. Academic Press, New York, pp 47–72

Albini FA (1984) Wildland fires. Am Sci 72: 590–597

Alexander M (1977) Introduction to soil microbiology, 2nd edn. John Wiley, New York

Allen ON (1957) Experiments in soil bacteriology, 3rd edn. Burgess, Minneapolis

Andersen AN, Yen AL (1985) Immediate effects of fire on ants in the semi-arid mallee region of north-western Victoria. Aust J Ecol 10: 25–30

Anderson HW (1976) Fire effects on water supply, floods, and sedimentation. Proc Annu Tall Timbers Fire Ecology Conf 15: 249–260

Anderson HW, Coleman GB, Zinke PJ (1959) Summer slides and winter scour – dry-wet erosion in southern California mountains. USDA For Serv Pac Southwest For Range Exp Stn Tech Pap 36

Anderson HW, Hoover MD, Reinhart KG (1976) Forests and water: effects of forest management on floods, sedimentation and water supply. USDA For Serv Gen Tech Rep PSW-18

Anderson JPE, Domsch KH (1975) Measurement of bacterial and fungal contributions to respiration of selected agricultural and forest soils. Can J Microbiol 21: 314–322

Andrews PL (1986) BEHAVE: fire behaviour prediction and fuel modelling system – BURN subsystem. USDA For Serv Gen Tech Rep INT-194

Anon (1979) Automatic sensing wet/dry precipitation collector. Aerochem Metrics, Florida

Archer S (1989) Have southern Texas savannas been converted to woodlands in recent history? Am Nat 134: 545–561

Arianoutsou M, Margaris NS (1981) Fire induced nutrient losses in a phryganic (East Mediterranean) ecosystem. Int J Biometeorol 25: 341–347

Arianoutsou-Faraggitaki M, Margaris NS (1982) Microbial activity after fire in a phryganic east mediterranean ecosystem. Proc Symp on Dynamics and management of mediterranean-type ecosystems, USDA For Serv Gen Tech Rep PSW-58, Pacific Southwest For Range Exp Stn, pp 321–324

Aschmann H (1973) Distribution and peculiarity of mediterranean ecosystems. In: di Castri F, Mooney HA (eds) Mediterranean-type ecosystems, origin and structure. Ecological Studies 7. Chapman & Hall, London, pp 11–19

Atkinson G (1984) Erosion damage following bushfires. J Soil Conserv Serv NSW 40: 4–9
Augspurger CK (1981) Reproductive synchrony of a tropical shrub: experimental studies on effects of pollinators and seed predators on *Hybanthus prunifolius* (Violaceae). Ecology 62: 775–788
Bååth E, Söderström B (1982) Seasonal and spatial variation in fungal biomass in a forest soil. Soil Biol Biochem 14: 353–358
Bailey SW, Hornbeck JW, Martin CW, Buso DC (1987) Watershed factors affecting stream acidification in the White Mountains of New Hampshire, USA. Environ Manage 11: 53–60
Baker HG (1986) Patterns of plant invasion in North America. In: Mooney HA, Drake JA (eds) Ecology of biological invasions of North America and Hawaii. Ecological Studies 58. Springer, Berlin Heidelberg New York, pp 4–26
Baker HG, Stebbins GL (eds) (1965) The genetics of colonizing species. Academic Press, New York
Bands DP, Bosch JM, Lamb, AJ, Richardson DM, van Wilgen BW, van Wyk DB, Versfeld DB (1987) Jonkershoek Forestry Research Centre. Pamphlet 384, Dept Environment Affairs, Pretoria
Banks CH (1964) Further notes on the effect of autumnal veld burning on storm flow in the Abdolskloof catchment, Jonkershoek. For S Afr 4: 79–84
Bazzaz FA (1986) Life history of colonizing plants: some demographic, genetic, and physiological features. In: Mooney HA, Drake JA (eds) Ecology of biological invasions of North America and Hawaii. Ecological Studies 58, Springer, Berlin Heidelberg New York, pp 96–110
Bean PA (1962) An enquiry into the effect of veldfires on certain geophytes. MSc Thesis, University of Cape Town
Beard JS (1976) Vegetation survey of Western Australia, 1: 1000000 series sheet 6, Murchison. University of Western Australia Press, Nedlands
Beard JS (1981) Vegetation survey of Western Australia, 1: 1000000 series sheet 7, Swan. University of Western Australia Press, Nedlands
Beard JS (1983) Ecological control of the vegetation of southwestern Australia: moisture versus nutrients. In: Kruger FJ, Mitchell DT, Jarvis JUM (eds) Mediterranean-type ecosystems: the role of nutrients. Ecological Studies 43, Springer, Berlin Heidelberg New York, pp 66–73
Beard JS (1984) Biogeography of the kwongan. In: Pate JS, Beard JS (eds) Kwongan: plant life of the sandplain. University of Western Australia Press, Nedlands, pp 1–25
Beard JS (1990) Plant life of Western Australia. Kangaroo Press, Kenthurst
Beattie AJ (1985) The evolutionary ecology of ant-plant mutualisms. Cambridge University Press, Cambridge
Becking JH (1959) Nitrogen-fixing bacteria of the genus *Beijerinckia* in South African soils. Plant Soil 11: 193–206
Bell DT, Hopkins AJM, Pate JS (1984) Fire in the kwongan. In: Pate JS, Beard JS (eds) Kwongan: plant life of the sandplain. University of Western Australia Press, Nedlands, pp 178–204
Bergl SM, Lamont BB (1988) The water relations, rooting patterns and phenologies of two sclerophyllous shrubs. In: Di Castri F, Floret Ch, Rambal S, Roy J (eds) Time scales and water stress. Proc 5th Int Conf on Mediterranean ecosystems, International Union of Biological Sciences, Paris, pp 569–573
Bigalke RC (1979a) Fauna. In: Smuts F (ed) Stellenbosch, 300 years. Town Council of Stellenbosch, Stellenbosch, pp 39–50
Bigalke RC (1979b) Aspects of vertebrate life in fynbos, South Africa. In: Specht RL (ed) Heathlands and related shrublands. Descriptive Studies. Elsevier, Amsterdam, pp 81–95
Binkley D, Kimmins JP, Feller MC (1982) Water chemistry profiles in an early- and a mid-successional forest in coastal British Columbia. Can J For Res 12: 240–248
Blackburn WH, Tueller PT (1970) Pinyon and juniper invasion in black sagebrush communities in east-central Nevada. Ecology 51: 841–848

Blake TJ (1983) Coppice systems for short-rotation intensive forestry: the influence of cultural, seasonal and plant factors. Aust For Res 13: 279–291

Blong RJ, Riley SJ, Crozier PJ (1982) Sediment yield from runoff plots following bushfire near Narrabeen Lagoon, NSW. Search 13: 36–38

Bond P, Goldblatt P (1984) Plants of the Cape Flora – a descriptive catalogue. J S Afr Bot Suppl Vol 13: 1–455

Bond WJ (1980) Fire and senescent fynbos in the Swartberg, Southern Cape. S Afr For J 114: 68–74

Bond WJ (1984) Fire survival of Cape Proteaceae – influence of fire season and seed predators. Vegetatio 56: 65–74

Bond WJ (1985) Canopy-stored seed reserves (serotiny) in Cape Proteaceae. S Afr J Bot 51: 181–6

Bond WJ (1987) Regeneration and its importance in the distribution of woody plants. PhD Thesis, University of California, Los Angeles

Bond WJ (1989a) Describing and conserving biotic diversity. In: Huntley BJ (ed) Biotic diversity in southern Africa: concepts and conservation. Oxford University Press, Cape Town, pp 2–18

Bond WJ (1989b) The tortoise and the hare: ecology of angiosperm dominance and gymnosperm persistence. Biol J Linn Soc 36: 227–249

Bond WJ, Breytenbach GJ (1985) Ants, rodents and seed predation in Proteaceae. S Afr J Zool 20: 150–154

Bond WJ, Slingsby P (1983) Seed dispersal by ants in shublands of the Cape Province and its evolutionary implications. S Afr J Sci 79: 231–233

Bond WJ, Slingsby P (1984) Collapse of an ant-plant mutualism: the Argentine ant (*Iridomyrmex humilis*) and myrmecochorous Proteaceae. Ecology 65: 1031–1037

Bond WJ, Ferguson M, Forsyth G (1980) Small mammals and habitat structure along altitudinal gradients in the Southern Cape Mountains. S Afr J Zool 15: 34–43

Bond WJ, Vlok J, Viviers M (1984) Variation in seedling recruitment of Cape Proteaceae after fire. J Ecol 72: 209–221

Bond WJ, Le Roux D, Erntzen R (1990) Fire intensity and regeneration of myrmecochorous Proteaceae. S Afr J Bot 56: 326–331

Booysen PdeV, Tainton NM (eds) (1984) Ecological effects of fire in South African ecosystems. Ecological Studies 48. Springer, Berlin Heidelberg New York

Bosch JM, Hewlett JD (1982) A review of catchment experiments to determine the effect of vegetation changes on water yield and evapotranspiration. J Hydrol 55: 3–23

Bosch JM, Schulze RE, Kruger FJ (1984) The effects of fire on water yield. In: Booysen PdeV, Tainton NM (eds) Ecological effects of fire in South African ecosystems. Ecological Studies 48. Springer, Berlin Heidelberg New York, pp 327–348

Bosch JM, van Wilgen BW, Bands DP (1986) A model for comparing water yields from fynbos catchments burnt at different intervals. Water SA 12: 191–196

Botha SA (1988) Die invloed van diere op die populasiebiologie van die Clanwilliam seder (*Widdringtonia cederbergensis* Marsh) met spesiale verwysing na die rol van klein soogdiere. MSc Thesis, University of Stellenbosch

Botha SA (1989) Effect of sowing season and granivory on the establishment of the Clanwilliam cedar. S Afr J Wildl Res 19: 112–117

Boucher C (1972) The vegetation of the Cape Hangklip area. MSc Thesis, University of Cape Town

Boucher DH, Vandermeer JH, Yih K, Zamora N (1990) Contrasting hurricane damage in tropical rain forest and pine forest. Ecology 71: 2022–2024

Bowen AJM (1979) Environmental chemistry of the elements. Academic Press, London

Bradstock RA (1990) Demography of woody plants in relation to fire: *Banksia serrata* L.f. and *Isopogon anemonifolius* (Salisb.) Knight. Aust J Ecol 15: 117–132

Bradstock RA, Myerscough PJ (1988) The survival and population responses to frequent fires of two woody resprouters: *Banksia serrata* and *Isopogon anemonifolius*. Aust J Bot 36: 415–31

Bradstock RA, O'Connel MA (1988) Demography of woody plants in relation to fire: *Banksia ericifolia* L.f. and *Petrophile pulchella* (Schrad) R.Br. Aust J Ecol 13: 505–18

Breytenbach GJ (1982) Small mammal responses to environmental gradients in the Groot Swartberg of the southern Cape. MSc Thesis, University of Pretoria
Breytenbach GJ (1984) Single agedness in fynbos: a predation hypothesis. In: Dell B (ed) Proc 4th Int Conf on Mediterranean ecosystems. Botany Department, University of Western Australia, Nedlands, pp 14–15
Breytenbach GJ (1986) Impacts of alien organisms on terrestrial communities with emphasis on communities of the south-western Cape. In: Macdonald IAW, Kruger FJ, Ferrar AA (eds) The ecology and management of biological invasions in southern Africa. Oxford University Press, Cape Town, pp 229–238
Breytenbach GJ (1987) Small mammal dynamics in relation to fire. In: Cowling RM, Le Maitre DC, McKenzie B, Prys-Jones RP, van Wilgen BW (eds) Disturbance and the dynamics of fynbos biome communities. S Afr Nat Sci Prog Rep 135, CSIR, Pretoria, pp 56–58
Breytenbach GJ (1989) Alien control: can we afford to slash and burn hakea in fynbos ecosystems? S Afr For J 151: 6–16
Bricker OP, Rice KC (1989) Acidic deposition to streams. Environ Sci Tech 23: 379–385
Britton DL (submitted) The benthic macroinvertebrate fauna of a South African mountain stream, and its response to fire. J Zool
Brookes PC, Tate KR, Jenkinson DS (1983) The adenylate energy charge of the soil microbial biomass. Soil Biol Biochem 15: 9–16
Brown G, Mitchell DT (1986) Influence of fire on the soil phosphorus status in sand-plain lowland fynbos, south-western Cape. S Afr J Bot 52: 67–72
Brown G, Mitchell DT, Stock WD (1984) Atmospheric deposition of phosphorus in a coastal fynbos ecosystem of the southwestern Cape, South Africa. J Ecol 72: 547–551
Brown JAH (1972) Hydrologic effects of a bushfire in a catchment in south-eastern New South Wales. J Hydrol 15: 77–96
Brown PJ, Le Maitre DC (1990) Fire-lilies, the firebirds of the fynbos. Veld Flora 76: 22
Brown PJ, Manders PT, Bands DP, Kruger FJ, Andrag RH (1991) Prescribed burning as a conservation management practice: a case history from the Cederberg mountains, Cape Province, South Africa. Biol Conserv 56: 133–150
Brownlie SF (1982) The effects of recent land use on a fynbos site: Final report. Unpublished report. School of Environmental Studies, University of Cape Town, Cape Town
Burch GJ, Moore ID, Burns J (1989) Soil hydrophobic effects on infiltration and catchment runoff. Hydrol Process 3: 211–222
Burgan RE (1987) A comparison of procedures to estimate fine dead fuel moisture for fire behaviour predictions. S Afr For J 142: 34–40
Burgan RE, Rothermel RC (1984) BEHAVE: fire behaviour prediction and fuel modelling system. Fuel subsystem. USDA For Serv Gen Tech Rep INT-167
Burtt BL (1970) The evolution and taxonomic significance of a subterranean ovary in certain monocotyledons. Isr J Bot 19: 77–90
Byram GM (1959) Combustion of forest fuels. In: Davis KP (ed) Forest fire: control and use. McGraw-Hill, New York, pp 155–182
Byram GM (1963) An analysis of the drying process in forest fuel material. Paper presented at the 1960 Int Symp on Humidity and moisture, Washington, DC
Campbell BM, van der Meulen F (1980) Patterns of plant species diversity in fynbos vegetation, South Africa. Vegetatio 43: 43–47
Campbell BM, Werger MJA (1988) Plant form in the mountains of the Cape, South Africa. J Ecol 76: 637–653
Campbell BM, MacKenzie B, Moll EJ (1979) Should there be more tree vegetation in the mediterranean climate region of South Africa? J S Afr Bot 45: 543–557
Campbell BM (1985) A classification of the mountain vegetation of the fynbos biome. Mem Bot Surv S Afr 50: 1–121
Campbell RE, Baker MBJ, Folliott PF, Larson LR, Avery CC (1977) Wildfire effects on a ponderosa pine ecosystem: an Arizona case study. USDA For Serv Res Pap RM-191
Carpenter FL, Recher HF (1979) Pollination, reproduction and fire. Am Nat 113: 871–9

Chan HT (1980) Reproductive biology of some Malaysian dipterocarps. II. Fruiting biology and seedling studies. Malaysian For 43: 438–51
Charnov EL, Shaffer WM (1973) Life-history consequences of natural selection: Cole's result revisited. Am Nat 109: 791–93
Chesson PL, Warner RR (1981) Environmental variability promotes coexistence in lottery competitive systems. Am Nat 117: 923–943
Christensen NL, Muller CH (1975) Effects of fire on factors controlling plant growth in *Adenostoma* chaparral. Ecol Monog 45: 29–55
Clarholm M, Rosswall T (1980) Biomass and turnover of bacteria in a forest soil and a peat. Soil Biol Biochem 12: 49–57
Coetzee JH, Giliomee JH (1985) Insects in association with the infloresence of *Protea repens* (L) (Proteaceae) and their role in pollination. J Entomol Soc South Afr 48: 303–314
Coetzee JH, Giliomee JH (1987) Seed predation and survival in the infructescences of *Protea repens* (Proteaceae). S Afr J Bot 53: 61–4
Collins BG, Rebelo AG (1987) Pollination biology of the Proteaceae in Australia and southern Africa. Aust J Ecol 12: 387–421
Colman EA (1951) Fire and water in southern California's mountains. USDA For Serv Calif For Range Exp Stn, Misc Pap 3
Comins HN, Noble IR (1985) Dispersal, variability and transient niches: species coexistence in a uniformly variable environment. Am Nat 126: 706–723
Cone GB (1973) The floras of the south western part of South Africa. Tuatara 20: 160–164
Connell JH (1978) Diversity in tropical rain forests and coral reefs. Science 199: 1302–1310
Connell JH, Slatyer RO (1977) Mechanisms of succession in natural communities and their role in community stability and organization. Am Nat 111: 1119–1144
Conrad CE, Oechel WC (eds) (1982) Dynamics and management of mediterranean-type ecosystems. USDA For Sev Gen Tech Rep PSW-58
Countryman CM, Philpot CW (1970) Physical characteristics of chemise as a wildland fuel. USDA For Serv Res Pap PSW-66
Cowling RM (1983a) Diversity relations in Cape shrublands and other vegetation in the south-eastern Cape, South Africa. Vegetatio 54: 103–127
Cowling RM (1983b) The occurrence of $C_3$ and $C_4$ grasses in fynbos and allied shrublands in the south eastern Cape. Oecologia (Berlin) 58: 121–127
Cowling RM (1984) A syntaxonomic and synecological study in the Humansdorp region of the fynbos biome. Bothalia 15: 175–227
Cowling RM (1987) Fire and its role in coexistence and speciation in Gondwanan shrublands. S Afr J Sci 83: 106–112
Cowling RM (ed) (1991) The ecology of fynbos: Nutrients, Fire and Diversity. Oxford University Press, Cape Town
Cowling RM, Campbell BM (1980) Convergence in vegetation structure in the mediterranean communities of California, Chile and South Africa. Vegetatio 43: 191–197
Cowling RM, Gxaba T (1990) Effects of a fynbos overstorey shrub on understorey community structure: implications for the maintenance of community-wide species richness. S Afr J Ecol 1: 1–7
Cowling RM, Lamont BB (1987) Post-fire recruitment of four co-occurring *Banksia* species. J Appl Ecol 24: 645–658
Cowling RM, Moll EJ, Campbell BM (1976) The ecological status of the understorey communities of pine forests on Table Mountain. S Afr For J 99: 13–23
Cowling RM, Campbell BM, Mustart P, McDonald DJ, Jarman ML, Moll EJ (1988) Vegetation classification in a floristically complex area: the Agulhas Plain. S Afr J Bot 54: 290–300.
Crawley MJ (1986) The population biology of invaders. Philos Trans R Soc Lond B 314: 711–731

Crawley MJ (1987) What makes a community invasible? In: Gray AJ, Crawley MJ, Edwards PJ (eds) Colonization, succession and stability. Blackwell, Oxford, pp 429–453

Crowell KL (1968) Rates of competitive exclusion by the Argentine ant in Bermuda. Ecology 49: 551–555

Cryer JD (1986) Time series analysis. Duxbury, Boston

Dafni A, Cohen D, Noy-Meir I (1981a) Life cycle variation in geophytes. Ann Mo Bot Gard 68: 653–60

Dafni A, Schmida A, Avishai M (1981b) Leafless autumnal-flowering geophytes in the Mediterranean region – phytogeographical, ecological and evolutionary aspects. Plant Syst Evol 137: 118–93

Danin A, Orshan G (1990) The distribution of Raunkiaer life forms in relation to the environment. J Veg Sci 1: 41–48

Daubenmire R (1968) Ecology of fire in grasslands. Adv Ecol Res 5: 209–266

Davidson DW, Morton SR (1981) Myrmecochory in some plants (F. Chenopodiaceae) of the Australian arid zone. Oecologia 50: 357–366

Davis SD (1989) Patterns in mixed chaparral stands: differential water status and seedling survival during summer drought. In: Keeley SC (ed) The Californian chaparral: paradigms re-examined. Sci Ser No 34, Natural History Museum, Los Angeles, pp 97–105

Davis SD, Mooney HA (1985) Comparative water relations of adjacent California shrub and grassland communities. Oecologia (Berlin) 66: 522–529

Davis SD, Mooney HA (1986) Water use patterns of four co-occurring chaparral shrubs. Oecologia (Berlin) 70: 172–177

Day J, Siegfreid WR, Louw GN, Jarman ML (eds) (1979) Fynbos ecology: a preliminary synthesis. S Afr Nat Sci Prog Rep 40, CSIR, Pretoria

Deacon HJ (1972) A review of the Post-Pleistecene in South Africa. S Afr Archaeol Soc Goodwin Ser 1: 26–45

Deacon HJ (1983a) An introduction to the Fynbos region, time scales and palaeoenvironments. In: Deacon HJ, Hendey QB, Lambrechts JJN (eds) Fynbos palaeoecology: a preliminary synthesis. S Afr Nat Sci Prog Rep 75, CSIR, Pretoria, pp 1–20

Deacon HJ (1983b) The peopling of the fynbos region. In: Deacon HJ, Hendey QB, Lambrechts JJN (eds) Fynbos palaeoecology: a preliminary synthesis. S Afr Nat Sci Prog Rep 75, CSIR, Pretoria, pp 181–204

Deacon HJ (1983c) The comparative evolution of mediterranean-type ecosystems: a southern perspective. In: Kruger FJ, Mitchell DT, Jarvis JUM (eds) Mediterranean-type ecosystems – the role of nutrients. Ecological Studies 43. Springer, Berlin Heidelberg New York, pp 3–40

DeBano LF (1966) Formation of non-wettable soils involves heat transfer mechanism. USDA For Serv Res Note PSW-132

DeBano LF (1971) The effect of hydrological substances on water movement in soil during infiltration. Soil Sci Soc Am Proc 35: 340–343

DeBano LF (1981) Water repellent soils: a state of the art. USDA For Serv Gen Tech Rep PSW-46

DeBano LF, Conrad CE (1976) Nutrients lost in debris and runoff water from a burned chaparral watershed. In: Proc Federal Inter-Agency Sedimentation Conf 3: 13–27

DeBano LF, Conrad CE (1978) The effect of fire on nutrients in a chaparral ecosystem. Ecology 59: 489–497

DeBano LF, Dunn PH (1982) Soil and nutrient cycling in mediterranean-type ecosystems: a summary and synthesis. Proc Symp on Dynamics and management of mediterranean-type ecosystems, USDA For Serv Gen Tech Rep PSW-58, Pac Southwest For Range Exp Stn, Berkeley, pp 358–364

DeBano LF, Krammes JS (1966) Water repellent soils and their relation to wildfire temperatures. Int Assoc Sci Hydrol Bull XI Annee 2: 14–19

DeBano LF, Schmidt LJ (1990) Potential for enhancing riparian habitats in the southwestern United States with watershed practices. Forest Ecol Manage 33/34: 385–403

DeBano LF, Osborn JF, Krammes JS, Letey J (1967) Soil wettability and wetting agents . . . our current knowledge of the problem. USDA For Serv Res Pap PSW-43
DeBano LF, Mann LD, Hamilton DA (1970) Translocation of hydrophobic substances into soil by burning organic litter. Soil Sci Soc Am Proc 34: 130–133
DeBano LF, Dunn PH, Conrad CE (1977) Fire's effect on physical and chemical properties of chaparral soils. In: Mooney HA, Conrad CE (eds) Proc Symp on Environmental consequences of fire and fuel management in mediterranean ecosystems. USDA For Serv Gen Tech Rep WO-3, pp 65–74
DeBano LF, Eberlein GE, Dunn PH (1979a) Effects of burning on chaparral soils I. Soil nitrogen. Soil Sci Soc Am J 43: 504–509
DeBano LF, Rice RM, Conrad CE (1979b) Soil heating in chaparral fires: effects on soil properties, plant nutrients, erosion and runoff. USDA For Serv Res Pap PSW-145
Debussche M, Escarré J, Lepart J (1980) Changes in mediterranean shrub communities with *Cytisus purgans* and *Genista scorpius*. Vegetatio 43: 73–82
Deeming JE, Brown JK (1975) Fuel models in the National Fire Danger Rating System. J For 73: 347–380
Deeming JE, Burgan RE, Cohen JD (1978) The National Fire Danger Rating System – 1978. USDA For Serv Gen Tech Rep INT 39
De Kock AE (1983) Argentynse mier (*Iridomyrmex humilis*) indringing in die fynbos van die Jonkershoekvallei. Rep 84/14, Jonkershoek Forestry Research Centre, Stellenbosch
De Kock AE (1990a) Interactions between the introduced Argentine ant, *Iridomyrmex humilis* Mayr, and two indigenous fynbos ant species. J Entomol Soc South Afr 53: 107–108
De Kock AE (1990b) The Argentine ant (*Irydomyrmex humilis*) (Mayr), indigenous ants and other epigaeic invertebrates of the fynbos in relation to fire and vegetation age. MSc Thesis, University of Stellenbosch
De Kock AE, Giliomee JH (1989) A survey of the Argentine ant, *Iridomyrmex humilis* (Mayr), (Hymenoptera: Formicidae) in South African fynbos. J Entomol Soc South Afr 52: 157–164
Dell B, Hopkins AJM, Lamont BB (eds) (1986) Resilience in mediterranean-type ecosystems. Junk, Dordrecht
Dell JD, Philpot CW (1965) Variations in the moisture content of several fuel size components of live and dead chamise. USDA For Serv Res Note PSW-83
Dennis B, Patil GP, Rossi O (1979) The sensitivity of ecological diversity indices to the presence of pollutants in aquatic communities. In: Calms J, Patil GP, Waters WE (eds) Statistical ecology II: environmental biomonitoring, assessment, prediction and management – certain case studies and related quantitative issues. International Co-operative Publishing House, Fairland, Maryland, pp 379–413
De Vos MP (1977) Knolontwikkeling by sommige genera van die Iridaceae en die sistematiese posisie van *Micranthus*. Tydsk Natuurwetenskappe 17: 5–19
Diamond J, Case TJ (1986) Overview: introductions, extinctions, exterminations and invasions. In: Diamond J, Case TJ (eds) Community ecology. Harper & Row, New York, pp 65–79
Di Castri F (1981) Mediterranean-type shrublands of the world. In: Di Castri F, Goodall DW, Specht RL (eds) Ecosystems of the world 11. Mediterranean-type shrublands. Elsevier, Amsterdam, pp 1–52
Di Castri F, Mooney HA (eds) (1973) Mediterranean-type ecosystems – origin and structure. Chapman & Hall, London
Digby PGN, Kempton RA (1987) Multivariate analysis of ecological communities. Chapman & Hall, London
Dobson AJ (1983) An introduction to statistical modelling. Chapman & Hall, London
Dodd J, Heddle EM, Pate JS, Dixon KW (1984) Rooting patterns of sandplain plants and their functional significance. In: Pate JS, Beard JS (eds) Kwongan: plant life of the sandplain. University of Western Australia Press, Nedlands, pp 146–177
Donnelly D (1983) Comparison of community structure of the epigaeic invertebrate fauna, with particular reference to ants and beetles, between fynbos sites under different management practices. MSc Thesis, University of Stellenbosch

Donnelly D, Giliomee JH (1985a) Community structure of epigaeic ants (Hymenoptera: Formicidae) in fynbos vegetation in the Jonkershoek Valley. J Entomol Soc South Afr 48: 247–257
Donnelly D, Giliomee JH (1985b) Community structure of epigaeic ants in a pine plantation and in newly burnt fynbos. J Entomol Soc S Afr 48: 259–265
Donnelly D, Giliomee JH (1988) The epigaeic invertebrate fauna in South African fynbos of different ages after fire. Contribution to Chapter 5: Soil and litter invertebrates, coordinated by JD Majer and P Greenslade. In: Specht RL (ed) Mediterranean-type ecosystems: a data source book. Kluwer, Dordrecht, pp 197–226
Doyle AT (1990) Use of riparian and upland habitats by small mammals. J Mammal 71: 14–23
Drake JA, Mooney HA, Di Castri F, Groves RH, Kruger FJ, Rejmanek M, Williamson M (eds) (1989) Biological invasions: a global perspective. Wiley, Chichester
Dunn PH, DeBano LF (1977) Fire's effect on biological and chemical properties of chaparral soils. In: Mooney HA, Conrad CE (eds) Proc Symp on Environmental consequences of fire and fuel management in mediterranean ecosystems, Palo Alto, California, pp 75–84
Dunn PH, DeBano LF, Eberlein GE (1979) Effects of burning on chaparral soils II. Soil microbes and nitrogen mineralization. Soil Sci Soc Am J 43: 509–514
Dunn PH, Wells WG, Dickey J, Wohlgemuth PM (1982) Role of fungi in postfire stabilization of chaparral ash beds. Proc Symp on Dynamics and management of mediterranean-type ecosystems, USDA For Serv Gen Tech Rep PSW-58, pp 378–381
Dunn PH, Barro SC, Poth M (1985) Soil moisture affects survival of microorganisms in heated chaparral soil. Soil Biol Biochem 17: 143–148
Dunne T, Leopold LB (1978) Water in environmental planning. WH Freeman, San Francisco
Dyer RA (1976) The genera of southern African flowering plants, vol 2. Gymnosperms and monocotyledons. Department of Agricultural Services, Pretoria
Dyrness CT (1976) Effect of wildfire on soil wettability in the High Cascades of Oregon. USDA For Serv Res Pap PNW-202
Eaton JS Likens GE, Bormann FH (1973) Throughfall and stemflow chemistry in a northern hardwood forest. J Ecol 61: 495–508
Edwards D (1984) Fire regimes in the biomes of South Africa. In: Booysen PdeV, Tainton NM (eds) Ecological effects of fire in South African ecosystems. Ecological Studies 48. Springer, Berlin Heidelberg New York, pp 19–38
Egler FE (1954) Vegetation science concepts 1. Initial floristic composition – a factor in old-field vegetation development. Vegetatio 4: 412–417
Ehleringer J, Mooney HA (1983) Productivity of desert and mediterranean-climate plants. In: Lange OL, Nobel PS, Osmond CB, Ziegler H (eds) Encyclopedia of plant physiology, vol 12D. Springer, Berlin Heidelberg New York, pp 205–231
Ehrlich PR (1986) Which animal will invade? In: Mooney HA, Drake JA (eds) Ecology of biological invasions of North America and Hawaii. Ecological Studies 58. Springer, Berlin Heidelberg New York, pp 79–95
Ellis RC, Graley AM (1987) Soil chemical properties as related to forest succession in a highland area in north-east Tasmania. Aust J Ecol 12: 307–317
Elton CS (1958) The ecology of invasions by animals and plants. Methuen, London
Emberger L (1955) Project d'une classification biogeographique des climats. In: Les divisions ecologiques du monde, Centre Nationale de la Recherche Scientifique (CNRS), Paris, pp 5–11
Erickson JM (1971) The displacement of native ant species by the introduced Argentine ant *Iridomyrmex humilis* Mayr. Psyche 78: 257–266
Esler KJ, Cowling RM (1990) Effects of density on the reproductive output of *Protea lepidocarpodendron*. S Afr J Bot 56: 29–33
Esterhuizen AD, van de Venter HA, Robbertse PJ (1986) A preliminary study of seed germination of *Watsonia fourcadei*. S Afr J Bot 52: 221–225
Everson TM, van Wilgen BW, Everson CS (1988) Adaptation of a model for rating fire danger in the Natal Drakensberg. S Afr J Sci 84: 44–49

Fagerstrom T (1988) Lotteries in communities of sessile organisms. Trends Ecol Evol 3: 303–6

Fairbanks BC, Woods LE, Bryant RJ, Elliott ET, Cole CV, Coleman DC (1984) Limitations of ATP estimates of microbial biomass. Soil Biol Biochem 16: 549–558

Felsenstein J (1976) The theoretical population genetics of variable selection and migration. Annu Rev Genet 10: 253–280

Finegan B (1984) Forest succession. Nature 312: 109–114

Fluker SS, Beardsley JW (1970) Sympatric associations of three ants: *Iridomyrmex humilis*, *Pheidole megacephala* and *Anoplolepis longipes* in Hawaii. Ann Entomol Soc Am 63: 1290–1296

Fox BJ (1981) The influence of disturbance (fire, mining) on ant and small mammal species diversity in Australian heathland. In: Conrad CE, Oechel WC (eds) Proc Symp on Dynamics and management of mediterranean-type ecosystems, San Diego State University, San Diego, California, pp 213–219

Fox BJ, Quinn RD, Breytenbach GJ (1985) A comparison of small mammal succession following fire in shrublands of Australia, California, and South Africa. Proc Ecol Soc Aust 14: 179–198

Fox MD, Fox BJ (1986) The susceptibility of natural communities to invasion. In: Groves RH, Burdon JJ (eds) Ecology of biological invasions. Cambridge University Press, Cambridge, pp 57–66

Fraser MW (1990) Small mammals, birds and ants as seed predators in post-fire mountain fynbos. S Afr J Wildl Res 20: 52–56

Frost PGH (1984) The responses and survival of organisms in fire-prone environments. In: Booysen PdeV, Tainton NM (eds) Ecological effects of fire in South African ecosystems. Ecological Studies 48. Springer, Berlin Heidelberg New York, pp 273–309

Fry M St L (1987) A detailed characterisation of soils under different fynbos-climate-geology combinations in the southwestern Cape. MSc Thesis, University of Stellenbosch

Fuentes ER, Etchegaray J (1983) Defoliation patterns in matorral ecosystems. In: Kruger FJ, Mitchell DT, Jarvis JUM (eds) Mediterranean-type ecosystems: the role of nutrients. Ecological Studies 43, Springer, Berlin Heidelberg New York, pp 525–542

Fugler SR (1983) The control of silky hakea in South Africa. Bothalia 14: 977–980

Gambell AW, Fisher DW (1966) Chemical composition of rainfall in eastern North Carolina and southeastern Virginia. Geological Survey Water-Supply Pap 1535-K, USA

GENSTAT 5 Committee of the Statistics Department, Rothamsted Experiment Station (1987) Genstat 5 Reference manual. Oxford University Press, Oxford

George AS (1981) The genus *Banksia*. Nuytsia 3: 239–473

George AS, Hopkins AJM, Marchant NG (1979) The heathlands of Western Australia. In: Specht RL (ed) Heathlands and related shrublands of the world: descriptive studies. Elsevier, Amsterdam, pp 211–230

Gill AM (1975) Fire and the Australia flora: a review. Aust For 38: 4–25

Gill AM (1981a) Adaptive responses of Australian vascular plants to fires. In: Gill AM, Groves RH, Noble IR (eds) Fire and the Australian biota. Australian Academy of Science, Canberra, pp 243–271

Gill AM (1981b) Fire: hazard and tool. In: Haigh C (ed) Heaths of New South Wales. National Parks and Wildlife Service, Sydney, pp 50–56

Gill AM (1981c) Fire adaptive traits of vascular plants. In: Mooney HA, Bonnicksen TM, Christensen NL, Lotan JE, Reiners WA (eds) Proc Conf on Fire regimes and ecosystem properties. USDA For Serv Gen Tech Rep WO-86, pp 208–230

Gill AM, Groves RH (1981) Fire regimes in heathlands and their plant ecological effects. In: Specht RL (ed) Heathlands and related shrublands of the world: analytical studies. Elsevier, Amsterdam, pp 61–84

Gill AM, Ingwersen F (1976) Growth of *Xanthorrhoea australis* R.Br. in relation to fire. J Appl Ecol 13: 195–203

Gill AM, Neser S, (1984) *Acacia cyclops* and *Hakea sericea* at home and abroad. In: Dell B (ed) Proc 4th Int Conf on Mediterranean ecosystems. Botany Department, University of Western Australia, Nedlands, pp 57–58

Gill AM, Groves RH, Noble IR (eds) (1981) Fire and the Australian biota. Australian Academy of Science, Canberra
Gilmour DA (1968) Water repellence of soils related to surface dryness. Aust For 32: 143–148
Gilmour DA, Cheney NP (1968) Experimental prescribed burn in Radiata pine. Aust For 32: 171–178
Giovannini G, Lucchesi S (1983) Effect of fire on hydrophobic and cementing substances of soil aggregates. Soil Sci 136: 231–236
Goldblatt P (1978) An analysis of the flora of southern Africa: its characteristics, relationships, and origins. Ann Mo Bot Gard 65: 369–436
Goldblatt P (1983) Geography of Iridaceae in Africa. Bothalia 14: 559–64
Goldblatt P (1989) The genus *Watsonia*. Ann Kirstenbosch Bot Gard 19: 1–148
Gómez-Pompa A, Vázquez-Yanes C (1981) Successional studies of a rain forest in Mexico. In: West DC, Shugart HH, Botkin DB (eds) Forest succession, concepts and application. Springer, Berlin Heidelberg New York, pp 246–266
Good RB (1973) A preliminary assessment of erosion following wildfires in Kosciusko National Park, NSW in 1973. J Soil Conserv Serv NSW 29: 191–199
Gordon-Gray KD, Wright FB (1969) *Cyrtanthus breviflorus* and *Cyrtanthus luteus* (Amaryllidaceae): observations with particular reference to Natal populations. J S Afr Bot 35: 35–62
Gray JT (1982) Community structure and productivity in *Ceanothus* chaparral and coastal sage scrub of southern California. Ecol Monogr 52: 415–435
Gray JT, Schlesinger WH (1981) Nutrient cycling in mediterranean type ecosystems. In: Miller PC (ed) Resource use by chaparral and matorral: a comparison of vegetation function in two mediterranean type ecosystems. Springer, Berlin Heidelberg New York, pp 259–285
Green LR (1981) Burning by prescription in chaparral. USDA For Serv Gen Tech Rep PSW-51
Greenacre MJ (1984) Theory and applications of correspondence analysis. Academic Press, London
Greenslade PJM, Thompson CH (1981) Ant distribution, vegetation and soil relationships in the Cooloola-Noosa River area, Queensland. In: Gillison AN, Anderson DJ (eds) Vegetation classification in Australia. CSIRO and Australian National University Press, Canberra, pp 192–207
Grime JP (1973) Competitive exclusion in herbaceous vegetation. Nature 242: 344–347
Groves RH (1983) Nutrient cycling in Australian heath and South African Fynbos. In: Kruger FJ, Mitchell DT, Jarvis JUM (eds) Mediterranean-type ecosystems. The role of nutrients. Ecological Studies 43. Springer, Berlin Heidelberg New York, pp 179–191
Groves RH (1986) Invasion of mediterranean ecosystems by weeds. In: Dell B, Hopkins AJM, Lamont BB (eds) Resilience in mediterranean-type ecosystems. Junk, Dordrecht, pp 129–145
Groves RH, Burdon JJ (eds) (1986) Ecology of biological invasions. Cambridge University Press, Cambridge
Groves RH, Di Castri F (eds) (1991) Biogeography of mediterranean invasions. Cambridge University Press, Cambridge
Grubb PJ (1977) The maintenance of species-richness in plant communities: the importance of the regeneration niche. Biol Rev 52: 107–145
Grubb PJ (1986) Problems posed by sparse and patchily distributed species in species-rich plant communities. In: Diamond J, Case TJ (eds) Community ecology. Harper & Row, New York, pp 207–225
Grubb PJ (1988) The uncoupling of disturbance and recruitment, two kinds of seed bank, and persistence of plant populations at the regional and local scales. Ann Zool Fenn 25: 23–36
Grubb PJ, Hopkins AJM (1986) Resilience at the level of the plant community. In: Dell B, Hopkins AJM, Lamont BB (eds) Resilience in mediterranean-type ecosystems. Junk, Dordrecht, pp 21–38

Grut M (1977) Notes on the history of forestry in the western Cape 1652–1872. S Afr For J 100: 32–37
Hall AV (1959) Observations on the distribution and ecology of Orchidaceae in the Muizenberg mountains, Cape Peninsula. J S Afr Bot 25: 265–278
Hall AV (1979) Invasive weeds. In: Day J, Siegfried WR, Louw GN, Jarman ML (eds) Fynbos ecology: a preliminary systhesis, S Afr Nat Sci Prog Rep 40, CSIR, Pretoria, pp 133–147
Hanes TL (1971) Succession after fire in the chaparral of southern California. Ecol Monogr 41: 27–52
Harper JL (1965) Establishment, aggression, and cohabitation in weedy species. In: Baker HG, Stebbins GL (eds) The genetics of colonizing species. Academic Press, New York, pp 243–265
Harper JL (1977) Population biology of plants. Academic Press, London
Hart JJ, Radosevich SR (1987) Water relations of two California chaparral shrubs. Am J Bot 74: 371–384
Haskins, CP, Haskins EF (1965) *Pheidole megacephala* and *Iridomyrmex humilis* in Bermuda – equilibrium or slow replacement? Ecology 46: 736–740
Hastings SJ, Oechel WC, Sionit N (1989) Water relations and photosynthesis of chaparral resprouts and seedlings following fire and hand clearing. In: Keeley SC (ed) The Californian chaparral: paradigms re-examined. Science Ser No 34. Natural History Museum of Los Angeles County, Los Angeles, California, pp 107–113
Hattingh CC (1945) Argentine ant versus indigenous ants. J Entomol Soc South Afr 8: 164
Haynes RA (1976) Aspects of the ecology and life-history of *Protea arborea* Houtt. Project Report, Field Ecology Course, University of Rhodesia, Salisbury
Heede BH, Harvey MD, Laird JR (1988) Sediment delivery linkages in a chaparral watershed following a wildfire. Environ Manage 12: 349–358
Hellmers H, Horton JS, Juhren G, O'Keefe J (1955) Root systems of some chaparral plants in southern California. Ecology 36: 667–678
Helvey JD, Tiedemann AR, Fowler WB (1976) Some climatic and hydrological effects of wildfire in Washington State. Proc Annu Tall Timbers Fire Ecol Conf 15: 201–222
Hewlett JD, Hibbert AR (1967) Factors affecting the response of small watersheds to precipitation in humid areas. In: Sopper WL, Lull HW (eds) Int Symp on Forest Hydrology. Pergamon Press, Oxford, pp 275–290
Hewlett JD, Nutter WL (1969) An outline of forest hydrology. University of Georgia Press, Athens
Higgins KB, Lamb AJ, van Wilgen BW (1987) Root systems of selected plant species in mesic fynbos in the Jonkershoek Valley, South Western Cape Province. S Afr J Bot 53: 249–257
Hill AR (1986) Stream nitrate-N in relation to variations in annual and seasonal runoff regimes. Water Resour Bull 22: 829–839
Hillel D (1980) Fundamentals of soil physics. Academic Press, New York
Hobbs RJ (1989) The nature and effects of disturbance relative to invasions. In: Drake JA, Mooney HA, Di Castri F, Groves RH, Kruger FJ, Rejmanek M, Williamson MH (eds) Ecology of biological invasions: a global perspective. Wiley, Chichester, pp 389–405
Hobbs RJ, Atkins L (1988) Spatial variability of experimental fires in south-west Western Australia. Aust J Ecol 13: 295–299
Hobbs RJ, Mallik AU, Gimingham CH (1984) Studies on fire in Scottish heathland communities. III. Vital attributes of the species. J Ecol 72: 963–976
Hoffman MT, Moll EJ, Boucher C (1987) Post-fire succession at Pella, a South African lowland fynbos site. S Afr J Bot 53: 370–374
Hölldobler B, Wilson EO (1990) The ants. Springer, Berlin Heidelberg New York
Holmes PM (1989) A comparative study of the seed bank dynamics of two congeneric alien invasive species. PhD Thesis, University of Cape Town
Holmes PM (1990) Dispersal and predation in alien *Acacia*. Oecologia 83: 288–290

Horne IP (1981) The frequency of veld fires in the Groot Swartberg mountain catchment area, Cape Province. S Afr For J 118: 56–60
Hoyt WG, Troxell HC (1934) Forests and stream flow. Tran Am Soc Civ Eng 99: 1–30
Hubbell SP (1979) Tree dispersion, abundance, and diversity in a tropical dry forest. Science 203: 1299–1309
Hughes L, Westoby M (1990) Removal rates of seeds adapted for dispersal by ants. Ecology 71: 138–148
Hunt GA, Fogel R (1983) Fungal hyphal dynamics in a western Oregon Douglas-fir stand. Soil Biol Biochem 15: 641–649
Huntley BJ (1987) Ten years of cooperative ecological research in South Africa. S Afr J Sci 83: 72–79
Ims RA (1990) The ecology and evolution of reproductive synchrony. Trends Ecol Evol 5: 135–40
Jackson ME (1980) Occurrence of the Argentine ant on the Witwatersrand. S Afr J Sci 76: 31–32
Jackson ML (1958) Soil chemical analysis. Prentice-Hall, Englewood Cliffs
Jacobs TC, Gilliam JW (1985) Riparian losses of nitrate from agricultural drainage waters. J Environ Qual 14: 472–478
James S (1984) Lignotubers and burls – their structure, function and ecological significance in mediterranean ecosystems. Bot Rev 50: 225–266
Janzen DH (1971) Seed predation by animals. Annu Rev Ecol Syst 2: 465–492
Janzen DH (1988) Management of habitat fragments in a tropical dry forest: growth. Ann Mo Bot Gard 75: 105–116
Jarman ML (1982) A look at the littlest floral kingdom. Scientiae 23: 9–19
Jarman ML (1988) A description of the fynbos biome project intensive study site at Pella. Ecosystems Programmes, Occas Rep 33, CSIR, Pretoria
Jefferey DJ, Holmes PM, Rebelo AG (1988) Effects of dry heat on seed germination in selected indigenous and alien legume species in South Africa. S Afr J Bot 54: 28–34
Jenkinson DS, Ladd JN (1981) Microbial biomass in soil: measurement and turnover. In: Paul EA, Ladd JN (eds) Soil biochemistry, vol 5. Marcel Dekker, New York, pp 415–471
Jenkinson DS, Powlson DS (1976) The effects of biocidal treatments on metabolism in soil-V. A method for measuring soil biomass. Soil Biol Biochem 8: 209–213
Jenkinson DS, Davidson SA, Powlson DS (1979) Adenosine triphosphate and microbial biomass in soil. Soil Biol Biochem 11: 521–527
Johnson C (1978) Australian myrtle. In: Stirton CH (ed) Plant invaders: beautiful but dangerous. Dept of Nature and Environmental Conservation of the Cape Prov Admin, Cape Town, pp 92–95
Johnstone IM (1986) Plant invasion windows: a time-based classification of invasion potential. Biol Rev 61: 369–394
Jones PCT, Mollison JE (1948) A technique for the quantitative estimation of soil microorganisms. J Gen Microbiol 2: 54–69
Jordaan PG (1965) Die invloed van 'n winterbrand op die voortplanting van vier soorte van die Proteaceae. Tydskr Natuurwet 5: 27–31
Jorgenson JR, Wells CG (1971) Apparent nitrogen fixation in soil influenced by prescribed burning. Proc Soil Sci Soc Am 35: 806–810
Karr JR, Schlosser IJ (1978) Water resources and the land-water interface. Science 201: 229–234
Kauffman JB, Uhl C, Cummings DL (1988) Fire in the Venezuelan Amazon 1: fuel biomass and fire chemistry in the evergreen rainforest of Venezuela. Oikos 53: 167–175
Keeley JE (1977) Seed production, seed populations in soil, and seedling production after fire for two congeneric pairs of sprouting and non-sprouting chaparral shrubs. Ecology 58: 820–829
Keeley JE (1981) Reproductive cycles and fire regimes. In: Mooney HA, Bonnicksen TM, Christensen NL, Lotan JE, Reiners WA (eds) Proc Conf on Fire regimes and ecosystem properties. USDA For Serv Gen Tech Report WO-86, pp 231–277

Keeley JE (1986) Resilience of mediterranean shrub communities to fire. In Dell B, Hopkins AJM, Lamont BB (eds) Resilience in mediterranean-type ecosystems. Junk, Doordrecht, pp 95–112
Keeley JE (1987) Role of fire in seed germination of woody taxa in California chaparral. Ecology 68: 434–443
Keeley JE, Keeley SC (1988) Chaparral. In: Barbour MG, Billings WD (eds) North American terrestrial vegetation. Cambridge University Press, Cambridge, pp 165–207
Keeley JE, Zedler PH (1978) Reproduction of chaparral shrubs after fire: a comparison of sprouting and seeding strategies. Am Midl Nat 97: 120–132
Keeley S, Johnson AW (1976) A comparison of the pattern of herb and shrub growth in comparable sites in Chile and California. Am Midl Nat 97: 120–132
Keller HM, Foster F, Weibel P (1986) Factors affecting stream water quality: results of a 15 year monitoring study in the Swiss prealps. In: Proc Symp on Monitoring to detect changes in water quality series, Budapest, pp 215–225
Kessell SR, Good RB, Potter MW (1982) Computer modelling in natural area management. Spec Publ 9, Australian National Parks and Wildlife Service
Kessell SR, Good RB, Hopkins AJM (1984) Implementation of two new resource management information systems in Australia. Environ Manage 8: 251–270
Kessell SR, Potter MW, Bevins CD, Bradshaw L, Jeske BW (1978) Analysis and application of forest fuels data. Environ Manage 2: 347–363
King JM, Day JA, Davies BR, Henshall-Howard M-P (1987) Particulate organic matter in a mountain stream in the south-western Cape, South Africa. Hydrobiologia 154: 165–187
Klein RG (1974) Environment and subsistence of prehistoric Man in the Southern Cape Province, South Africa. World Archaeol 5: 249–284
Kleinbaum DG, Kupper LL (1978) Applied regression analysis and other multivariable methods. Duxbury, North Scituate, Mass
Kluge RL (1983) The hakea fruit weevil, *Erytenna consputa* (Coleoptera: Curculionidae), and the biological control of *Hakea sericea* Schrader in South Africa. PhD Thesis, Rhodes University, Grahamstown
Knight RS (1988) Aspects of plant dispersal in the southwestern Cape with particular reference to the roles of birds as dispersal agents. PhD Thesis, University of Cape Town
Koen JH, Breytenbach W (1988) Ant species richness of fynbos and forest ecosystems in the southern Cape. S Afr J Zool 23: 184–188
Köppen W (1931) Grundriss der Klimakunde. De Gruyter, Berlin
Krammes JS, DeBano LF (1965) Soil wettability: a neglected factor in watershed management. Water Resour Res 1: 283–286
Krammes JS, Osborn J (1969) Water-repellent soils and wetting agents as factors influencing erosion. In: DeBano LF, Letey J (eds) Proc Symp on Water repellent soils. University of California, Riverside, pp 177–186
Kruger FJ (1977a) A preliminary account of aerial plant biomass in fynbos communities of the mediterranean-type climate zone of the Cape Province. Bothalia 12: 301–307
Kruger FJ (1977b) Ecology of Cape fynbos in relation to fire. In: Mooney HA, Conrad CE (eds) Proc Symp on Environmental consequences of fire and fuel management in mediterranean ecosystems, Palo Alto, California, pp 230–244
Kruger FJ (1977c) Invasive woody plants in the Cape fynbos with special reference to the biology and control of *Pinus pinaster*. In: Proc 2nd Natl Weeds Conf S Afr, Balkema, Cape Town, pp 57–74
Kruger FJ (1978) Some aspects of the demography of *Watsonia pyramidata* (Andr.) in relation to fire. Paper presented at the Annu Congr S Afr Asso Botany, January 1978
Kruger FJ (1979) South African heathlands. In: Specht RL (ed) Heathlands of the world. 9A. Heathlands and related shrublands: descriptive studies. Elsevier, Amsterdam, pp 19–80
Kruger FJ (1981) Seasonal growth and flowering rhythms: South African heathlands. In: Specht RL (ed) Ecosystems of the world, vol 9B. Heathlands and related shrublands. Analytical studies. Elsevier, Amsterdam, pp 1–4

Kruger FJ (1982) Prescribing fire frequencies in Cape fynbos in relation to plant demography. In: Conrad CE, Oechel WE (eds) Dynamics and management of mediterranean-type ecosystems. USDA For Serv Gen Tech Rep PSW-58, pp 118–122
Kruger FJ (1983) Plant community diversity and dynamics in relation to fire. In: Kruger FJ, Mitchell DT, Jarvis JUM (eds) Mediterranean-type ecosystems: the role of nutrients. Ecological Studies 43. Springer, Berlin Heidelberg New York, pp 466–472
Kruger FJ (1984) Effects of fire on vegetation structure and dynamics. In: Booysen PdeV, Tainton NM (eds) Ecological effects of fire in South African ecosystems. Ecological Studies 48. Springer, Berlin Heidelberg New York, pp 219–244
Kruger FJ (1987) Succession after fire in selected fynbos communities of the South-Western Cape. PhD Thesis, University of the Witwatersrand, Johannesburg
Kruger FJ, Bigalke RC (1984) Fire in fynbos. In: Booysen P deV, Tainton NM (eds) Ecological effects of fire in South African ecosystems. Ecological Studies 48. Springer, Berlin Heidelberg New York, pp 67–114
Kruger FJ, Lamb AJ (1979) Conservation of the Kogelberg State Forest. Preliminary assessment of the effects of management from 1967 to 1978. Rep 79-02, Jonkershoek Forestry Research Centre, Stellenbosch
Kruger FJ, Mitchell DT, Jarvis JUM (eds) (1983) Mediterranean-type ecosystems: the role of nutrients. Ecological Studies 43. Springer, Berlin Heidelberg New York
Kruger FJ, Richardson DM, van Wilgen BW (1986) Processes of invasion by alien plants. In: Macdonald IAW, Kruger FJ, Ferrar AA (eds) The ecology and management of biological invasions in southern Africa. Oxford University Press, Cape Town, pp 145–155
Kruger FJ, Richardson DM, Smith RE (1988) Plant water relations in sclerophyllous trees and shrubs of riparian and hillslope habitats at Swartboskloof, southwestern Cape Province, South Africa. In: Di Castri F, Floret Ch, Rambal S, Roy J (eds) Time scales and water stress. Proc 5th Int Conf on Mediterranean ecosystems, International Union of Biological Sciences, Paris, pp 575–581
Kruger FJ, Breytenbach GJ, Macdonald IAW, Richardson DM (1989) Characteristics of invaded mediterranean-climate regions. In: Drake JA, Mooney HA, Di Castri F, Groves RH, Kruger FJ, Rejmanek M, Williamson MH (eds) Ecology of biological invasions: a global perspective. Wiley, Chichester, pp 181–213
Kummerow J (1981) Structure of roots and root systems. In: Di Castri F, Goodall DW, Specht RL (eds) Ecosystems of the world 11: mediterranean-type shrublands. Elsevier, Amsterdam, pp 269–288
Kummerow J, Krause J, Jow W (1977) Root systems of chaparral shrubs. Oecologia (Berlin) 29: 163–177
Kummerow J, Montenegro G, Krause D (1981) Biomass, phenology, and growth. In: Miller PC (ed) Resource use by chaparral and matorral, Ecological Studies 39. Springer, Berlin Heidelberg New York, pp 69–96
Lambrechts JJN (1979) Geology, geomorphology and soils. In: Day J, Siegfried WR, Louw GN, Jarman ML (eds) Fynbos ecology: a preliminary synthesis. S Afr Nat Sci Prog Rep 40, CSIR, Pretoria, pp 16–26
Lambrechts JJN (1986) Phosphorus levels and adsorption parameters of podzolized fynbos and indigenous forest soils. Paper read at the combined congress of the S Afr Soc Crop Production, S Afr Weed Science Society and the Soil Science Society of S Afr, Pretoria, January 1986
Lambrechts JJN, Theron AA, Fry M (1986) Detail characterization of soils under different fynbos-climate-geology combinations in South and Southwestern Cape. Proj Rep Ser, Ecosystem Programmes, CSIR, Pretoria
Lamont B (1982) Mechanisms for enhancing nutrient uptake in plants, with particular reference to mediterranean South Africa and Western Australia. Bot Rev 48: 597–689
Lamont BB (1983) Strategies for maximizing nutrient uptake in two mediterranean ecosystems of low nutrient status. In: Kruger FJ, Mitchell DT, Jarvis JUM (eds) Mediterranean-type ecosystems. The role of nutrients. Springer, Berlin Heidelberg New York, pp 246–273

Lamont BB (1985) The comparative reproductive biology of three *Leucospermum* species (Proteaceae) in relation to fire responses and breeding system. Aust J Bot 33: 139–45
Lamont BB, Collins BG, Cowling RM (1985) Reproductive biology of the Proteaceae in Australia and South Africa. Proc Ecol Soc Aust 14: 213–224
Lamont BB, Le Maitre DC, Cowling RM, Enright NJ (in press) Canopy seed storage in woody plants. Bot Rev
Langford KJ (1976) Change in yield of water following a bushfire in a forest of *Eucalyptus regnans*. J Hydrol 29: 87–114
Lawton JH, Brown KC (1986) The population and community ecology of invading insects. Philos Trans R Soc Lond B 314: 607–617
Le Houérou HN (1981) Impact of man and his animals on mediterranean vegetation. In: Di Castri F, Goodall DW, Specht RL (eds) Ecosystems of the world 11: mediterranean-type shrublands. Elsevier, Amsterdam, pp 63–86
Leitch CJ, Flinn DW, van De Graaff RHM (1983) Erosion and nutrient loss resulting from Ash Wednesday (February 1983) wildfires: a case study. Aust For 46: 173–180
Le Maitre DC (1984) A short note on seed predation in *Watsonia pyramidata* (Andr.) Stapf in relation to season of burn. J S Afr Bot 50: 407–15
Le Maitre DC (1987a) Effects of season of burn on species populations and composition of fynbos in the Jonkershoek Valley. S Afr J Bot 53: 284–292
Le Maitre DC (1987b) Dynamics of canopy-stored seed in relation to fire. In: Cowling RM, Le Maitre DC, McKenzie B, Prŷs-Jones RP, van Wilgen BW (eds) Disturbance and the dynamics of fynbos biome communities. S Afr Nat Sci Prog Rep 135, CSIR, Pretoria, pp 24–45
Le Maitre DC (1988a) Effects of season of burn on the regeneration of two Proteaceae with soil-stored seed. S Afr J Bot 54: 575–580
Le Maitre DC (1988b) The effects of parent density and season of burn on the regeneration of *Leucadendron laureolum* (Proteaceae) in the Kogelberg. S Afr J Bot 54: 581–584
Le Maitre DC (1989) Seed regeneration of the Proteaceae in Langrivier. In: Richardson DM (ed) The short-term effects of fire in Langrivier, Jonkershoek State Forest, Stellenbosch. Rep J2/89, S Afr For Res Inst, Stellenbosch, pp 39–50
Le Maitre DC (1990) The influence of seed ageing on the plant on seed germination *Protea neriifolia* R.Br. (Proteaceae). S Afr J Bot 56: 49–53
Le Maitre DC, Midgely JJ (1991) Plant reproductive ecology. In: Cowling RM (ed) The ecology of fynbos: Nutrients, Fire and Diversity. Oxford University Press, Cape Town (in press)
Leopold AC (1961) Senescence in plant development. Science 134: 1727–32
Lepart J, Debussche M (1991) Invasion processes as related to succession and disturbance. In: Groves RH, Di Castri F (eds) Biogeography of mediterranean invasions. Cambridge University Press, Cambridge, pp 159–177
Leverich WJ, Levin DA (1979) Age-specific survivorship and reproduction in *Phlox drummondii*. Am Nat 113: 881–903
Levyns MR (1929) Veld burning experiments at Ida's valley, Stellenbosch. Trans R Soc S Afr 17: 61–92
Levyns MR (1962) Past plant migrations in South Africa. Ann Cape Prov Mus 2: 7–10
Levyns MR (1966) *Haemanthus canaliculatus*, a new fire-lily from the Western Cape Province. J S Afr Bot 32: 73–5
Lieberburg I, Kranz PM, Siep A (1975) Bermudan ants revisited: the status and interaction of *Pheidole megacephala* and *Iridomyrmex humilis*. Ecology 56: 473–478
Likens GE, Borman FH, Pierce RS, Eaton JS, Johnson NM (1977) Biogeochemistry of a forested ecosystem. Springer, Berlin Heidelberg New York
Lindley AJ, Bosch JM, van Wyk DB (1988) Changes in water yield after fire in fynbos catchments. Water S A 14: 7–12
Linhart H, Zucchini W (1986) Model selection. Wiley, New York
Lombaard, HB (1971) 'n Ekologiese studie van aspekte van die voortplanting van Protea mellifera en Protea pulchella. PhD Thesis, University of Stellenbosch

Long G, Fay F, Thiault M, Trabaud L (1967) Essais de determination experimentale de la productivité d'une garrigue à *Quercus coccifera*. Document 39, CNRS (CEPE), Montpellier

Loos MA, Schlosser IF, Mapham WR (1979) Phenoxy herbicide degradation in soils: quantitative studies of 2,4-D- and MCPA-degrading microbial populations. Soil Biol Biochem 11: 377–385

Lossaint P (1973) Soil-vegetation relationships in mediterranean ecosystems of southern France. In: Di Castri F, Mooney HA (eds) Mediterranean type ecosystems. Origin and structure. Springer, Berlin Heidelberg New York, pp 199–210

Louw GN (1969) The nutritive value of natural grazings in South Africa. Proc S Afr Soc Anim Prod 8: 57–61

Louw HA, Webley DM (1959) The bacteriology of the root region of the oat plant grown under controlled pot culture conditions. J Appl Bacteriol 22: 216–226

Low AB (1983) Phytomass and major nutrient pools in an 11-year post-fire coastal fynbos community. S Afr J Bot 2: 98–104

Lowrance R, Todd R, Fail J, Hendrickson O, Leonard R, Asmussen L (1984) Riparian forests as nutrient filters in agricultural watersheds. BioScience 34: 374–377

Lowrance R, Leonard R, Scheridan J (1985) Managing riparian ecosystems to control nonpoint pollution. J Soil Water Conserv 40: 87–91

Lumac Systems AG (1979) Environmental applications. ATP method for microbial biomass in soil (Lumit application 508). Lumac Systems AG, Basel

Lundgren B, Söderström B (1983) Bacterial numbers in a pine forest soil in relation to environmental factors. Soil Biol Biochem 15: 625–630

Mac Arthur RH, Wilson EO (1967) The theory of island biogeography. Princeton University Press, Princeton, New Jersey

Macdonald IAW (1984) Is the fynbos biome especially susceptible to invasion by alien plants? A re-analysis of available data. S Afr J Sci 80: 369–377

Macdonald IAW, Jarman ML (1984) Invasive alien organisms in the terrestrial ecosystems of the fynbos biome, South Africa. S Afr Nat Sci Prog Rep 85, CSIR, Pretoria

Macdonald IAW, Richardson DM (1986) Alien species in terrestrial ecosystems of the fynbos biome. In: Macdonald IAW, Kruger FJ, Ferrar AA (eds) The ecology and management of biological invasions in southern Africa. Oxford University Press, Cape Town, pp 77–91

Macdonald IAW, Kruger FJ, Ferrar AA (eds) (1986) The ecology and control of biological invasions in southern Africa. Oxford University Press, Cape Town

MacGillivray DB (1990) The insect pollination of selected *Erica* species. MSc Thesis, University of Stellenbosch

MacKay SM, Cornish PM (1982) Effects of wildfire and logging on the hydrology of small catchments near Eden, NSW. Proc 1st Natl Symp on Forest hydrology, Melbourne, pp 112–117

Maclean GL (1985) Roberts' birds of southern Africa, 5th edn. John Voelcker Bird Book Fund, Cape Town

MacVicar CN, de Villiers JM, Loxton RF, Verster E, Lambrechts JJN, Merryweather FR, le Roux J, van Rooyen TH, von Harmse HJ M (1977) Soil classification. A binomial system for South Africa. Dept Agric Tech Serv, Pretoria

Magurran AE (1988) Ecological diversity and its measurement. Croom Helm, London

Majer JD (1978) An improved pitfall trap for sampling ants and other epigaeic invertebrates. J Aust Entomol Soc 17: 261–262

Majer JD (1983) Ants: Bio-indicators of minesite rehabilitation, land-use and land conservation. Environ Manage 7: 375–383

Majer JD, Greenslade P (1988) Soil and litter invertebrates. In: Specht RL (ed) Mediterranean-type ecosystems: a data source book. Kluwer, Dordrecht, pp 197–226

Majer JD, Day JE, Kabay ED, Perriman WS (1984) Recolonization by ants in bauxite mines rehabilitated by a number of different methods. J Appl Ecol 21: 355–375

Manders PT (1987) A transition matrix model of the population dynamics of the Clanwilliam cedar (*Widdringtonia cedarbergensis*) in natural stands subject to fire. For Ecol Manage 20: 171–186
Manders PT (1990a) Fire and other variables as determinants of forest/fynbos boundaries in the Cape Province. J Veg Sci 1: 483–490
Manders PT (1990b) Soil seed banks and post-fire seed deposition across a forest-fynbos ecotone in the Cape Province. J Veg Sci 1: 491–498
Manders PT (1990c) Quantifying soil seed banks. A comparison of physical separation and seedling emergence techniques in Cape fynbos and forest vegetation. S Afr J Ecol 1: 27–30
Manders PT (1991) The relationships between forest and mountain fynbos in the south-western Cape Province of South Africa. PhD Thesis, University of Cape Town.
Manders PT, Richardson DM (1992) Colonization of Cape fynbos communities by forest species. For Ecol Manage (in press)
Margaris NS (1977) Decomposers and the fire cycle in mediterranean-type ecosystems. In: Mooney HA, Conrad CE (eds) Proc Symp on Environmental consequences of fire and fuel management in mediterranean ecosystems, USDA For Serv Gen Tech Rep WO-3, Washington, DC, pp 37–45
Mark AF (1965) Flowering, seeding, and seedling establishment of narrow-leaved snow tussock, *Chionochloa rigida*. N Z J Bot 3: 180–193
Mark AF (1969) Ecology of snow tussocks in the mountain grasslands of New Zealand. Vegetatio 18: 289–306
Markotter ED (1936) Die lewensgeskiedenis van sekere geslagte van die Amaryllidaceae. Ann Univ Stellenbosch 14, Ser A (2)
Martens R (1985) Estimation of the adenylate energy charge in unamended and amended agricultural soils. Soil Biol Biochem 17: 765–772
Martin ARH (1966) The plant ecology of the Grahamstown Nature Reserve: II. Some effects of burning. J S Afr Bot 32: 1–39
Martin CW, Harr RD (1988) Precipitation and streamwater chemistry from undisturbed watersheds in the Cascade mountains of Oregon. Water, Air Soil Poll 42: 203–219
Masson PH, Moll EJ (1987) The factors affecting forest colonisation of fynbos in the absence of fire at Orange kloof, Cape Province, South Africa. S Afr For J 143: 5–10
Mazurek H, Romane F (1986) Dynamics of young *Pinus pinaster* vegetation in a mediterranean area: diversity and niche strategy. Vegetatio 66: 27–40
McArthur AG (1966) Weather and grassland fire behaviour. Comm Aust For Timb Bur Leaflet 100, 23 pp
McArthur AG (1967) Fire behaviour in eucalypt forest. Comm Aust For Timb Bur Leaflet 107, 25 pp
McClanahan TR, Wolfe RW (1987) Dispersal of ornithochorous seeds from forest edges in central Florida. Vegetatio 107: 107–112
McDonald DJ (1985) The plant communities of Swartboschkloof, Jonkershoek. S Afr Nat Sci Prog Rep 104, CSIR, Pretoria
McDonald DJ (1987) Ordination by detrended correspondence analysis (DCA) of the vegetation of Swartboschkloof, Jonkershoek, Cape Province. Bothalia 17: 121–129
McDonald DJ (1988) A synopsis of the plant communities of Swartboschkloof, Jonkershoek, Cape Province. Bothalia 18: 233–260
McDonald DJ, Morley M (1988) A checklist of the flowering plants and ferns of Swartboschkloof, Jonkershoek, Cape Province. Bothalia 18: 261–270
McKenzie B (1978) A quantitative and qualitative study of the indigenous forests of the southwestern Cape. MSc Thesis, University of Cape Town
McKenzie B, Moll EJ, Campbell BM (1977) A phytosociological study of Orange Kloof, Table Mountain, South Africa. Vegetatio 34: 41–53
Merino O, Martin MP, Martin A, Merino J (1990) Successional and temporal changes in primary productivity in two mediterranean scrub ecosystems. Acta Oecol Plant 11: 103–112
Mersereau RC, Dyrness CT (1972) Accelerated mass wasting after logging and slash burning in western Oregon. J Soil Water Conserv 27: 112–114

Michell MR (1922) Some observations on the effects of a bush fire on the vegetation of Signal Hill. Trans R Soc S Afr 10: 213–232
Middlemiss E (1963) The distribution of *Acacia cyclops* in the Cape Peninsula by birds and other animals. S Afr J Sci 59: 419–420
Midgley JJ (1987a) The derivation, utility and implications of a divergence index for the genus *Leucadendron* (Proteaceae). Bot J Linn Soc 95: 137–52
Midgley JJ (1987b) Aspects of the evolutionary biology of the Proteaceae, with emphasis on the genus *Leucadendron* and its phylogeny. PhD Thesis, University of Cape Town
Midgley JJ (1989) Season of burn of serotinous Proteaceae: a critical review and further data. S Afr J Bot 55: 165–70
Midgley JJ, Clayton P (1990) Short -term effects of an autumn fire on small mammal populations in Southern Cape coastal mountain fynbos. S Afr For J 153: 27–30
Midgley JJ, Hoekstra TT, Bartholemew R (1989) The effect of date of planting on field establishment of serotinous Cape Proteaceae. Vegetatio 79: 185–192
Milewski AV (1981) A comparison of vegetation height in relation to the effectiveness of rainfall in the mediterranean and adjacent arid parts of Australia and South Africa. J Biogeogr 8: 107–116
Milewski AV, Bond WJ (1982) Convergence of myrmecochory in mediterranean Australia and South Africa. In: Buckley RC (ed) Ant-plant interactions in Austriala. Junk, Den Haag, pp 89–96
Milewski AV, Cowling RM (1985) Anomalies in the plant and animal communities in similar environments at the Barrens, Western Australia and the Caledon coast, South Africa. Proc Aust Ecol Soc 14: 199–212
Miller J (1985) Plant water relations along a rainfall gradient between the succulent Karoo and mesic mountain fynbos in the Cedarberg mountains near Clanwilliam, South Africa. MSc Thesis, University of Cape Town
Miller JM, Miller PC, Miller PM (1984) Leaf conductance and xylem pressure potentials in fynbos plant species. S Afr J Sci 80: 381–385
Miller PC (ed) (1981a) Resource use by chaparral and matorral. A comparison of vegetation function in two mediterranean-type ecosystems. Ecological Studies 39. Springer, Berlin Heidelberg New York
Miller PC (1981b) Similarities and limitations of resource utilization in mediterranean type ecosystems. In: Miller PC (ed) Resource use by chaparral and matorral. Ecological Studies 39. Springer, Berlin Heidelberg New York, pp 369–407
Miller PC (1983) Canopy structure of mediterranean-type shrubs in relation to heat and moisture. In: Kruger FJ, Mitchell DT, Jarvis JUM (eds) Mediterranean-type ecosystems – the role of nutrients. Ecological Studies 43. Springer, Berlin Heidelberg New York, pp 133–166
Miller PC, Hajek E (1981) Resource availability and environmental characteristics of mediterranean type ecosystems. In: Miller PC (ed) Resource use by chaparral and matorral: a comparison of vegetation function in two mediterranean type ecosystems. Ecological Studies 39. Springer, Berlin Heidelberg New York, pp 17–41
Miller PC, Poole DK (1979) Patterns of water use by shrubs in southern California. For Sci 25: 84–98
Miller PC, Bradbury DE, Hajek E, LaMarche V, Thrower NJW (1977) Past and present environment. In: Mooney HA (ed) Convergent evolution in Chile and California. Dowden, Hutchinson & Ross, Stroudsburg, pp 27–72
Miller PC, Miller J M, Miller PC (1983) Seasonal progression of plant water relations in fynbos in the western Cape Province, South Africa. Oecologia (Berlin) 56: 392–396
Miller TB, Johnson FD (1986) Sampling and data analyses of narrow, variable width gallery forests over environmental gradients. Trop Ecol 27: 132–142
Mills JN (1986) Herbivores and early postfire succession in southern Californian chaparral. Ecology 67: 1637–1649
Mills JN, Kummerow J (1989) Herbivores, seed predators and chaparral succession. In: Keeley SC (ed) The Californian chaparral: paradigms re-examined. Sci Ser No 34. Natural History Museum, Los Angeles, pp 49–55

Milton SJ (1979) Australian acacias in the S.W. Cape: pre-adaptation, predation and success. In: Neser S, Cairns ALP (eds) Proc 3rd Natl Weeds Conf S Afr, Balkema, Cape Town, pp 69–78

Milton SJ, Siegfried WR (1981) Above-ground biomass of Australian Acacias in the southern Cape, South Africa. J S Afr Bot 47: 701–716

Mirov NT (1967) The genus *Pinus*. Ronald Press, New York

Mitchell DT (1987) Fire and fynbos ecosystem nutrient dynamics. In: Cowling RM, Le Maitre DC, McKenzie B, Prys-Jones RP, van Wilgen BW (eds) Disturbance and the dynamics of fynbos biome communities. S Afr Nat Sci Prog Rep 135, CSIR, Pretoria, pp 69–73

Mitchell DT, Allsopp N (1984) Changes in the phosphorus composition of seeds of *Hakea sericea* (Proteaceae) during germination under low phosphorus conditions. New Phytol 96: 239–247

Mitchell DT, Brown G (1986) Influence of fire on the soil phosphorus status in sand plain lowland fynbos, south-western Cape. S Afr J Bot 52: 67–72

Mitchell DT, Coley PGF (1987) Litter production and decomposition from shrubs of *Protea repens* growing in sand plain lowland and mountain fynbos, southwestern Cape. S Afr J Bot 53: 25–31

Mitchell DT, Brown G, Jongens-Roberts SM (1984) Variation of forms of phosphorus in the sandy soils of coastal fynbos, southwestern Cape. J Ecol 72: 575–584

Mitchell DT, Coley PGF Webb S, Allsopp N (1986) Litterfall and decomposition processes in the coastal fynbos vegetation, southwestern Cape, South Africa. J Ecol 74: 977–993

Mitchell DT, Stock WD, Jongens-Roberts SM (1987) Nitrogen and phosphorus cycling in the fynbos biome. Occas Rep 18, Terrestrial Ecosystem Section, CSIR, Pretoria

Moll EJ, Campbell BM (1976) The ecological status of Table Mountain. Unpublished report, Botany Department, University of Cape Town

Moll EJ, Sommerville JEM (1985) Seasonal xylem pressure potentials of two South African coastal fynbos species in three soil types. S Afr J Bot 51: 187–193

Moll EJ, McKenzie B, McLachlan D (1980) A possible explanation for the lack of trees in the fynbos, Cape Province, South Africa. Biol Conserv 17: 221–228

Moll EJ, Campbell BM, Cowling RM, Bossi L, Jarman ML, Boucher C (1984) A description of the major vegetation categories in and adjacent to the fynbos biome. S Afr Nat Sci Prog Rep 83, CSIR, Pretoria

Mooney HA (ed) (1977) Convergent evolution in Chile and California. Mediterranean climate ecosystems. Dowden, Hutchinson & Ross, Stroudsburg

Mooney HA, Conrad CE (eds) (1977) Symposium on the environmental consequences of fire and fuel management in mediterranean ecosystems. USDA For Serv Gen Tech Rep WO-3

Mooney HA, Drake JA (eds) (1986) Ecology of biological invasions of North America and Hawaii. Ecological Studies 58. Springer, Berlin Heidelberg New York

Mooney HA, Parsons DJ (1973) Structure and function of the California chaparral – an example from San Dimas. In: DiCastri F, Mooney HA (eds) Mediterranean-type ecosystems – origin and structure. Chapman & Hall, London, pp 83–112

Mooney HA, Field C, Gulmon SL, Rundel P, Kruger FJ (1983) Photosynthetic characteristics of South African sclerophylls. Oecologia (Berlin) 58: 398–401

Mooney HA, Hamburg SP, Drake JA (1986) The invasions of plants and animals into California. In: Mooney HA, Drake JA (eds) Ecology of biological invasions of North America and Hawaii. Ecological Studies 58. Springer, Berlin Heidelberg New York, pp 250–272

Morrow PA (1983) The role of sclerophyllous leaves in determining insect grazing damage. In: Kruger FJ, Mitchell DT, Jarvis JUM (eds) Mediterranean-type ecosystems: the role of nutrients. Springer, Berlin Heidelberg New York, pp 509–524

Moulton MP, Pimm SL (1986a) Species introductions to Hawaii. In: Mooney HA, Drake JA (eds) Ecology of biological invasions of North America and Hawaii. Ecological Studies 58. Springer, Berlin Heidelberg New York, pp 231–249

Moulton MP, Pimm SL (1986b) The extent of competition in shaping an introduced avifauna. In: Diamond J, Case TJ (eds) Community ecology. Harper & Row, New York, pp 80–97

Mueller-Dombois D (1964) Effect of depth to water table on height growth of tree seedlings in a greenhouse. For Sci 10: 306–316

Mutch RW (1970) Wildland fires and ecosystems – a hypothesis. Ecology 51: 1046–1051

Myers RL (1983) Site susceptibility to invasion by the exotic tree *Melaleuca guinguenervia* in southern Florida. J Appl Ecol 20: 645–658

Nahal I (1981) The mediterranean climate from a biological viewpoint. In: Di Castri F, Goodall DW, Specht RL (eds) Ecosystems of the world 11, mediterranean-type shrublands. Elsevier, Amsterdam, pp 63–86

Nannipieri P, Johnson RL, Paul EA (1978) Criteria for measurement of microbial growth and activity in soil. Soil Biol Biochem 10: 223–229

Nash AJ (1963) A method of evaluating the effects of topography on the soil water balance. For Sci 9: 413–422

Naveh Z (1974) The ecology of fire in Israel. In: Proc Annu Tall Timbers Fire Ecol Conf 13: 131–170

Naveh Z (1982) Mediterranean landscape evolution and degradation as multivariate biofunctions: theoretical and practical implications. Landscape Plann 9: 125–146

Naveh Z, Whittaker RH (1979) Structural and floristic diversity of shrublands in northern Israel and other mediterranean areas. Vegetatio 41: 171–190

Neethling D-M, Boucher C (1986) 'n Strukturele en floristiese opname van die kleinsoogdiernavorsings terrein te Swartboskloof, Jonkershoek. Final Rep to the Fynbos Biome Project, National Programme Environmental Science, CSIR, Pretoria

Nel JAJ, Rautenbach IL, Breytenbach GJ (1980) Mammals of the Kammanassie Mountains, southern Cape Province. S Afr J Zool 15: 255–261

Neser S (1968) Studies on some potentially useful insect enemies of the needle bush (*Hakea* spp. – Proteaceae). PhD Thesis, Australian National University, Canberra

Noble IR, Slatyer RO (1980) The use of vital attributes to predict successional changes in plant communities subject to recurrent disturbances. Vegetatio 43: 5–21

O'Dowd DJ, Gill AM (1984) Predator satiation and site alteration following fire: mass reproduction of mountain ash (*Eucalyptus delegatensis*) in southeastern Australia. Ecology 65: 1052–1066

O'Dowd DJ, Hay ME (1980) Mutualism between harvester ants and a desert ephemeral: seed escape from rodents. Ecology 61: 531–540

Oades JM, Jenkinson DS (1979) Adenosine triphosphate content of the soil microbial biomass. Soil Biol Biochem 11: 201–204

Oechel WC (1988) Seedling establishment and water relations after fire in a mediterranean ecosystem. In: Allen EB (ed) The reconstruction of disturbed arid ecosystems: an ecological approach. Westview Press, pp 34–45

Old SM (1969) Microclimate, fire, and plant production in an Illonois prairie. Ecol Monogr 39: 355–384

Olivier W, Werner W (1980) The genus *Cyrtanthus* Ait. Veld & Flora 68: 78–81

O'Loughlin, EM, Cheney NP, Burns J (1982) The Bushrangers experiment: hydrological response of a eucalypt catchment to fire. Proc 1st Nat Symp on Forest hydrology, I. E. Aust Natl Conf Publ 82/6 Melbourne, pp 132–138

Orians GH (1986) Site characteristics favoring invasion. In: Mooney HA, Drake JA (eds) Ecology of biological invasions of North America and Hawaii. Ecological Studies 58. Springer, Berlin Heidelberg New York, pp 133–148

Orshan G (1983) Approaches to the definition of mediterranean growth forms. In: Kruger FJ, Mitchell DT, Jarvis JUM (eds) Mediterranean-type ecosystems – the role of nutrients. Ecological Studies 43. Springer, Berlin Heidelberg New York, pp 86–102

Orshan G (1989) Plant pheno-morphological studies in mediterranean-type ecosystems. Kluwer, Dordrecht

Osborn JF, Pelishek RE, Krammes JS, Letey J (1964) Soil wettability as a factor in erodibility. Soil Sci Soc Am Proc 28: 294–295

Pack DH (1980) Precipitation chemistry patterns: a two-network data set. Science 208: 1143
Parker VT (1984) Correlation of physiological divergence with reproductive mode in chaparral shrubs. Madrono 31: 231–242
Pase CP, Ingebo PA (1965) Burned chaparral to grass: early effects on water and sediment yields from two granitic soil watersheds in Arizona. Proc Ann Ariz Watershed Symp, pp 8–11
Pate JJ, Dixon KW (1982) Tuberous, cormous and bulbous plants – biology of an adaptive strategy in Western Australia. University of Western Australia Press, Nedlands
Pate JS, Dixon KW, Orshan G (1984) Growth and life form characteristics of kwongan species. In: Pate JS, Beard JS (eds) Kwongan: plant life of the sandplain. University of Western Australia Press, Nedlands, pp 84–100
Patil GP, Taille C (1979) An overview of diversity. In: Grassle JF, Patil GP, Smith W, Taillie C (eds) Statistical ecology 6: ecological diversity in theory and practice. International Co-operative Publishing House, Fairland, Maryland, pp 3–27
Paul EA, Johnson RL (1977) Microscopic counting and adenosine-5'-triphosphate measurement in determining microbial growth in soils. Appl Environ Microbiol 34: 263–269
Peet RK, Glenn-Lewin DC, Wolf JW (1983) Prediction of man's impact on plant species diversity: a challenge for vegetation science. In: Holzner W, Werger MJA, Ikusima I (eds) Man's impact on vegetation. Junk, The Hague, pp 41–54
Penman HL (1948) Natural evaporation from open water, bare soil and grass. Proc R Soc Lond Ser A 193: 120–146
Petal J (1978) The role of ants in exosystems. In: Brian MV (ed) Production ecology of ants and termites. IBP No 13. Cambridge University Press, Cambridge, pp 293–325
Peterjohn WT, Correll DL (1984) Nutrient dynamics in an agricultural watershed: observations on the role of a riparian forest. Ecology 65: 1466–75
Peters NE (1984) Evaluation of environmental factors affecting yields of major dissolved ions of streams in the United States. US Geological Survey, Watersupply Pap 228
Petrusewicz K, Andrzejewski R (1962) Natural history of a free living population of house mice (*Mus musculus* L.) with particular reference to groupings within the population. Ekologia Polska Ser A 10: 85–122
Phillips JD (1989) An evaluation of the factors determining the effectiveness of water quality buffer zones. J Hydrol 107: 133–145
Philpot CW (1963) The moisture content of ponderosa pine and whiteleaf manzanita foliage in the central Sierra Nevada. USDA For Serv Res Note PSW-39
Philpot CW (1969) Seasonal changes in heat content and ether extractive content of chamise. USDA For Serv Res Pap INT 61
Philpot CW, Mutch RW (1971) The seasonal trends in moisture content, either extractives and energy of ponderosa pine and Douglas fir needles. USDA For Serv Res Pap INT-102
Pimm SL (1986) Community stability and structure. In: Soulé ME (ed) Conservation biology: the science of scarcity and diversity. Sinauer Associates, Sunderland, pp 309–329
Pimm SL (1989) Theories of predicting success and impact of introduced species. In: Drake JA, Mooney HA, Di Castri F, Groves RH, Kruger FJ, Rejmanek M, Williamson MH (eds) Ecology of biological invasions: a global perspective. Wiley, Chichester, pp 351–367
Poole DK, Miller PC (1975) Water relations of selected species of chaparral and coastal sage communities. Ecology 56: 1118–1128
Poole DK, Roberts SW, Miller PC (1981) Water utilization. In: Miller PC (ed) Resource use by chaparral and matorral. Ecological Studies 39. Springer, Berlin Heidelberg New York, pp 123–149
Postgate JR (1982) The fundamentals of nitrogen fixation. Cambridge University Press, Cambridge

Prins AJ (1978) Hymenoptera. In: Werger MJA (ed) Biogeography and ecology of southern Africa. Junk, The Hague, pp 825–875

Putz FE, Brokaw NVL (1989) Sprouting of broken trees on Basso Colorado Island, Panama. Ecology 70: 508–512

Pyne SJ (1984) Introduction to wildland fire – fire management in the United States. Wiley, New York

Raison RJ (1980) A review of the role of fire in nutrient cycling in Australian native forests, and of methodology for studying the fire-nutrient interaction. Aust J Ecol 5: 15–21

Raunkiaer C (1934) The life-forms of plants and statistical plant geography. Clarendon Press, Oxford

Raven PH, Axelrod DI (1978) Origin and relationships of the California flora. Univ Calif Pub Bot 72: 1–134

Read J, Hill RS (1983) Rainforest invasion onto Tasmanian old-fields. Aust J Ecol 8: 149–161

Rees AR (1972) The growth of bulbs. Academic Press, London

Rehfuess KE (1989) Acid deposition – extent and impact on forest soils, nutrition, growth and disease phenomena in central Europe: a review. Water, Air Soil Poll 48: 1–20.

Reid C, Dyer RA (1984) A review of southern African species of *Cyrtanthus*. Am Plant Life Soc, La Jolla, California

Rejmanek M (1989) Invasibility of plant communities. In: Drake JA, Mooney HA, Di Castri F, Groves RH, Kruger FJ, Rejmanek M, Williamson MH (eds) Ecology of biological invasions: a global perspective. Wiley, Chichester, pp 369–388

Reynolds B, Hornung M, S Hughes S (1989) Chemistry of streams draining grassland and forest catchments at Plynlimon, mid-Wales. Hydrol Sci J 34: 667–686

Richardson DM (1985) Height growth of seedlings of *Protea repens Hakea sericea*, *Pinus pinaster* and *Acacia longifolia*. Rep 85/26, Jonkershoek Forestry Research Centre, Stellenbosch

Richardson DM (1988) Age structure and regeneration after fire in a self-sown *Pinus halepensis* forest on the Cape Peninsula, South Africa. S Afr J Bot 54: 140–144

Richardson DM (1989) The ecology of invasions by *Pinus* (Pinaceae) and *Hakea* (Proteaceae) species, with special emphasis on patterns, processes and consequences of invasion in mountain fynbos of the southwestern Cape Province, South Africa. PhD Thesis, University of Cape Town

Richardson DM (1990) Pines on the move: invasions in perspective. Afr Wildl 44: 162–167

Richardson DM, Bond WJ (1991) Determinants of plant distribution: evidence from pine invasions. Am Nat 137: 639–668

Richardson DM, Brink MP (1985) Notes on *Pittosporum undulatum* in the south-western Cape. Veld & Flora 71: 75–77

Richardson DM, Brown PJ (1986) Invasion of mesic mountain fynbos by *Pinus radiata*. S Afr J Bot 52: 529–536

Richardson DM, Fraser M (in press) The birds of Swartboskloof, Jonkershoek Valley, Stellenbosch. Southern Birds

Richardson DM, Kruger FJ (1990) Water relations and photosynthetic characteristics of selected trees and shrubs of riparian and hillslope habitats in the southwestern Cape Province, South Africa. S Afr J Bot 56: 214–225

Richardson DM, van Wilgen BW (1984) Factors affecting the regeneration success of *Hakea sericea*. S Afr For J 131: 63–68

Richardson DM, van Wilgen BW (1986a) Effects of thirty-five years of afforestation with *Pinus radiata* on the composition of mesic mountain fynbos near Stellenbosch. S Afr J Bot 52: 309–315

Richardson DM, van Wilgen BW (1986b) The effects of fire in felled *Hakea sericea* and natural fynbos and the implications for weed control in mountain catchments. S Afr For J 139: 4–14

Richardson DM, van Wilgen BW, Mitchell DT (1987) Aspects of the reproductive ecology of four Australian *Hakea* species (Proteaceae) in South Africa. Oecologia (Berlin) 71: 345–354

Richardson DM, Macdonald IAW, Forsyth GG (1989) Reductions in plant species richness under stands of alien trees and shrubs in the fynbos biome. S Afr For J 149: 1–8

Richardson DM, Cowling RM, Le Maitre DC (1990) Assessing the risk of invasive success in *Pinus* and *Banksia* in South African mountain fynbos. J Veg Sci 1: 629–642

Richardson DM, Macdonald IAW, Holmes PM, Cowling RM (1991) Plant and animal invasions. In: Cowling RM (ed) The ecology of fynbos: Nutrients, Fire and Diversity. Oxford University Press, Cape Town (in press)

Richardson GR, Lubke RA, Jacot Guillarmod A (1984) Regeneration of grassy fynbos near Grahamstown (eastern Cape) after fire. S Afr J Bot 3: 153–162

Ronan NM (1986) The hydrological effects of fuel reduction burning and wildfire at Wallaby Creek. Melbourne and Metropolitan Board of Works, Rep No MMBW-W-0015, Melbourne, Australia

Rothermel RC (1972) A mathematical model for predicting fire spread in wildland fuels. USDA For Serv Res Pap INT-115

Rothermel RC, Philpot CW (1973) Predicting changes in chaparral flammability. J For 71: 640–643

Roughgarden J, Diamond J (1986) Overview: the role of species interactions in community ecology. In: Diamond J, Case TJ (eds) Community ecology. Harper & Row, New York, pp 333–343

Rourke JP (1980) The proteas of southern Africa. Purnell, Cape Town

Rourke JP (1982) The proteas of southern Africa, 2nd edn. Centaur, Johannesburg

Roux ER, Warren JL (1963) Studies in the autecology of Australian acacias in South Africa. 2. Symbiotic nitrogen fixation in *Acacia cyclops*. A. Cunn. S Afr J Sci 59: 294–295

Rowe PB, Storey H, Hamilton EL (1951) Some results of hydrologic research. In: Some aspects of watershed management in southern California. USDA For Serv Calif For Range Expt Stn Misc Pap 1, pp 19–29

Rowe PB, Countryman CM, Storey HC (1954) Hydrologic analysis used to determine effects of fire on peak discharge and erosion rates in southern California watersheds. USDA For Serv Calif For Range Exp Stn

Rowley J (1970) Effects of burning and clipping on temperature, growth, and flowering of narrow-leaved snow tussock. N Z J Bot 8: 264–282

Rundel PW (1981a) Fire as an ecological factor. In: Lange OL, Nobel PS, Osmond CB, Ziegler H (eds) Physiological plant ecology I: response to the physical environment. Encyclopedia of plant physiology, vol 12A. Springer, Berlin Heidelberg New York, pp 501–538

Rundel PW (1981b) Structural and chemical components of flammability. In: Mooney HA, Bonnicksen TM, Christensen NL, Lotan JE, Reiners WA (eds) Proc Conf on Fire regimes and ecosystem properties. USDA For Serv Gen Tech Rep WO-86, pp 183–207

Rundel PW (1983) Impact of fire on nutrient cycles in mediterranean-type ecosystems with reference to chaparral. In: Kruger FJ, Mitchell DT, Jarvis JUM (eds) Mediterranean-type ecosystems. The role of nutrients. Springer, Berlin Heidelberg New York, pp 192–207

Russell WB (1983) Conservation of arable land in the USA: report on service tour. Unpublished report, S Afr Dept of Agriculture and Fisheries, Natal Region, Cedara

Rutherford MC (1978) Karoo-fynbos biomass along an elevational gradient in the Western Cape. Bothalia 12: 555–560

Rutherford MC (1981) Survival, regeneration and leaf biomass changes in woody plants following spring burns in *Burkea africana – Ochna pulchra* savanna. Bothalia 13: 531–552

Rutherford MC, Pressinger FM, Musil CF (1986) Standing crops, growth rates and

resource use efficiency in alien plant invaded ecosystems. In: Macdonald IAW, Kruger FJ, Ferrar AA (eds) The ecology and management of biological invasions in southern Africa. Oxford University Press, Cape Town, pp 189–199
Rycroft HB (1947) A note on the immediate effects of veld burning on storm flow in a Jonkershoek stream catchment. J S Afr For Assoc 15: 80–88
Samways MJ (1981) Comparison of ant community structure (Hymenoptera: Formicidae) in citrus orchards under chemical and biological control of red scale, *Aonidiella aurantii* (Maskell) (Hemiptera: Diaspididae). Bull Entomol Res 71: 58–60
SAS Institute Inc. (1985a) SAS users guide: basics, version 5 edn. SAS Inst Inc, Cary, North Carolina
SAS Institute Inc. (1985b) SAS user's guide: statistics, version 5 edn. SAS Inst Inc, Cary, North Carolina
Savage SM, Osborn J, Letey J, Heaton C (1972) Substances contributing to fire-induced water repellency in soils. Soil Sci Soc Am Proc 36: 674–678
Schlegel HG (1985) General microbiology, 6th edn. Cambridge University Press, Cambridge
Schlesinger WH, Gill DS (1980) Biomass, production and changes in the availability of light, water and nutrients during their development of pure stands of the chaparral shrub, *Ceanothus megacarpus*, after fire. Ecology 61: 781–789
Schlesinger WH, Hasey MM (1981) Decomposition of chaparral shrub foliage: losses of organic and inorganic constituents from deciduous and evergreen leaves. Ecology 62: 762–774
Schlettwein CHG, Giliomee JH (1987) Comparison of insect biomass and community structure between fynbos sites of different ages after fire, with particular reference to ants, leafhoppers and grasshoppers. Ann Univ Stellenbosch Ser A32(2): 1–76
Schnürer JM, Clarholm M, Rosswall T (1985) Microbial biomass and activity in an agricultural soil with different organic matter contents. Soil Biol Biochem 17: 611–618
Scholtz A (1986) Palynological and paleobotanical studies in the southern Cape. MA Thesis, University of Stellenbosch
Schulze RE, Lambson C (1986) RADSLOPE: a model to simulate radiant energy fluxes an potential evaporation on sloping terrain. Unpublished paper, Dept Agric Eng, Univ of Natal, Pietermaritzburg, South Africa
Schutte KH (1960) Trace element deficiencies in Cape vegetation. J S Afr Bot 26: 45–49
Schweitzer FR, Scott KD (1973) Early occurrence of domestic sheep in sub-Saharan Africa. Nature 241: 547
Scott DF (1989) The effects of the Langrivier fire on overland flow and soil erosion. In: Richardson DM (ed) The short-term effects of fire in Langrivier, Jonkershoek State Forest, Stellenbosch. Rep J2/89, S Afr For Res Inst, Stellenbosch, pp 69–77
Scott DF, van Wyk DB (1990) The effects of wildfire on soil wettability and hydrological behaviour of an afforested catchment. J Hydrol 121: 239–256
Shaughnessy GL (1980) Historical ecology of alien woody plants in the vicinity of Cape Town, South Africa. PhD Thesis, University of Cape Town
Shaughnessy GL (1986) A case study of some woody plant introductions to the Cape Town area. In: Macdonald IAW, Kruger FJ, Ferrar AA (eds) The ecology and management of biological invasions in southern Africa. Oxford University Press, Cape Town, pp 37–43
Shmida A (1981) Mediterranean vegetation in California and Israel: similarities and differences. Isr J Bot 30: 105–123
Shmida A, Ellner SP (1984) Coexistence of plant species with similar niches. Vegetatio 58: 29–55
Siegfried WR (1983) Trophic structure of some communities of fynbos birds. J S Afr Bot 49: 1–43
Silvertown J (1980) The evolutionary ecology of mast seeding in trees. Biol J Linn Soc 14: 235–250
Sim TR (1907) The forests and forest flora of the Cape Colony. Taylor & Henderson, Aberdeen
Simberloff D (1986) Introduced insects: a biogeographic and systematic perspective. In: Mooney HA, Drake JA (eds) Ecology of biological invasions of North America and

Hawaii. Ecological Studies 58. Springer, Berlin Heidelberg New Youk, pp 4–26
Skaife SH (1961) The study of ants. Longmans, London
Skaife SH (undated) South African nature notes. Maskew Miller, Cape Town
Skead CJ (1980) Historical mammal incidence in the Cape Province, vol 1. The Western and Northern Cape. Cape Department of Nature and Environmental Conservation, Cape Town
Slingsby P (1982) The Argentine ant – how much of a threat? Veld & Flora 68: 102–104
Slingsby P, Bond WJ (1981) Ants – friends of the fynbos. Veld & Flora 67: 39–45
Slingsby P, Bond WJ (1982) Of ants and proteas. Afr Wildl 36: 104–107
Smith RE (1990) Comparative water relations of *Protea nitida* seedlings and sprouters after fire. MSc Thesis, University of Natal, Pietermaritzburg
Smith RE, Higgins KB (1990) The root systems of three species of *Protea* shrubs in the Jonkershoek Valley, Cape Province. S Afr J Plant Soil 7: 238–240
Smith RE, Richardson DM (1990) Comparative post-fire water relations of selected re-seeding and resprouting fynbos plants in the Jonkershoek Valley, Cape Province, South Africa. S Afr J Bot 56: 683–694
Smith RM, Stamey WL (1965) Determining the range of tolerable erosion. Soil Sci 100: 414–424
Smithers RHN (1983) The mammals of the southern African subregion. University of Pretoria, Pretoria
Sneeuwjagt RJ, Frandsen WH (1977) Behaviour of experimental grass fires vs predictions based on Rothermel's fire model. Can J For Res 7: 357–367
Snijman D (1984) A revision of the genus *Haemanthus* (Amaryllidaceae). J S Afr Bot Suppl Vol 12: 1–139
Söderström BE (1979) Seasonal fluctuations of active fungal biomass in horizons of a podzolized pine-forest soil in central Sweden. Soil Biol Biochem 11: 149–154
Söhnge APG (1988) The geology of Swartboskloof. Rep J3/88, Jonkershoek Forestry Research Centre, Stellenbosch
Sommerville JEM (1981) A comparison of the seasonal growth of indigenous and non-indigenous species in the South Western Cape. J S Afr Bot 47: 797–805
Southwood TRE (1978) Ecological methods, 2nd edn. Chapman & Hall, London
Specht RL (1969) A comparison of the sclerophyllous vegetation characteristic of mediterranean type climates in France, California, and southern Australia. II. Dry matter, energy, and nutrient accumulation. Aust J Bot 17: 293–308
Specht RL (1981a) Primary production in mediterranean-climate ecosystems regenerating after fire. In: Di Castri F, Goodall DW, Specht RL (eds) Ecosystems of the world 11. Mediterranean-type shrublands. Elsevier, Amsterdam, pp 257–267
Specht RL (1981b) Responses to fires in heathlands and related shrublands. In: Gill AM, Groves RH, Noble IR (eds) Fire and the Australian biota. Australian Academy of Science, Canberra, ACT, pp 391–415
Specht RL (ed) (1988) Mediterranean-type ecosystems – a data source book. Kluwer, Dordrecht
Specht RL, Moll EJ (1983) Mediterranean-type heathlands and sclerophyllous shrublands of the world: an overview. In: Kruger FJ, Mitchell DT, Jarvis JUM (eds) Mediterranean-type ecosystems. The role of nutrients. Springer, Berlin Heidelberg New York, pp 41–65
Specht RL, Rayson P (1957) Dark Island Heath (Ninety-mile Plain, South Australia). I. Definitions of the ecosystem. Aust J Bot 5: 52–75
Specht RL, Rayson P, Jackman ME (1958) Dark Island Heath (Ninety-mile Plain, South Australia). VI. Pyric succession: changes in composition, coverage, dry weight and mineral nutrient status. Aust J Bot 6: 59–88
Stearns SC (1976) Life-history tactics: a review of the ideas. Q Rev Biol 51: 3–47
Stock WD, Lewis OAM (1986a) Soil nitrogen and the role of fire as a mineralizing agent in a South African coastal fynbos ecosystem. J Ecol 74: 317–328
Stock WD, Lewis OAM (1986b) Atmospheric input of nitrogen to a coastal fynbos ecosystem of the south-western Cape Province, South Africa. S Afr J Bot 52: 273–276
Stock WD, Lewis OAM, Allsopp N (1988) Soil nitrogen mineralization in a coastal fynbos succession. Plant Soil 106: 295–298

Stohlgren TJ, Rundel PW (1986) A population model for a long-lived, resprouting chaparral shrub: *Adenostoma fasciculatum*. Ecol Model 34: 245–257
Stone EC (1951) The stimulative effect of fire on the flowering of the golden brodiaea (*Brodiaea ixioides* Wats. var *lugens* Jeps.). Ecology 32: 534–537
Stork NE (1988) Insect diversity: facts, fiction and speculation. Biol J Linn Soc 35: 321–337
Strid AK (1974) A taxonomic revision of the genus *Bobartia* L. (Iridaceae). Opera Bot 37: 1–45
Sutherland JP (1990) Perturbations, resistance, and alternative views of the existence of multiple stable points in nature. Am Nat 136:270–275
Swank WT, Waide JB (1988) Characterization of baseline precipitation and stream chemistry and nutrient budgets for control watersheds. In: Swank WT, Crossley DA (eds) Forest hydrology and ecology at Coweeta. Ecological Studies 66. Springer, Berlin Heidelberg New York, pp 57–79
Taylor HC (1969) A vegetation survey of the Cape of Good Hope Nature Reserve. MSc Thesis, University of Cape Town
Taylor HC (1977) The Cape floral kingdom, an ecological view. Proc 2nd Natl Weeds Conf S Afr, Balkema, Cape Town, pp 19–33
Taylor HC (1978) Capensis. In: Werger MJA (ed) Biogeography and ecology of southern Africa. Junk, The Hague, pp 171–229
Tenhunen JD, Catarino FM, Lange OL, Oechel WC (eds) (1987) Plant response to stress. Functional analysis in mediterranean ecosystems. Springer, Berlin Heidelberg New York
Teramura AH (1980) Relationship between stand age and water repellency of chaparral soils. Bull Torrey Bot Club 107: 42–46
Thomas CM, Davis SD (1989) Recovery patterns of three chaparral shrub species after wildfire. Oecologia (Berlin) 80: 309–320
Thompson K (1978) The occurrence of buried viable seed in relation to environmental gradients. J Biogeogr 5: 425–430
Thompson K, Grime PH, Mason G (1977) Seed germination in response to diurnal fluctuations in temperature. Nature 267: 147–149
Thompson MW (1990) Mapping of the December 1988 – January 1989 South Western Cape fires using multitemporal Landsat multispectral scanner data. Rep FOR 16, CSIR Division of Forest Science and Technology, Pretoria
Thrower NJW, Bradbury DE (1977) Chile-California mediterranean scrub atlas. Dowden, Hutchinson & Ross, Stroudsburg
Tiedemann AR, Conrad CE, Dieterich JH, Hornbeck JW, Megahan WF, Viereck LA, Wade DD (1979) Effects of fire on water – a state-of-knowledge review. USDA For Serv Gen Tech Rep WO-10
Trabaud L (1981) Man and fire: impacts on mediterranean vegetation. In: Di Castri F, Goodall DW, Specht RL (eds) Mediterranean-type shrublands of the World. Elsevier, Amsterdam, pp 523–537
Trabaud L (1991) Is fire an agent favouring invasions? In: Groves RH, Di Castri F (eds) Biogeography of mediterranean invasions. Cambridge University Press, Cambridge, pp 179–190
Trujillo DP (1976) Chemical properties of chaparral fuels change during preheating before planning. USDA For Serv Res Note RM-320
Tutin TG, Heywood VH, Burges NA, Moore DM, Valentine DH, Walters SM, Webb DA (1980) Flora Europaea, vol 5. Cambridge University Press, Cambridge
Tyson PD, Kruger FJ, Louw CW (1988) Atmospheric pollution and its implications in the Eastern Transvaal Highveld. S Afr Nat Sci Prog Rep 150, CSIR, Pretoria
van Daalen JC (1984) Distinguishing features of forest species on nutrient-poor soils in the southern Cape. Bothalia 15: 229–239
van der Heyden F, Lewis OAM (1989) Seasonal variation in photosynthetic capacity with respect to plant water stress of five species of the mediterranean climate region of South Africa. S Afr J Bot 55: 509–515

van der Merwe P (1966) Die flora van Swartboschkloof, Stellenbosch, en die herstel van soorte na 'n brand. Ann Univ Stellenbosch Ser A 14: 691–736
van der Zel DW, Kruger FJ (1975) Results of the multiple catchment experiments at the Jonkershoek Research Station, South Africa. II. Influence of protection of fynbos on stream discharge in Langrivier. For S Afr 16: 13–18
van Hensbergen HJ (1984) Ecological separation between four species of carnivore in the western Pyrenees, Spain. PhD Thesis, University of Cambridge
van Laar A (1984) The topographical distribution of rainfall within the Jonkershoek Valley. S Afr For J 128: 30–32
van Staden J (1978) Seed viability in *Protea neriifolia*. I. The effects of time of harvesting on seed viability. Agroplantae 10: 65–7
van Vliet HR, Kempster DP, Satory DP, Gerber FA, Schoonraad IJ (1988) Analytical methods manual. Hydrological Research Institute, Department of Water Affairs, Pretoria
van Wagner CE (1977) Conditions for the start and spread of crown fire. Can J For Res 7: 23–34
van Wilgen BW (1981) Some effects of fire frequency on fynbos plant community composition and structure at Jonkershoek, Stellenbosch. S Afr For J 118: 42–55
van Wilgen BW (1982) Some effects of post-fire age on the above ground plant biomass of fynbos (macchia) vegetation in South Africa. J Ecol 70: 217–225
van Wilgen BW (1984a) Adaptation of the United States Fire Danger Rating System to fynbos conditions. I. A fuel model for fire danger rating in the fynbos biome. S Afr For J 129: 61–65
van Wilgen BW (1984b) Fire climates in the southern and western Cape Province and their potential use in fire control and management. S Afr J Sci 80: 358–362
van Wilgen BW (1985) The derivation of fire hazard indices and burning prescriptions from climatic and ecological features of the fynbos biome. PhD Thesis, University of Cape Town
van Wilgen BW (1986) An outline of proposed research at the Swartboskloof study site. Occas Rep 7, Ecosystem Programmes, CSIR, Pretoria
van Wilgen BW (1987) Prŷs Fire regimes in the fynbos biome. In: Cowling RM, Le Maitre DC, McKenzie B, Prys-Jones RP, van Wilgen BW (eds) Disturbance and the dynamics of fynbos biome communities. S Afr Nat Sci Prog Rep 135, CSIR, Pretoria, pp 6–14
van Wilgen BW, Burgan RE (1984) Adaptation of the United States Fire Danger Rating system to fynbos conditions. Part II. Historic fire danger in the fynbos biome. S Afr For J 129:66–78
van Wilgen BW, Kruger FJ (1981) Observations on the effects of fire in mountain fynbos at Zachariashoek, Paarl. J S Afr Bot 47: 195–212
van Wilgen BW, Kruger FJ (1985) The physiography and fynbos vegetation communities of the Zachariashoek catchments, south-western Cape Province. S Afr J Bot 51: 379–399
van Wilgen BW, Le Maitre DC (1981) Preliminary estimates of nutrient levels in fynbos vegetation and the role of fire in nutrient cycling. S Afr For J 119: 24–28
van Wilgen BW, Richardson DM (1985) The effects of alien shrub invasions on vegetation structure and fire behaviour in South African fynbos shrublands: a simulation study. J Appl Ecol 22: 955–966
van Wilgen BW, Siegfried WR (1986) Seed dispersal properties of pine species as a determinant of invasive potential. S Afr J Bot 52: 546–548
van Wilgen BW, Viviers ML (1985) The effects of season of fire on serotinous Proteaceae in the Western Cape and the implications for catchment management. S Afr For J 133: 49–53
van Wilgen BW, Wills AJ (1988) Fire behaviour prediction in savanna vegetation. S Afr J Wildl Res 18: 41–46
van Wilgen BW, Le Maitre DC, Kruger FJ (1985) Fire behaviour in South African fynbos (macchia) vegetation and predictions from Rothermel's fire model. J Appl Ecol 22: 207–216

van Wilgen BW, Everson CS, Trollope WSW (1990a) Fire management in southern Africa: some examples of current objectives, practices and problems. In: Goldammer JG (ed) Fire in the tropical biota. Ecological Studies 84. Springer, Berlin Heidelberg New York, Heidelberg, pp 177–215
van Wilgen BW, Higgins KB, Bellstedt DU (1990b) The role of vegetation structure and fuel chemistry in excluding fire from forest patches in the fire-prone fynbos shrublands of South Africa. J Ecol 78: 210–222
van Wilgen BW, Bond WJ, Richardson DM (1991) Ecosystem management. In: Cowling RM (ed) The ecology of fynbos: Nutrients, Fire and Diversity. Oxford University Press, Cape Town (in press)
van Wyk DB (1981a) Influence of prescribed burning on nutrient budgets of mountain fynbos catchments in the south-western Cape, Republic of South Africa. In: Conrad CE, Oechel WC (eds) Proc Symp on Dynamics and management of mediterranean-type ecosystems. USDA For Serv Gen Tech Rep PSW-58, pp 390–396
van Wyk DB (1981b) Water quality of manipulated mountain catchments in the western Cape. In: Maaren H (ed) Proc Worksh on the Effect of rural land use and catchment management on water resources, Department of Water Affairs, Forestry and Environmental Conservation Rep TR 113, pp 76–89
van Wyk DB (1983) Apparatus for sampling of streams for chemical quality and sediment. Water S A 9: 88–92
van Wyk DB (1987) Some effects of afforestation on streamflow in the western Cape Province, South Africa. Water S A 13: 31–36
van Wyk DB (1990) Atmospheric deposition at selected sites in the mountain catchments of South Africa. In: Proc 1st IUAPPA Regional Conf on Air pollution: towards the 21st century, CSIR, Pretoria, pp 1–19
van Wyk DB, Lesch W (1989) The effect of the Langrivier fire on water quality and sediment yield. In: Richardson DM (ed) The short-term effects of fire in Langrivier, Jonkershoek State Forest, Stellenbosch. Rep J2/89, South African Forestry Research Institute, Stellenbosch, pp 78–88
van Wyk DB, Bosch JM, Braune E, Bruwer CA, Hart RC, King JM, Mitchell DS, Rogers KH, Simpson DE, Twinch AJ (1984) Catchment management. In: Hart RC, Allanson BR (eds) Limnological criteria for management of water quality in the southern hemisphere. S Afr Nat Sci Prog Rep 93, CSIR, Pretoria, pp 1–31
Versfeld DB (1981) Overland flow on small plots at the Jonkershoek Forestry Research Station. S Afr For J 119: 35–40
Versfeld DB (1985) Rainfall interception in *Protea neriifolia* – measurement and prediction. MSc Thesis, University of Stellenbosch
Versfeld DB, van Wilgen BW (1986) Impacts of woody aliens on ecosystem properties. In: Macdonald IAW, Kruger FJ, Ferrar AA (eds) The ecology and control of biological invasions in South Africa. Oxford University Press, Cape Town, pp 239–246
Viessman W, Knapp JW, Lewis GL, Harbaugh TE (1977) Introduction to hydrology, 2nd edn. Harper & Row, New York
Viro PJ (1974) Effects of forest fire on soil. In: Kozlowski TT, Ahlgren CE (eds) Fire and ecosystems. Academic Press, New York, pp 7–45
Virzo De Santo A, Alfani A, Sapio S (1978) Microbial populations of two soils in the beech forests of Monte Taburno (Campania Appenines). Pedobiologia 18: 48–56
Vitousek PM (1986) Biological invasions and ecosystem properties: can species make a difference? In: Mooney HA, Drake JA (eds) Ecology of biological invasions in North America and Hawaii. Ecological Studies 58. Springer, Berlin Heidelberg New York, pp 163–176
Vitousek PM (1990) Biological invasions and ecosystem processes: towards an integration of population biology and ecosystem studies. Oikos 57: 7–13
Vogel J, Fuls A, Ellis R (1978) The geographic distribution of Kranz grasses in South Africa. S Afr J Sci 74: 209–215
Vogt KA, Edmonds RL, Antos GC, Vogt DJ (1980) Relationships between $CO_2$ evolution, ATP concentrations and decomposition in four forest ecosystems in western Washington. Oikos 35: 72–79

Vose PB, Ruschel AP (1981) Associative $N_2$-fixation, vols I, II. CRC Press, Boca Raton

Waher IG (1974) Certain aspects of the biology of *Watsonia* spp. MSc Thesis, University of Cape Town

Walker J (1981) Fuel dynamics in Australian vegetation. In: Gill AM, Groves RH, Noble IR (eds) Fire and the Australian biota. Aust Acad Sci. Canberra, pp 101–128

Waller DM (1979) Models of mast fruiting in tress. J Theor Biol 80:223–32

Warcup JH (1981) Effect of fire on the soil microflora and other non-vascular plants. In: Gill AM, Groves RH, Noble IR (eds) Fire and the Australian biota. Aust Acad Sci, Canberra, pp 203–214

Ward PS (1987) Distribution of the introduced Argentine ant (*Iridomyrmex humilis*) in natural habitats of the Lower Sacramento Valley and its effects on the indigenous ant fauna. Hilgardia 55: 1–16

Warner RR, Chesson PL (1985) Coexistence mediated by recruitment fluctuations: a field guide to the storage effect. Am Nat 125: 769–787

Wells CG, Campbell RE, DeBano LF, Lewis CE, Fredriksen RL, Franklin EC, Froelich RC, Dunn PH (1979) Effect of fire on soil: a state-of-knowledge review. USDA For Serv Gen Tech Rep WO-7

Wells PV (1969) The relation between mode of reproduction and extent of speciation in woody genera of the California chaparral. Evolution 23: 264–267

Werger MJA, Kruger FJ, Taylor HC (1972) A phytosociological study of the Cape fynbos and other vegetation at Jonkershoek, Stellenbosch. Bothalia 10: 599–614

White F (1978) The Afromontane region. In: Werger MJA (ed) The biogeography and ecology of southern Africa. Junk, The Hague, pp 463–513

Wicht CL (1939) Forest influences research technique at Jonkershoek. J S Afr For Assoc 3: 65–80

Wicht CL (1945) Preservation of the vegetation of the south-western Cape. Spec Publ R Soc S Afr, Cape Town

Wicht CL (1948a) Hydrological research in South African forestry. J S Afr For Assoc 6: 4–21

Wicht CL (1948b) A statistically designed experiment to test the effects of burning on a sclerophyll scrub community. 1. Preliminary account. Trans R Soc S Afr 31: 479–501

Wicht CL (1967a) Forest hydrology research in the South African Republic. In: Sopper WE, Lull W (eds) Int Symp on Forest hydrology. Pergamon, Oxford, pp 75–84

Wicht CL (1967b) The validity of conclusions from South African multiple watershed experiments. In: Sopper WL, Lull HW (eds) Int Symp on Forest hydrology. Pergamon, Oxford, pp 749–760

Wicht CL, Meyburgh JC, Boustead P (1969) Rainfall at the Jonkershoek Forest Hydrological Research Station. Ann Univ Stellenbosch Ser A 44 (1): 1–66

Wiens D, Rourke JP, Casper BB, Rickardt EA, Lapine TR, Peterson CJ, Channing A (1983) Nonflying mammal pollination of southern African proteas: a noncoevolved system. Ann Missouri Bot Gard 70: 1–31

Willan K (1979) Design and field tests of a modified small mammal livetrap. S Afr J Zool 14: 81–84

Willan K (1986) Bait selection in laminate toothed rats and other southern African small mammals. Acta Theriol 31: 359–363

Willan K, Bigalke RC (1982) The effects of fire regime on small mammals in S W Cape montane fynbos (Cape macchia). In: Conrad CE, Oechel WC (eds) Dynamics and management of mediterranean-type ecosystems. USDA For Serv Gen Tech Rep PSW-58, pp 207–212

Wilson SD, Keddy PA (1986) Species competitive ability and position along a natural stress/disturbance gradient. Ecology 67: 1236–1242

Witkowski ETF (1991) Growth and competition between seedlings of *Protea repens* and the alien invasive, *Acacia saligna*, in relation to nutrient availability. Func Ecol 5: 101–110

Wright EW, Bollen WB (1961) Microflora of Douglas-fir forest soil. Ecology 42: 825–828

Yarranton GA, Morrison RG (1974) Spatial dynamics of a primary succession: nucleation. J Ecol 62: 417–428

Yeaton RI, Bond WJ (1991) Competition between two shrub species: dispersal differences and fire promote co-existence. Am Nat 138: 328–341

Yodzis P (1986) Competition, mortality and community structure. In: Diamond J, Case TJ (eds) Community ecology. Harper & Row, New York, pp 480–491

Young KR, Ewel JJ, Brown BJ (1987) Seed dynamics during forest succession in Costa Rica. Vegetatio 71: 157–173

Zammit C, Westoby M (1987) Seedling recruitment strategies in obligate-seeding and resprouting *Banksia* shrubs. Ecology 68: 1984–1992

Zar JH (1974) Biostatistical analysis. Prentice-Hall, Englewood Cliffs

Zar JH (1984) Biostatistical analysis, 2nd edn. Prentice-Hall, Englewood Cliffs

Zedler PH, Zammit CA (1989) A population-based critique of concepts of change in the chaparral. In: Keeley SC (ed) The California chaparral. Paradigms reexamined. Science Series No 34. Natural History Museum of Los Angeles County, Los Angeles, pp 73–83

Zedler PH, Gautier CR, McMaster GS (1983) Vegetation change and response to extreme events: the effect of a short interval between fires in California chaparral and coastal sage. Ecology 64: 809–818

Zinke PJ (1977) Mineral cycling in fire-type ecosystems. In: Mooney HA, Conrad CE (eds) Proc Symp on Environmental consequences of fire and fuel management in mediterranean ecosystems, Palo Alto, California, pp 85–92

# Subject Index


*Acacia*
*cyclops* 161, 166
*longifolia* 162, 166, 170
*mearnsii* 166
*melanoxylon* 166
*saligna* 161, 166, 170
spp. 169
*Acantholepis capensis* 209, 214
Acidity (*see* pH)
*Acomys subspinosus* 188–191, 193–194
*Acrolophia capensis* 74
Actinomycetes (*see* Soil)
*Adenanthos cygnorum* 119
*Adenostoma fasciculatum* 43–45, 61, 111–112, 117, 123–124, 141, 201
*Aeropetes tulbaghia* 153
*Aethomys namaquensis* 188–191, 193–195, 200
Afforestation 5, 8–9, 179, 233
*Agapanthus africanus* 72
*Agathelpis dubia* 80
*Agathosma giftbergensis* 39, 116
Age at maturity (*see* Vital attributes)
Air pollution 244–245, 257
Albedo 29
*Albuca*
*canadensis* 73
sp. 97
Aliens (*see* Invasive animals, Invasive plants)
*Aloe mitriformis* 16
*Amyryllis* sp. 158
Ammonification 265, 271
Ammonium 242
budget 250–253
in atmospheric deposition 244, 246
in stream water 244–247, 249, 253–254
*Anapalina triticea* 73
*Anaxeton asperum* 75
*Annesorrhiza inebrians* 75
Annual plants 55, 58, 62, 69
*Anoplolepis custodiens* 17, 207–209
Antecedent precipitation index 218
Antecedent wetness 221, 225–226, 228
*Anthospermum*
*aethiopicum* 79
*galioides* 79
Ants 17, 86, 182, 201, 203–215
community composition 208–214
species occurring in Swartboskloof 209–210
*Apis* spp. 149
*Arctostaphylos*
*glandulosa* 124
*glauca* 111, 117, 124
*pungens* 44
spp. 45, 123–125
*Arctotis semipapposa* 75, 97
Argentine ant (*see Iridomyrmex humilis*)
*Aristea*
*africana* 73
*major* 14, 73, 266
*spiralis* 73
*Aspalathus*
*ciliaris* 62, 65, 77
*spicata* 90
*Athrixia heterophylla* 75
Atmospheric deposition 241–242, 244–246, 250–253, 255, 257
ATP (*see* Soil microbial ATP)
*Aulax pallasia* 79
Australia (*see also* Kwongan) 23–24, 41, 48–49, 52, 70, 109, 118–119, 125, 141–142, 146, 166–171, 180, 182, 199–202, 215, 235

Bacteria (*see also* Soil, Nitrogen-fixing bacteria)
rhizosphere 259, 264–266, 268–269, 272
*Banksia*
*attenuata* 123
*hookeriana* 119
*menziesii* 119

*Banksia* (cont.)
potential invaders of mountain fynbos 179
spp. 125–126, 169, 179, 202
Bedload 221–222, 234, 240, 242, 253–254
Bedrock 248, 255
*Berkheya herbacea* 75
*Berzelia lanuginosa* 11, 76, 192
Bicarbonate 242
budget 250–253
in atmospheric deposition 244, 246
in stream water 244, 246–247, 249, 253–254
Biogeochemical cycles (see Nutrient cycling)
Biomass (*see also* Phytomass)
annual increments 170, 243
as fuel 38–39
bacterial (see Soil)
comparison of fynbos and other mediterranean-type shrublands 167–171
effect of invasive plants on 167–171
effect of post-fire age on 41
steady-state 40–41, 170–171, 174, 279, 281
the role of soil moisture on 120–121
Birds
as seed dispersers 56, 62, 81, 83, 86, 88, 90, 93, 105–106, 109, 123, 166
species occurring in Swartboskloof 17
*Blaeria dumosa* 58, 62, 77
*Blechnum australe* 15, 16
*Bobartia*
*gladiata* 73
*indica* 73
spp. 145, 148
*Boophane*
spp. 158
*Brabejum stellatifolium* 15, 44, 46, 97
*Brachylaena neriifolia* 111
*Bradypterus victorini* 17
*Briza maxima* 161–162
*Brodiaea* spp. 145
*Brunia nodiflora* 13–15, 44, 46–47, 76
*Brunsvigia* spp. 158
*Bulbine tuberosa* 65, 72

*Caesia contorta* 72
Calcium 242
budget 250–253
in atmospheric deposition 244–246
in bedload 254
in phytomass 243
in stream water 244–247, 249, 253–254
volatilized 250, 255–256
*Calopsis membranacea* 74
*Camponotus*
*maculatus* 209, 211
*niveosetosus* 209
*Cannomois virgata* 74
Carbon
in bedload 254
in soil 248
*Carpacoce vaginellata* 79
*Cassytha ciliolata* 78
*Casuarina* spp. 169, 272
Catchment
experiments 234–237, 252–253
management (see Management)
*Ceanothus*
*crassifolius* 124
*greggii* 111–112, 117, 124, 201
*leucodermis* 124
*megacarpus* 111–112
*spinosus* 111–112
spp. 123, 125, 169, 272
Cellulose decomposition 271
*Centella glabrata* 75
*Chaetops frenatus* 17
Chamise (*see Adenostoma fasciculatum*)
Chaparral, USA,
biomass 169–171
ecophysiology 109–112, 116–118
fire frequency in 58
fire intensity in 51–52
fire regime 52
fuel properties of 41, 44–45, 47, 49
fire-regeneration strategies in 69, 123–126
geophytes in 146
hydrology 216–217, 230, 236–237
nutrient budgets 252, 255–256
seed predation in 201
soil microorganisms in 270–271
Chemistry
budgets (*see* Nutrient budgets)
constituents 240–256
of atmospheric deposition 244–245
of bedload 254
of phytomass 243
of rainfall 244–246, 249
of soils 248
of stream water 244–247, 250
*Chenopodium ambrosioides* 97
Chile (*see also* Fundo Santa Laura, Matorral) 22, 24, 28, 69–70, 146, 169–171, 202
*Chionochloa rigida* 148
*Chironia baccifera* 78
Chlorine 242
budget 250–253
in atmospheric deposition 244–246
in stream water 244–247, 249, 253–254

*Chrysanthemoides monilifera* 75, 62, 97
*Cliffortia*
  *cuneata* 14–15, 23, 44, 46, 79, 84, 90, 111, 113–115
  *polygonifolia* 79, 192
  *ruscifolia* 14, 79, 90, 97
  spp. 58, 60
Climate
  classification 22–24
  drought 22–23, 31–32, 168, 171–172
  evaporation 27, 29–30
  evapotranspiration 31
  influence on fire 32–35
  lightning 32, 42
  mediterranean
    definition 22–23
    in Swartboskloof 21–24, 35
  pluviometric quotient 23–24
  radiation 27–29
  rainfall (see also Chemistry) 21, 25–27, 29, 31–32, 169–171, 220–221, 225
  snow 27
  temperature 21, 26–27
  vapour pressure deficit 26–27
  wind 27, 30–31
*Clutia*
  *alaternoides* 77
  *rubicaulis* 77
$CO_2$ evolution (*see* Soil microorganisms)
Coexistence 279, 281
  of seeders and sprouters 71, 108–122, 126–127, 142
    implications for management 143–144, 276, 283
*Coleonema juniperinum* 79
*Colletotrichum gloeosporioides* 198
Colonization 62, 65, 85, 86–95, 105–106, 173, 175, 256, 273, 278–279, 281
Community boundaries (*see* also Forest/fynbos boundary) 52, 66–69, 86, 89, 95, 105, 277–280
Conductance, leaf (*see* Ecophysiology)
*Conyza bonariensis* 75, 162
Correspondence analysis (see Ordination)
*Corymbium*
  *glabrum* 75
  *scabrum* 75
  *villosum* 75
Cover (*see* Plant cover)
Coweeta, USA 246, 252
*Crassula*
  *capensis* 77
  *fascicularis* 77
*Crematogaster*
  *liengmei* 209
  *peringueyi* 209, 211
Critical surface tension (CST) 226–227
*Crocidura flavescens* 188–191, 193–195
*Cunonia capensis* 11, 16, 44, 46, 84–85, 87–90, 92, 97, 99–104, 106, 110–111, 113
*Cyanella hyacinthoides* 75
*Cyclopia genistoides* 77
*Cymbopogon marginatus* 14, 74
*Cyphia*
  *bulbosa* 76
  *volubilis* 76
*Cyrtanthus*
  *ventricosus*
    life cycle 65, 152–153
    pollination (*see* Pollination)
    *Watsonia borbonica*, comparison with 157–160
    vital attributes 72
  spp. 158

*Dendromus mesomelas* 189
*Dilatris pillansii* 73
*Diosma hirsuta* 79, 97, 116
*Diospyros*
  *glabra* 15, 77, 93, 97, 192
  *whyteana* 97
*Disparago ericoides* 76
Dispersal (*see* Seed dispersal)
Disturbance
  fire as 81–84, 86, 89–92, 105, 164–165, 172–175, 180
  role in plant invasions 164, 168, 172–175
*Dorylus helvolus* 209
*Drosera*
  *aliciae* 77
  *capensis* 77
Dummy variable analysis 219, 221–223

Echo Valley, USA 23–24, 28
Ecophysiology
  comparison between forest and fynbos species 100–102
  comparison between seeders and sprouters 109–118
  diallel experiment 101–103
  leaf conductance 99
    seeders vs sprouters 109–114, 116, 119
  photosynthetic capacity 109, 117, 170
  shade tolerance
    seeders vs sprouters 114–116
  transpiration 108, 110, 113, 115, 228
  water relations 23, 98–103
    seeders vs sprouters 110–119
  water stress 98–103
  xylem pressure potential 23, 171, 98, 102–103

Ecophysiology xylem ... (cont.)
seeders vs sprouters 1102–112, 116–119
Ecotone (*see* Community boundaries)
*Ehrharta*
*bulbosa* 74
*dura* 74
*ramosa* 74
Elaiosomes (*see also* Myrmecochory) 182, 203
*Elegia*
*asperifolia* 11, 74
*capensis* 11
*juncea* 74
*Elytropappus*
*glandulosus* 76
*gnaphaloides* 116
*rhinocerotis* 169
Emmental, Switzerland 9
Epicormic buds 127
*Eremia totta* 77
*Erepsia anceps* 78
*Erica*
*articularis* 77
*coccinea* 15, 77
*hispidula* 14–15, 77
*longifolia* 77
*plukenetii* 44, 46–47, 111
*sphaeroidea* 77
Erodibility (*see also* Soil erosion) 232
Erosion (*see* Soil erosion, Sediment yields)
*Eucalyptus* spp. 48–49, 125, 169, 216, 226, 229–230, 235
*Euphorbia genistoides* 77
Exclosure experiments 164, 187, 196–198
Exotics (*see* Invasive animals, Invasive plants)

*Festuca scabra* 74
*Ficinia*
*deusta* 72
*filiformis* 72
*nigrescens* 73
Fire
as disturbance 172–175
as a management practice 18, 143, 257, 283–284
details of the March 1987 fire 18–20
effects on animal habitat 194, 200, 215
effects on catchment hydrology 216–257
effects on soil microorganisms 259, 261–272
effects on soil moisture 261, 268
effects on soil nitrogen 269, 271–272
effects on soil pH 259–260, 268, 271
frequency 6–7, 42–43, 52
effects on biomass and fuel loads 41–42, 281–283
effects on species composition 69, 277–283
effects on streamflow 281–283
effects on vegetation structure 279–283
role in plant invasions 279–280
hazard 31, 33–35
history in Swartboskloof 6–7, 42–43
ignition sources 33, 41–42
intensity 19–20, 49, 51
effect on plant mortality 83, 133–134
effect on plant recruitment 141–142
effect on soils 230, 233
prescribed burning (*see* Fire as a management practice)
probability of 6–7, 40–43
rate of spread 19, 49, 51
role in plant invasions 170–175, 177–180
season 33–35
stimulated flowering of geophytes 145–160
weather 19–20, 30, 32–35
Fire response types (*see also* Vital attributes)
autoregenerating long-lived sprouters 55, 57, 65, 67, 70
facultative seeder-sprouters 55, 57, 65, 67, 70
fire ephemerals 55, 57, 65, 67, 70
list for Swartboskloof 72–80
obligate seeders 55, 57, 65, 67, 70
obligate vegetative regenerating sprouters 55, 57, 65, 67, 70
relative advantages of seeding and sprouting 122–124
spectra for Swartboskloof 57
Flowering
fire-stimulated flowering 145–160, 277
mass flowering 148, 156–157, 159–160
Fluoride 242
budget 250–253
in atmospheric deposition 244, 246
in stream water 244, 246–247, 249, 253–254
Forest (see also Riparian zones)
as refugia for small mammals 187, 194
community description 15–17
distribution in fynbos 12, 81–82, 167, 169–170
development in the absence of fire 81, 280
fire intensity in 51

fuel properties of 48–50, 278
Forest/fynbos boundary
conceptual model 81–84
dynamics 81–107
edaphic factors 96–97
historical factors 103–104
stability of 104–107, 278
the role of fire 48–50, 52, 84–86, 278
the role of nutrients 96–98
the role of soil moisture 81, 98–104
Formicidae (*see* Ants)
Fuel
arrangement 38, 43, 48–50, 52
buildup/decay 40
classes 38
comparison between vegetation types 48–52
crude fat content 44–48
energy content 38–39, 44–47
load 223–230
in forest 48–50
in fynbos 40–41, 48–50
models 39–40
moisture content 20, 43–48
packing ratio 44
Fundo Santa Laura, Chile 24, 28
Fungi
in soil (*see* Soil)
pathogenic (*see Colletotrichum*)
Fynbos Biome Project, vi 1

Garrigue, France 170
*Gazania serrata* 76
*Geissorhiza aspera* 73
Geology (*see* Swartboskloof)
Geophytes (*see also Cyrthanthus, Watsonia*) 60, 65–66, 145–160
*Gerbera crocea* 76
Germination 55,' 60, 85–86, 89–92, 114–115, 124, 126, 138–139, 156, 164, 168, 178, 196–198, 277
*Gethyllis* sp. 157–158
*Gladiolus*
*blommesteinii* 72
*carneus* 72
*Gnidia oppositifolia* 80
Granivory (*see* Seed predation)
Grassland
fire intensity in 51
Grasses 70–74, 171–172

Habitat
for small mammals 185–186, 189–196, 200
*Haemanthus*
*canaliculatus* 148
*coccineus* 72
spp. 158
Haemorodaceae 158
*Hakea*
*gibbosa* 166, 177
*salicifolia* (*H. saligna*) 177
*sericea* (*H. acicularis, H. tenuifolia*) 161–163, 166, 170, 177, 180, 275, 279, 283
spp. 123, 161, 175, 177–178, 180
*suaveolens* (*H. drupacea*) 177
*Halleria*
*lucida* 16, 97
*Haplocarpha lanata* 76
*Hartogiella schinoides* 16, 77, 94, 96–97
*Heeria argentea* 16, 75, 97
*Helichrysum*
*cymosum* 62, 76
*nudifolium* 76
*teretifolium* 76
*zeyheri* 76
Herbivory 198
Herpetofauna 17
*Hessea stellaris* 156
*Heteromeles arbutifolia* 111, 117, 124
*Hibbertia aurea* 118
History
of fire 6–7, 18–20
of land-use 7–9
Hubbard Brook, USA 246, 252
Hydrology
hydrological experiments 234–239
hydrological response 5, 216, 229, 233–239
infiltration 216–217, 226, 229
interception 228
overland flow 216–217, 223–225, 229, 238–239, 280
paired catchment experiments 9, 217
peak discharge 216, 218, 220
quickflow 218–220
response ratio 220
spates 220
stormflow 220
streamflow 217–221, 228–229, 234–237
surface runoff 222–226, 229–230
Hygrophilous communities 10–12, 64–65, 69, 192
*Hypochoeris radicata* 76, 161–162
*Hypodiscus*
*albo-aristatus* 15, 74
*aristatus* 75
*willdenowia* 75
*Hypoponera spei* 209

*Ilex mitis* 16, 44, 46, 84, 87–89, 92, 97

*Indigofera alopecuroides* 77
Insects 17–18, 203–215
Invasion (*see* Invasive animals, Invasive plants)
Invasive animals (*see Iridomyrmex humilis*)
Invasive plants
  alien flora of Swartboskloof 162
  determinants of invasibility 164–165
  disturbance, the role of 164, 168, 172–175
  effects on biomass 50, 170
  effects on fuel properties 50, 52
  effects on species diversity 280–282
  effects on streamflow 280–284
  history of introduction 175–179
  invasibility of fynbos 161–181
  invasion windows 172–175, 279
  life history attributes of 164, 166, 168, 174–179
  management of 283–284
  profile of a successful invader 174–179
  risk assessment 176–179
*Iridomyrmex humilis* 17, 203–215
  distribution in Swartboskloof 208
  effect on indigenous ants 208–214
  rate of invasion in Swartboskloof 207
*Ischyrolepis*
  *capensis* 75
  *gaudichaudiana* 14, 44, 46, 75, 114–115
  *sieberi* 15, 75

*Kiggelaria africana* 44, 46, 62, 66, 78, 90–92, 94, 96–97, 99–104, 106, 111–113, 124
*Knowltonia vesicatoria* 16, 97
Kosciusko National Park, Australia 235
Kwongan (Australia) 70, 118–119, 146, 169, 171, 202

*Lachenalia orchioides* 73
Land-use (*see* History)
Leaching 249, 251, 255
*Leptogenys nitida* 209
*Leptospermum laevigatum* 161, 166
*Leucadendron*
  *laureolum* 173, 279
  *salignum* 79, 110–111, 113–115, 124
  spp. 71, 123–125, 169, 172
*Lichtensteinia lacera* 75
Life forms (plants)
  list for Swartboskloof 72–80
  definitions 80
*Linum africanum* 78
Litter
  decomposition 40, 270–271
  – fall rates 40
  mass 204
*Lobelia coronopifolia* 77
Longevity (*see also* Vital attributes) 124
Lottery model (*see also* Non-equilibrium dynamics) 172

*Macrochaetium hexandrum* 73
Magnesium 242
  budget 250–253
  in atmospheric deposition 244–246
  in bedload 254
  in phytomass 243
  in soil 248
  in stream water 244–247, 249, 253–254
  volatilized 250, 255–256
*Mairea lasiocarpa* 76
Malee-broombush 170
Mammals
  as seed predators 196–198, 200–202
  effects on vegetation structure 200–202
  habitat (*see* Habitat)
  population densities 188–191, 199
  response to fire 191–196
  species occurring in Swartboskloof 17, 189
  survival of fire 189–191, 277
  trapping techniques 184–185
Management 9, 143–144, 257, 276, 279–280, 283–284
Mast fruiting 148
Mattoral (Chile) 69–70, 146, 169–170, 202
*Maytenus*
  *acuminata* 16, 62, 66–67, 77, 84, 94, 96
  *oleoides* 44, 59, 77, 89, 94, 96–97, 124
Mediterranean Basin 146, 180
Mediterranean climate (*see* Climate)
*Meranoplus peringueyi* 208, 209
*Merxmuellera*
  *cincta* 11, 13, 74, 192
  *rufa* 74
  *stricta* 14, 74
*Metalasia muricata* 76
Mice (*see* Mammals)
*Micranthus alopecuroides* 73
Microorganisms (*see* Soil microorganisms)
Mineralization 258–259, 265, 267, 269–272
*Monomorium*
  *braunsi* 209
  *lubricum* 209
  *musicum* 209, 211
  *nuptialis* 209
  *schultzei* 209
  spp. 208
  *springvalense* 209
  *torvicte* 209

*willowmorense* 208–209, 211
*Montinia caryophyllacea* 78
Montpellier, France 24
*Moraea*
*bituminosa* 73
*tricuspidata* 74
Multiple regression 219, 221, 225
*Muraltia*
*alopecuroides* 79
*heisteria* 79
*Mus minutoides* 188–191, 193–195, 200
*Myosorex varius* 188–191, 193–195
Myrmecochory 166, 201, 203–204, 277
*Myrsine africana* 62, 78, 97
*Myrsiphyllum scandens* 16

*Nebelia paleacea* 15, 76
*Nectarinia violacea* 17
*Neesenbeckia punctoria* 73
Nitrate 242
budget 250–253
in atmospheric deposition 244–246
in stream water 244–247, 249, 253–254
Nitrification 269, 271
Nitrogen 242
budget 250–253
– fixation (*see also* Nitrogen-fixing bacteria) 168, 255
in atmospheric deposition 244–246
in bedload 254
in phytomass 243
in soil 248
in stream water 244–247, 249, 253–254
volatilized 250, 255–256
Nitrogen-fixing bacteria 259, 264–266, 268–269, 271–272
Non-equilibrium dynamics 172–175, 275, 278–279
Nucleation 82–83, 92–96
Nutrient(s)
budgets 240–241, 250–253, 257
content in plants 243
cycling 97, 240–257, 258, 258–259, 274
role in forest development 96–98

*Ocymyrmex cilliei* 209
*Oftia africana* 80
*Olea*
*europaea* subsp. *africana* 14, 84, 90, 94, 96–97, 124
spp. 169
*Olinia ventosa* 16, 96–97
Orchidaceae 158
Ordination 10–11, 185, 192, 211–212, 214
*Ornithogalum hispidum* 65, 73
*Osmitopsis*
*afra* 76
*asteriscoides* 11, 76, 192
*Osteospermum*
*junceum* 76
*tomentosum* 76
*Otholobium*
*fruticans* 78, 110–111, 113
*oliquum* 78
*Otomys irroratus* 188–191, 193–195
*Oxalis*
*bifida* 79
*lanata* 79
*purpurea* 79
spp. 146, 158
*strigosa* 79
*tenuifolia* 79

*Paraserianthus lophantha* 166
*Passiflora quadrangularis* 161–162
*Pelargonium*
*cucullatum* 78
*elongatum* 62
*longifolium* 78
*myrrhifolium* 78
spp. 146, 158
*tabulare* 78
*Pellaea pteroides* 72
*Penaea mucronata* 79
*Pentameris*
*macrocalycina* 74
*thuarii* 74
spp. 70
*Pentaschistis*
*aristidoides* 74
*colorata* 74
*curvifolia* 74
*malouinensis* 74
*pallescens* 74
Perches, role in succession to forest 88–89
*Peucedanum sieberianum* 75
pH
in atmospheric deposition 244–246
of soil (*see* Soil)
in stream water 244, 246–247, 249, 253–254
*Pheidole*
*capensis* 17, 208
*foreli* 209, 211
*tenuinodes* 209
Phenology (*see also* Flowering) 150
Phosphate 242
budget 250–253
in atmospheric deposition 244, 246
in stream water 244–247, 249, 253–254

Phosphorus 242
budget 250–253
in atmospheric deposition 244–246
in bedload 254
in phytomass 243
in soil 248
in stream water 244–247, 249, 253–254
volatilized 250, 255–256
Photosynthetic capacity (see Ecophysiology)
*Phylica*
*imberbis* 79
*pubescens* 79, 90
spp. 58
Phytomass (*see also* Biomass) 242–243, 251
Phytosociology 10–17
*Pinus*
*halepensis* 162–163, 166, 177, 178, 180, 275
*pinaster* 161–163, 166, 170, 177, 180, 275
*pinea* 179
*ponderosa* 44
*radiata* 3, 5, 8, 66, 72, 161–162, 166, 170, 180, 275, 284
*Pittosporum undulatum* 94, 96, 161–162
*Plagiolepis*
*jouberti* 209
*pygmaea* 209
Plant-animal interactions 56, 62, 81, 83, 86, 88, 90, 93, 105–106, 109, 123, 156, 166, 181–204, 276–277, 279
Plant cover 11–16
Plant-water relations (*see* Ecophysiology)
*Platycaulos depauperatus* 75, 192
*Platylophus trifoliatus* 97
*Podalyria*
*calyptrata* 90, 97
*montana* 78
*Podocarpus*
*elongatus* 16, 84, 97
spp. 169
Pollination 18, 148–149, 152–153, 156–157
Pollution (*see* Air pollution)
*Polygala bracteolata* 79
Potassium
budget 250–253
in atmospheric deposition 244, 246
in bedload 254
in phytomass 243
in soil 248
in stream water 244–247, 249, 253–254
volatilized 250, 255, 256
*Praomys* (*Myomyscus*) *verreauxi* 188–189, 193–195
*Prismatocarpus diffusus* 77
*Promerops cafer* 77
*Protasparagus*
*rubicundus* 72
spp. 97, 148
*Protea*
*acaulos* 79
*glabra* 116
*neriifolia*
age at maturity 135
competitive ability 101
fire response pattern 62–63, 67–69
fuel properties 44
generalized life history 124
inter-fire recrutiment 139–141, 143, 183
leaf conductance 100, 111
moisture content of foliage 46
mortality rates 128–131, 143
population growth after fire 128–129, 143
regeneration in burnt and unburnt sites 90
reproductive development 134–135
roots 104, 111
seed germination 90–92, 138–139, 143
seed predation by mammals 187, 197–198
seed production 89, 135–137, 1432
seed release 88
seedling establishment in burnt and unburnt sites 90–91
seedling growth, effect of soil moisture 99–100, 103
seedling mortality 198
seedlings, location of 88–89
seedling: parent ratio 128
serotiny in 136
stand density 128
vital attributes 79
xylem pressure potential 111
*nitida*
effect of soil characteristics on distribution 97
competitive ability 101
generalized life history profile 124
inter-fire recruitment 140–141, 143
leaf conductance 100, 111, 114
lignotuber 134
maturity classes 131
mortality rates 128, 130–135, 140, 143
population growth after fire 128–129, 143
roots 104, 111

seed germination 89–92, 138–139, 143
seed production 136–138, 143
seedling establishment in burnt and unburnt sites 91
seedling growth, effect of soil moisture 99–100, 103
seedling : parent ratio 128
stand density 128
transpiration 115
vital attributes 79
xylem pressure potential 111
*repens* 14, 17, 40, 43, 47, 62, 67, 79, 90–92, 99, 100, 111, 113–115, 124, 138, 170, 192, 259, 266
spp. 17, 66, 67, 71, 102, 141, 169
*Pseudotsuga menziesii* 44–45
*Psoralea aculeata* 78
*Pteridium aquilinum* 14, 72
Pugnacious ant (*see Anaplolepis custodiens*)

*Quercus*
*dumosa* 111–112, 124

*Rafnia capensis* 78
*Rapanea melanophloeos* 16, 96–97
Raunkiaer's life forms (*see* Life forms)
Refugia 194
Regeneration strategies (*see* Fire response types)
Resilience 69–71, 124, 141, 144, 160, 179, 275–277, 282–283
Resource utilization
implications for invasibility 167–171
Respiration by soil microorganisms 261
*Restio*
*filiformis* 43, 75
*triticeus* 75
*Rhabdomys pumilio* 188–190, 193–195, 200–201
*Rhamnus californica* 117
Rhizosphere (*see* Bacteria, Soil, *Watsonia*)
*Rhus*
*angustifolia* 14, 15, 75, 84, 93, 97, 192
*laurina* (*Malosma laurina*) 111–112, 124
*ovata* 111, 124
*rosmarinifolia* 75
*tomentosa* 75, 93, 97
spp. 117
*Rhynchosia totta* 78
Riparian zones
as nutrient and sediment filters 256–257, 279–281
occurrence in Swartboskloof 10
susceptibility to fire 48–50, 278
Rodents (*see* Mammals)
*Roella ciliata* 58, 65, 77
Roots 91, 99–102, 104, 111–113, 117–119, 256, 268–269
*Rumohra adiantiformis* 16

San Diego, USA 24
San Dimas, USA 24
Savanna
fire intensity in 51
*Scabiosa columbaria* 77
*Schizaea pectinata* 72
*Schizodium obliquum* 74
*Sciurus carolinensis* 179
Scottish heathland 70
Season
effect on moisture content of fuel 43–47
effects on soil microorganisms 259, 261–272
effects on soil moisture 260–261, 265, 269–270
effects on soil nutrients 259, 265, 267, 272
effects on soil pH 259–260, 265, 269
role in fire weather and occurrence 33–35
*Sebaea exacoides* 58, 62, 78
*Secamone alpini* 16
Sediment yields (*see also* Soil erosion) 219, 221–225, 229, 234–239, 280–281
Seed
banks 58–59, 72–80, 85–87, 135–138, 201, 166
deposition 86–89
dispersal by ants (*see* Myrmecochory)
dispersal by birds (*see* Birds)
dispersal by wind 178, 180
germination requirements (*see also* Germination) 89–92, 168, 178
longevity 58–59, 68, 72–80
nutrient content 168
predation 156, 187, 196–198, 200–202, 276, 279
storage in the canopy (*see* Serotiny)
storage in the soil 85–86, 166
Seeders (*see also* Fire response types) 108–144
evolution of reseeding 125, 143, 275–276
*Selago spuria* 80
*Senecio*
*grandiflorus* 76
*paniculatus* 76
*pinifolius* 76
*pubigerus* 65, 76
Senescence 18, 40–41, 60–61, 108, 126, 140–141

*Serinus*
  *leucopterus* 17
  *totta* 17
Serotiny 136, 166, 168, 174–175, 178–179, 275–277
Shrews (*see also Crocidura, Myosorex*) 17
Silicon 242
  budget 250–253
  in atmospheric deposition 244, 246
  in stream water 244, 246, 247, 249, 253–254
*Smithustroma emarginata* 210
Sodium 242
  budget 250–253
  in atmospheric deposition 244–246
  in bedload 254
  in soil 248
  in stream water 244–247, 249, 253–254
Soil
  bacteria 259, 262–272
  biomass/populations of actinomycetes 259, 262–263, 267–268, 271–272
  chemical properties 248, 251, 258–259, 267
  erosion (*see also* Sediment yields) 216, 229–230, 232, 234–239
  fungi 259, 262–265, 267–268, 270–272
  heating 223, 231, 233, 255–256
  microbial ATP 259, 261–262, 265, 267–268, 270–271
  microorganisms 255, 258–272
  moisture 98–103, 204, 260–265, 267–268, 270
    deficit 23, 30
    gradients 119–121
  nutrients 258–259, 265, 267, 271–272
  pH 248, 258–260, 265, 267–268
  physical properties 258–259, 267
  rhizosphere populations of bacteria 259, 264–266, 268–269, 272
  texture 97, 227, 248, 258
  water-repellency (*see* Water-repellent soils))
  weathering 248–249, 251, 255
*Solanum tomentosum* 65, 80
*Spiloxene capensis* 73
Sprouters (*see also* Fire response types) 108–144, 277
*Staberoha cernua* 75
Stability of community boundaries 66–69, 277–278
*Stirlingia latifolia* 119
*Stoebe plumosa* 76
Stream water chemistry 244–250, 252–253
*Struthiola ciliata* 80
Succession
  fynbos to forest 81–107, 278
  phases in fynbos 60
  small mammal communities 189, 199–200
Sulphate 242
  budget 250–253
  in atmospheric deposition 244–246
  in stream water 244–247, 249, 253–254
Survival mechanisms (*see* Fire response types)
Suspended sediment (*see* Sediment yields)
Swartboskloof
  location 2
  climate (*see* Climate)
  fauna 17–18
  fire history (*see* History)
  geology 2–3
  hydrology (*see also* Hydrology) 5–7
  land-use history (*see* History)
  physical features 2–5
  research rationale, v-viii 18, 273–275
  soils, basic description 3–5
  vegetation 10–17

*Tetramorium*
  *bothae* 210
  *capense* 210
  *emeryi* 210
  *flaviceps* 210–211
  *grassi* 210
  *lobulicorne* 210
  *quadrispinosum* 210
  spp. 208
*Tetramyrma* sp.
*Tetraria*
  *bromoides* 73
  *burmannii* 73
  *capillacea* 15, 73
  *compar* 73
  *cuspidata* 43, 73
  *fasciata* 11, 73
  *fimbriolata* 73
  *involucrata* 15
  *ustulata* 73
*Thamnochortus*
  *fruticosus* 75
  *gracilis* 75
*Themeda triandra* 74
*Thereianthus spicatus* 74
*Thesium*
  *capitatum* 80
  spp. 58, 65, 67
  *strictum* 80
  *virgatum* 800

*Trachyandra*
  *hirsuta* 72
  *hirsutiflora* 72
  *tabulare* 72
Total dissolved solids 250–251
Transpiration (*see* Ecophysiology)
Trees (*see also* Afforestation, Forest, Invasive plants) 15–16, 24, 167–170
*Tritonia crispa* 74

*Urginea*
  *dregei* 65, 73
  spp. 158
*Ursinia*
  *paleacea* 76
  *pinnata* 65, 76

Vacant tree niche hypothesis 168, 170
Vegetation cover (*see* Plant cover)
Vital attributes of plant species (*see also* Fire response types)
  definition of terms 55–56
  juvenile periods 58, 72–80
  list for Swartboskloof species 72–80
  longevity 58–59, 72–80
  seed bank longevity 58–59, 72–80
  spectra for Swartboskloof 59
Volatilization 240–242, 250, 255–256

*Wachendorfia paniculata* 73
Wagon Wheel Gap, Colorado, USA 9
Water drop penetration time (WDPT) 226–227
Water quality (*see* Stream water chemistry)
Water relations (*see* Ecophysiology)
Water-repellent soils 226–228, 230–232
Watersheds (see Catchment)
Water stress (*see* Ecophysiology)
*Watsonia*
  *borbonica* (*W. pyramidata*)
    *Cyrthanthus ventricosus*, comparison with 157–160
    fire frequency, effect of 158–159
    fire-stimulated flowering 148, 153–157, 276
    flowering cues 154–156
    leaf conductance 114
    leaf harvesting, effect on flowering 153–154
    life cycle 149–152
    mass flowering 145, 146–147
    pollination (*see* Pollination)
    rhizosphere bacteria of 266
    transpiration 115
    vital attributes 74
  bulbs as food 7
  *fourcadei* 151
  *pyramidata* (*see W. borbonica*)
Weathering (*see* Soil)
*Widdringtonia nodiflora* 44, 72
*Willdenowia sulcata* 75

*Xanthorrhoea australis* 146
Xylem pressure potential (*see* Ecophysiology)

*Zaluzianskya* sp. 97
*Zantedeschia aethiopica* 16, 97

MIX
Papier aus verantwortungsvollen Quellen
Paper from responsible sources
FSC® C105338
FSC
www.fsc.org

If you have any concerns about our products,
you can contact us on
**ProductSafety@springernature.com**

In case Publisher is established outside the EU,
the EU authorized representative is:
**Springer Nature Customer Service Center GmbH**
**Europaplatz 3, 69115 Heidelberg, Germany**

Printed by Libri Plureos GmbH
in Hamburg, Germany